The Long March to the West

Geopolitical Affairs

Series Editor and Editor-in-Chief: Michel Korinman

Editor: John Laughland

This new books series will will cover global and regional geopolitics and current affairs. Each issue contains articles by commentators from all points of the political spectrum, the goal being to present as many different points of view as possible. Geopolitics studies how geography and politics interact; *Geopolitical Affairs* aims to be the last word in what is happening on our planet today.

International Advisory Board

Bakyt Alicheva-Himy, Michalis Attalides, Young-seo Baik, Bayram Balci, Sanjib Baruah, Dusan Batakovic, Dieter Bednarz, Peter Brown, Hamit Bozarslan, Hans-Jörg Brey, Michel Carmona, Steven David, Michael F. Davie, Bruno Drweski, Gérard-François Dumont, Mhamed Hassine Fantar, Luigi Vittorio Ferraris, Sir Andrew Green KCMG, Pedro José Garcia Sanchez, Manlio Graziano, William Gueraiche, Jean-Louis Guereña, Michel Gurfinkiel, Mezri Haddad, Simon Haddad, Xing Hua, Anthony Jones, Emil Kazakov, Thierry Kellner, Pierre Kipré, Paul Kratoska, Lord Lamont of Lerwick, Frank Lestringant, Dana Lindaman, Michel Makinsky, Yohanan Manor, Florence Mardirossian, Attila Melegh, Natalia Narotchnitskaya, Jean-Baptiste Onana, Thierry Pairault, Anna Papasavva, Michel Pérez, Sandro Provvisionato, Jean-Claude Redonnet, Michel Roux, Muhammed Siad Sahib, Luigino Scricciolo, Günter Seufert, Georges Tadonki, Comi Toulabor, Edward Vickers, Ludwig Watzal, Fabrizio Vielmini

1. Shia Power: Next Target Iran?
2. The Long March to the West: Twenty-First Century Migration in Europe and the Greater Mediterranean Area
3. Russia, the Caucasus and Eastern Europe
4. The Americas

Geopolitical Affairs is published by the Daedalos Institute of Geopolitics in Cyprus in cooperation with OGENI, the Geopolitical Observatory for Nations and the World (*Observatoire géopolitique des espaces nationaux et internationaux*) at the Sorbonne in Paris.

The Long March to the West

Twenty-First Century Migration in Europe and the Greater Mediterranean Area

Editors

Michel Korinman

John Laughland

VALLENTINE MITCHELL ACADEMIC

LONDON · PORTLAND, OR

Daedalos Institute of
Geopolitics, Cyprus

OGENI
(Observatoire géopolitique des espaces nationaux
et internationaux), Sorbonne, Paris

First published in 2007 in Great Britain by
VALLENTINE MITCHELL ACADEMIC
Suite 314, Premier House, 112–114 Station Road,
Edgware, Middlesex HA8 7BJ

and in the United States of America by
VALLENTINE MITCHELL ACADEMIC
c/o ISBS, 920 NE 58th Avenue, Suite 300
Portland, OR 97213-3786

www.vmbooks.com

Copyright © 2007

British Library Cataloguing in Publication Data
A catalogue record has been applied for

ISBN 978 0 85303 780 4 (cloth)
ISBN 978 0 85303 781 1 (paper)

Library of Congress Cataloging in Publication Data
A catalogue record has been applied for

This group of studies first appeared in a special issue of
Geopolitical Affairs, 1/2 [ISSN 1753-5417],
published by Vallentine Mitchell Academic.

All rights reserved. No part of this publication may be reproduced, stored in or introduced into a retrieval system, or transmitted, in any form or by any means, electronic, mechanical, photocopying, recording or otherwise, without the prior written permission of the publisher of this book.

Typeset in 10.5/12pt Garamond 3 by FiSH Books, Enfield, Middx
Printed in Great Britain by Antony Rowe Ltd, Chippenham, Wilts

Contents

Editorial: Migration: The Triple Constraint   ix
by Michel Korinman

PART I. THE PROBLEM

Migration and the New World Order   3
by Demetris Christofias

The New Logic of Migration in the Twenty-First Century   5
by Gérard-François Dumont

The Catholic Church and the 'Others': A Universalist Strategy   15
by Manlio Graziano

Farewell Italy: Demography and Immigration   24
by Giuseppe Sacco

Humanitarian NGOs and the Migration Policies of States: 56
A Financial and Strategic Analysis
by Marc-Antoine Pérouse de Montclos

PART II. FORTRESS EUROPE – THE SOLUTION?

What Migration Policy for the EU?   69
by Eberhard Rhein

Immigration: A French Taboo   78
by Ivan Rioufol

Why is the Vlaams Belang So Popular?   84
by Paul Belien

Turks in Germany: Problems and Perspectives   90
by Asiye Öztürk

A Foreign, Frightening World: The Germans' Attitude towards Islam by Elisabeth Noelle and Thomas Petersen	98
The End of the Ideology of Multiculturalism in the United Kingdom by John Laughland	102
Italy: From a Country of Illegal Emigrants to a Country of Illegal Immigrants by Valeria Palumbo	109
Immigrants in Italy by Luigino Scricciolo	116
Migrations as a Factor of International Cooperation and Stability by Guido Lenzi	126
Immigration, Radicalization and Islamic Discourse in Europe: An Examination and a Sufi Response by Nir Boms	129
The Migration Policy of the City of Athens by Kalliopi Bourdara	141
The Social Integration of Immigrants in Greece by Markos Papakonstantis	144
Are the Spanish For or Against Immigration? by Marisa Ortún Rubio	152
The Revival of Islam in Spain by Rosa María Rodríguez Magda	164
The Integration Strategies of Different Immigrant Groups in Spain by Rosa Aparicio Gómez	169
Immigration, Self-Government and Management of Identity: The Catalan Case by Ricard Zapata-Barrero	179

PART III. EUROPE'S SOUTHERN FRONTIER

Immigration and the Canary Islands by Javier Morales Febles	205

CONTENTS vii

Lampedusa, European Landfall 209
by Bruno Siragusa

From an Emigrant to an 'Immigrant' Country: A Malta Case Study 212
by Joe Inguanez

Immigration in Cyprus: New Phenomenon or Delayed 220
Responsiveness
by Anna Papasavva

Interview with Neoklis Sylikiotis, Minister of the Interior of the 228
Republic of Cyprus

Chinese versus Palestinians? 234
by Yohanan Manor

PART IV. EUROPE'S SOUTHERN MARCHES

Migrations and Algerian–French Relations: A Shared 251
Responsibility
by Sid Ahmed Ghozali

The Sudanese Population in Egypt 259
by Doreya Awny

PART V. SOUTH OF THE SOUTH

The Geopolitics of African Migrations to Europe: International 273
Hypocrisy About the Right and Need for Africans to Emigrate
by Georges Tadonki

Migrations and Nation Building in Black Africa: The Case of 281
Ivory Coast since the Mid-Twentieth Century
by Pierre-Aimeé Kipré

Is the Partition of Ivory Coast a Consequence of Migrations in 300
the Colonial Era?
by Christian Bouquet

Emigration from Burkina Faso and the Ivorian Crisis 308
by Augustin Loada

Experiences at a Consulate in Africa 320
by Alexandre Leitão

PART VI. WAR AND MIGRATION

Conflicts in Chad and Darfur   329
by Doual Mbainaissem

The Migration and Displacement of Assyro-Chaldeans in Iraq   343
by Françoise Brié

Displacement and Deportations in Iraq   354
by Françoise Brié

PART VII. EAST MEETS WEST

The Impact of East European Immigration on the UK   369
by Sir Andrew Green

Migration and Migration Narratives in the Era of Globalization   372
by Attila Melegh

East European Emigration to the West: A 'New South' or an 'Anti-South'?   382
by Bruno Drweski

Albanian Organized Crime   395
by Xavier Raufer

Transnational Networks: The Case of the Chinese of Zhejiang   405
by Véronique Poisson

PART VIII. PSYCHOANALYSIS AND GEOPOLITICS

Islam Wrestling with the 'West': Reflections on the Foundations   417
by Philippe Réfabert

Immigrants and the West's Culture of Dependency   423
by Theodore Dalrymple

Migration: The Triple Constraint

Editorial
MICHEL KORINMAN

The scenarios contained in the report of the United Nations Population Division from 2001 onwards are unambiguous: by 2005, it will be impossible to maintain either the ratio between active and inactive members of the population in Europe, or even the current level of the population, without mass immigration. There would need to be 161 million immigrants in Europe (including Russia) and 80 million in the European Union to maintain ratio between active and inactive members of the population, and 100 million in Europe and 47 million in the EU to maintain the demographic level. The need is particularly urgent in countries where the birth-rate has collapsed in the past twenty years, such as Spain and Italy. Italy, Germany and Russia respectively need 13, 18 and 28 million foreigners to maintain the demographic level and 19, 25 and 33 million foreigners to maintain the ratio between working and inactive members of the population.[1] Such a forecast (over a period of half a century!) is probably quite transitory, but clearly suggests that borders are being erased (deliberately?), that they are things of the past, and that there is a permanent movement of populations.

This indeed is one of the apocalyptic predictions in Jacques Attali's latest book, *A Short History of the Future*. He writes, 'In total, in twenty-five years, about fifty million people every year will move to a different country. Approximately one billion people will live in a country other than their country of origin or their parents' country of origin.'[2]

More modestly but in the same vein, the World Bank predicts a rise in migration but also calls for measures to encourage people to return to their countries of origin. It puts the current number of migrants in the world at 200 million, while recommending that states put in place better systems for managing flows of labour by encouraging 'circular migrations'. The idea is to abolish the obstacles to migration in all OECD countries and to encourage migrants to return home in order that their countries of origin can benefit from the skills they have acquired while abroad.[3] In this volume, Eberhard Rhein offers an implicit criticism of these proposals.

Europe is a paradoxical case in this regard: in 1993, western Europe admitted some 20 million immigrants in 1993, that is, a quarter of all migrants in the world, even though it represents less than 10 per cent of the world's population and even though its territory is but one-fifteenth of land on the planet.[4]

In 1990, Greece created its National Foundation for the Reception and Reinstallation of Repatriated Greeks (ΕΘΝΙΚΟ ΙΔΡΥΜΑ ΥΠΟΔΟΧΗΣ ΚΑΙ ΑΠΟΚΑΤΑΣΤΑΣΗΣ ΑΠΟΔΗΜΩΝ ΚΑΙ ΠΑΛΙΝΝΟΣΤΟΥΝΤΩΝ ΟΜΟΓΕΝΩΝ ΕΛΛΗΝΩΝ [ΕΙΥΑΠΟΕ]) in an attempt to encourage members of the Greek diaspora, rather than foreign immigrants, to settle in Greece. Opinion polls in Germany, Britain and France all show that the public is worried about immigration at the dawn of the new millennium. Gordon Brown, the British Chancellor of the Exchequer, has said Britain should have a special day to celebrate its national identity – and this in a country which has long been attached to the principle of multiculturalism.[5] In the most recent example of the same phenomenon in France, the presidential candidate Nicolas Sarkozy called for the creation of a Ministry of Immigration and National Identity. It mattered little that he later amended this to 'Republican Identity' in the face of furious attacks by his rivals – the figures are there: 55 per cent of French people support him on this, 58 and 56 per cent respectively for employees and workers. Sixty-five per cent of those polled said they thought immigrants should adhere to the collective identity of their country of residence.[6]

Moreover, the governments of western Europe are under a triple constraint. First, there has emerged uncontrollable pockets of violence, like the parts of Italy described in Luigino Scricciolo's article in this volume. The most famous case is the anti-delinquent wall put up by the left-wing city council of Padua to screen off the 'African ghetto' of dealers from the rest of the town.[7] This is in a country where immigration is a recent phenomenon: indeed, it is because Germany and France restricted immigration from Morocco, Tunisia and elsewhere in Africa from 1974 onwards that these people have been heading instead to Italy or Spain since the 1990s.[8] Should they be divided out across the territory and, if so, by what means? These are obviously very controversial issues, but the debate is being forced by the Mayors of Sassuolo (who has already tried out the method), Modena, Bovezzo (near Brescia) and the Minister of Social Solidarity, Paolo Ferrero.[9] Western European states are probably extremely concerned at the possible emergence of gangs of local inhabitants as has occurred in Italy. Worse still would be the creation of violent groups of local self-defence militias, that is to say, a real transition from the national right to the extreme right.

The same goes for the latent tensions between those islands where illegal immigrants arrive in order to enter the European paradise and the states to which they belong, like the Canary Islands, Lampedusa, or even between a new member state like Malta and the EU. Migrants chose the first of these routes from Senegal or Mauritania because of the creation of an electronic surveillance system in the Straits of Gibraltar and the reinforcement of the security barriers around the two Spanish enclaves of Ceuta and Melilla in Morocco. The nationalist autonomous government of the Canary Islands has lost control over the arrival of the *Cayucos*, large fishing boats overloaded with migrants, and people are now attacking the central government for having failed to predict the situation. The Canaries are asking for extra help from Madrid and even the United Nations. But the national government, which between January and August 2006 has moved more than 11,000 immigrants to Madrid, Barcelona, or Malaga, also must justify its acts to local and regional governments on the mainland who are attacking it for dumping so many illegal immigrants in their cities.[10] In 2005, 1,882 migrants landed on the coasts of Malta, which is the equivalent, in terms of population, to the arrival of 270,000 people in France, since Malta has only 402,000 inhabitants. The two major parties have declared that national interest overrides human rights, especially since the Maltese feel that other countries could well absorb groups of immigrants who have landed on their island, or else organize repatriations with them. There are 1.5 million people in Libya who would gladly follow the migrants to Europe.[11] The Mayor of Lampedusa, Bruno Siragusa of *Forza Italia*, makes the same point: his island is a mere 20.2 square km and it has 5,515 inhabitants. But 2,100 refugees landed there during the period 1–16 July 2006,[12] a situation which demanded the dispatch of a European Union task force.

The conference on migration held in Rabat in July 2006, on the initiative of Morocco and Spain, had three goals: to mark out the new routes being used by illegal immigrants after Morocco started to reinforce its border controls; to strengthen cooperation with African police forces who will have databases and alert systems, and to elaborate development programmes to encourage immigrants to return and settle in their countries of origin.[13] The ministerial conference between the European Union and the African Union on 22 and 23 November 2006 was held in parallel with negotiations between Italy and Libya which will provide for an exchange: the Italians will build a hospital or the first part of a motorway along the coast between Benghazi and Tripoli as compensation for damages incurred by Libya under colonialism and during the war, while Libya will award lucrative contracts to the *Ente nazionale idrocarburi* (ENI), the Italian national hydrocarbon producer. But if Libya, a country of 5,850,000 inhabitants which has between 500,000 and 1 million illegal immigrants on its territory and some 500,000 legal immigrants has accepted that the EU border control agency FRONTEX will patrol its coastline (under Italian and Maltese command), it has also asked for the EU to help control its 3,000 km of border in the desert. The Italian Minister for

Infrastructure, Antonio di Pietro, has accused Libya of manipulating the immigrants in order to force Rome to agree to build the motorway.[14] More generally, the North African countries who say that they are 'overrun'[15] are starting to say, on the fringes of 'Euro-Mediterranean' international conferences, that it is necessary to go beyond administrative measures and to adopt instead a geopolitical perspective on the issue. Morocco, after all, formally applied to join the European Community in 1987. Why not admit the states of North Africa, since they are closer to Europe than Turkey is? If they were admitted as EU member states, they would be much more inclined to operate as the southern 'marches' of the West. Both Morocco and Tunisia have also declared war on the Muslim veil. Naturally gas and oil are in the background of all this.

Then there is Sub-Saharan Africa. Kofi Annan can imply as much as he likes that the nation-state is a thing of the past: it is precisely within the context of the tradition of the nation-state inherited from France that numerous intellectuals from Ivory Coast explain the measures taken to restrict immigrants into their own country in the 1990s. In this volume, Pierre Kipré, Christian Bouquet and Augustin Loada debate this fascinating point. In Abidjan, people feel that it is precisely France which ought to have been more understanding.

There may also be a series of 'diagonal' shocks, at the intersection of the usual North–South and East–West axes of migration. On a visit to North Africa, the German Foreign Minister, Frank-Walter Steinmeier, was made aware of a major concern: was Europe preparing to extend enlargement to include Ukraine and Georgia, two Christian countries, to the detriment of enlargement to the southern Mediterranean? His Moroccan opposite number insisted that the EU ought to have equal relations with Rabat as with Kiev.[16] The North Africans are well acquainted with the theory of 'migratory relays'. There are supposed to be two million Romanians in western Europe at the moment, above all in Italy and Spain: this comprises about 20 per cent of the economically active population of Romania. Half of them are illegal. This means that the exodus has already taken place. But those parts of eastern Europe which have been abandoned, like North-East and East Central Romania – 35 per cent of the population of Vrancea has left – are now receiving immigrants from Asia because citizens of Moldova and Ukraine, traditional sources of cheap labour, prefer to try their luck further west.[17] In the same way, the Baltic states (though to a lesser extent in Estonia) have been emptied of their well-educated and dynamic young people in a way that recalls the historic emigration from those territories in the time of the pogroms and of Tsarist Russification. The population of Latvia declined by 200,000 people between 2004 and 2006, going from 2.3 to 2.1 million. 350,000 Lithuanians, or 10 per cent of the population, have emigrated since independence in 1991. The gaps left in the labour market have been filled by Belarussians, Russians and Ukrainians.[18] Russia is itself a transit country for some two million foreigners a year.[19] Just as intra-European integration between 1950 and 1972 played a positive role in the 'emergence of cultural proximity',[20] so population movements from the East will also widen mental horizons.

As the former Iraqi Minister for Migrants and Displaced Persons, Pascale Warda, put it at the conference on 'Peoples in Migration in the Twenty-First Century', organized by the Daedalos Institute of Geopolitics in Nicosia, Cyprus, if the 175 million migrants in the world in 2000 were gathered together into one country, it would be the fifth largest country in the world.

Michel Korinman is Professor of Geopolitics at the Sorbonne in Paris, Director General of the Daedalos Institute of Geopolitics in Cyprus, and Editor-in-Chief of *Geopolitical Affairs*.

NOTES

1 See Jacques Barou, *Europe, terre d'immigration Flux migratoires et intégration* (Grenoble, Presses universitaires de Grenoble, 2006, 2nd edn), pp. 217–18; United Nations Population Division, *World Population Prospects: the 2000 Revision* (New York:, United Nations, 2001); Kofi Annan, 'Les migrants font avancer l'humanité',"*Le Monde*, 9 June 2006.
2 Jacques Attali, *A Short History of the Future* (Paris, Fayard, 2006), p. 203.
3 Maurice Schiff and Çaglar Özden (eds), *International Migration, Remittances and the Brain Drain* (Research Department of the World Bank, 2005) <http://econ.worldbank.org/programs/migration>; Ali Mansoor and Bryce Quillin (eds), *Migration and Remittances: Eastern Europe and the Former Soviet Union*, 2007 <Cneal1@worldbank.org>; *Un nouveau rapport de la Banque mondiale en faveur de l'équité* <www.genreenaction.net/spip.php?article3699>.
4 See Barou, *Europe, terre d'immigration*, p. 53.
5 Ibid., pp. 74, 81, 144. See also Laure Mandeville, 'Londres s'interroge sur le multiculturalisme', *Le Figaro*, 6 September 2006.
6 See Guillaume Perrault, 'Politoscope: "l'identité nationale" approuvée', *Le Figaro*, 16 March 2007.
7 See Richard Heuzé, 'Padoue construit un mur antidélinquants', *Le Figaro*, 15 August 2006.
8 See Barou, *Europe, terre d'immigration*, p. 50.
9 See Grazia Maria Mottola, 'Trasferiamo gli immigrati ma lo Stato deve aiutarci', *Corriere della Sera*, 5 September 2006. See also, on Bovezzo: Marco Imarisio, 'Nel Senegal di Brescia, "È un ghetto autogestito"', *Corriere della Sera*, 8 September 2006. See also the anti-gypsy ditch dug by the town centre-left council of Schio in the Veneto: 'Veneto, dopo il Muro un fossato anti nomadi', *Corriere della Sera*, 1 November 2006.
10 See 'El presidente canario pide un gabinete de crisis ante la llegada masiva de inmigrantes', *El País*, agencies, 17 August 2006; J.A.R., 'Coalición Canaria acusa al Gobierno de actuar "tarde y sin sentido de anticipación"', *El País*, 1 September 2006; J. M. Pardellas, 'El presidente canario affirma que la islas "son un embalse a punto de reventar"', *El País*, 6 September 2006; Thierry Oberlé, 'Clandestins: crise entre les Canaries et Madrid', *Le Figaro*, 25 August 2006.
11 See Stéphane Kovacs, 'Malte se sent envahie par les immigrés', *Le Figaro*, 20 June 2006.
12 See Heinz-Joachim Fischer, 'Immer vor Lampedusa Afrikanische "Bootsflüchtlinge" machen auch der Regierung Prodi zu schaffen', *Frankfurter Allgemeine Zeitung*, 3 August 2006.

13 See Thierry Oberlé, 'La conférence sur les migrations s'ouvre à Rabat', *Le Figaro*, 10 July 2006.
14 See Hans-Christian Rössler, 'Libyens grösstes Problem ist nicht die Küste', *Frankfurter Allgemeine Zeitung*, 1 September 2006 and Dino Martirano, 'Libia-Italia, "nuove basi per il negoziato"' *Corriere della Sera*, 24 November 2006.
15 'Marokko sieht sich "überrannt"', *Frankfurter Allgemeine Zeitung*, 29 September 2005.
16 See Johannes Leithäuser, 'Ein Gürtel von Kiew bis Rabat', *Frankfurter Allgemeine Zeitung*, 20 November 2006.
17 Cf. Arielle Thedrel, 'La Roumanie aussi accueille des immigrants', *Le Figaro*, 29 December 2006.
18 Cf. Robert von Lucius, 'Wie zu Beginn der sowjetischen Besatzung Junge Menschen verlassen in Scharen das Baltikum, um im Ausland ihr Glück zu suchen – die heimische Wirtschaft leidet', *Frankfurter Allgemeine Zeitung*, 28 July 2006.
19 See Barou, *Europe, terre d'immigration*, p. 89.
20 Ibid., p. 43.

PART I
The Problem

Migration and the New World Order

DEMETRIS CHRISTOFIAS

Migration issues today constitute thorny questions for our world and call upon us to respond effectively and without delay. They are matters of a particularly urgent nature that demand systematic study, in order to face them in a just and fruitful manner. Developments triggered in the last few years at world level by the establishment of the so-called 'New World Order' have created situations of exploitation all over the universe, as well as areas of tension and confrontation – whether it be national, religious, racial or economic.

Military aggression against peoples, devastation, wretchedness and occupation create insecurity and push peoples to look for a safer place in order to survive. At the same time, the leaders of this 'new world order', in their excessive complacency and arrogance, along with their various economic, military and political interests, attempt to dominate world developments, ignoring the intensifying inequalities between richer and poorer countries.

The resurgence of sources of tension and poverty, as a result of the overexploitation of some countries, has led to a dramatic reappearance of the mass movement of peoples from poorer to richer countries. We in the Mediterranean region experience this migration in a particular way, in that it has unfortunately led to the reoccurrence of racist incidents among various peoples and societies.

Responding to such phenomena constitutes one of the greatest challenges faced by today's societies. In this framework, it is imperative to shape a new social culture, which will integrate and not exclude, protect and not marginalize, our fellow human beings, who become victims of political, racial, religious, or nationalistic persecution and cleansing.

To face these problems we need to go back to their prime causes and to try to cure these causes. Migration is the symptom of a sick society and racism is a dangerously contagious disease. It is our duty to look for the causes of racism, whether economic, social, or political, and to deal with them.

Countering the reflex feeling of fear against the 'other', the 'foreign', the 'different', we must instead project the principles of hospitality, solidarity, humanity. Instead of the 'frightened and secluded society', we must convince our citizens of the need for an open, democratically structured state.

At the basis of our tradition and civilization lies our solidarity and humanity. Securing equal opportunities for each and every man and woman and safeguarding their rights civilizes society and in turn promotes civilization itself. We must come up with satisfactory answers and effective solutions against the phenomenon of migration, the upheaval that is generated in economic activity and labour relations by these population movements. Only thus can we hold back feelings of racism, xenophobic incidents and the existence of intolerant behaviour. This is the field in which our efforts must concentrate in order to respond to the needs of the new era, which constitutes a confirmed reality.

The twenty-first century will evolve into a period of severe social problems and dramatic population movements if we do not attend with the utmost of our strength to the problems of those peoples and countries that are scourged by wars, poverty, exploitation, social inequality, unjust allocation of wealth and a one-sided enjoyment of productive resources.

We must look for answers to the problems of migration, mainly in the implementation of financial aid, debt cancellation, and the economic and social development of the migrants' countries of origin, in the full knowledge that the economic model of globalized neo-liberal capitalism produces new poverty and greater inequalities, pushing these populations to abandon their countries.

The century which has just begun risks being characterized as an era of darkness and intolerance, if we do not hasten to fight poverty, illiteracy, exploitation, racism and prejudice. Facing problems connected with peoples' migration is the field *par excellence*, where the principles of humanity, solidarity, civilization and democracy will be judged in all societies. It is a field of struggle for all of us, a ground for collective action and joint effort.

Demetris Christofias is the President of the House of Representatives of Cyprus.

The New Logic of Migration in the Twenty-First Century

GÉRARD-FRANÇOIS DUMONT

Migrations have been a constant issue throughout the history of humanity. To mention but a few examples from the past centuries, when France revoked the Edict of Nantes, some cities in other western European countries grew richer thanks to the migrations that followed. Migrations were also the basis for population settlements in today's Americas and Australia; conversely, they account for the fact that Ireland still has fewer inhabitants today than in 1840.

These world-wide migrations are constantly redrawing cultural maps. The factors behind them are similar to those in previous centuries. However, profoundly new factors have also appeared. I have called these the 'new logic' of migration.[1] Today as in the past, there are four classic factors influencing population movements: politico-religious factors, economic factors, demographic factors and mixed factors, which will be analysed in succession.

Politico-Religious Factors

Political and religious factors have a dual nature, depending on whether they repel or attract. If one examines factors of repulsion, several types appear: international wars, domestic civil wars or conflicts, political decisions and liberticidal regimes. First of all, wars very often lead to exodus, like the wars in the former Yugoslavia in the 1990s.

Domestic wars or civil conflicts represent the second type of factor for repulsion. One cannot understand Catalonia's desire to gain autonomy in a newly democratic country if one does not take into account memories from the exodus (*la Retirada*) of hundreds of thousands of Catalans fleeing civil war in early 1939 after Franco's forces had seized Barcelona. In the Kurdish zones in Turkey, internal military operations have driven populations out or obliged them to leave for Europe, particularly for Germany.

Political decisions, through which populations are driven out of their national territory, are a third type of cause. For instance, the Convention signed on 23 June 1946 between the Belgian and Italian governments stipulated that the Italian government would try to send two thousand workers a week to Belgium.

At the national level, the attitude of Soviet authorities in 1974 can illustrate a factor for repulsion. Eager to remove the political opponents who had gained too much of a reputation to be sent back into penal servitude, they decided to deprive Alexander Solzhenitsyn of his Soviet nationality and send him to the West.

As the fourth type of repulsion, dictatorships typically cause the emigration of their nationals, either when the regime is installed, or later, once the poor democratic credentials of the new regimes have become apparent. During the forty years of its existence, the GDR (German Democratic Republic, East Germany) was a regime which pushed its own citizens out. The same applies to Castro's long-standing regime in Cuba.

Four types of political repulsion thus force people to leave their territories, which modifies the cultural geography of the planet.

In opposition to the territories where these forces of repulsion are at work, other countries are attractive because of the political decisions or conditions they offer. For example, Turkish emigration to Germany, particularly in the 1960s, was stimulated by the political will of successive German governments which were keen to sign migration treaties with Turkey. Since the 1990s, immigration from Belarus, Ukraine, Russia, or Central Asia to Germany can be explained by the fact that the fundamental 1949 law allows anyone of German extraction to settle[2] and acquire nationality after six months. In other territories, economic, tax or financial laws can attract populations.

Other political decisions, like the policy of allowing family regrouping, can attract populations, all the more so when texts allow for a wide interpretation. Thus in France, following a judgment passed by the Council of State, it was legal for the families of polygamous workers to join them between 1980 and 1993; this became one of the reasons for the increase in immigration from Sub-Saharan Africa.

Russia is another interesting case. Since 1991, Russia first tried to contain the waves of Russians coming from the former USSR republics that had become independent. It thus attempted to appear reluctant to welcome them, with very moderate success. The idea was to preserve a Russian presence in the countries of the Community of Independent States (CIS) and in the Baltic states. Since 2002, Russia has realized how quickly it was losing its population (around 700,000 inhabitants a year) and it has radically changed its policies: it has now decided on an attractive policy towards twenty million potential migrants, improving its capacities as a host country and even devising 'credits for emigration'.

When one takes a closer look at politico-religious factors, one sees that international migrations often combine repulsion and attraction. For example, Jewish immigration to Israel is both due to repulsive political factors in the country of origin (like Nasser's decisions in Egypt in 1956) and attractive factors, given Israel's desire to increase its Jewish population.

The political factors behind migrations, which have been at work throughout the history of humanity, will continue into tomorrow, because the political decisions and situations are likely to generate repulsive or attractive factors. A second set of factors lies in the economic field.

Economic Factors

Notorious economic imbalances exist between countries. Some know how to maximize the value of their potential and human capital, or have energy resources or precious minerals that allow them a *rentier* economy. Conversely, others do not know how to create the conditions for economic expansion or they cannot maximize their assets. Some people, when they notice these gaps and have very little hope of improvement in their country, emigrate in search of better standards of living. Geographically, these international migrations are of two main kinds: South–North or South–South.

One of the most interesting examples of South–North migrations from the 1960s to the 1980s is the Ivory Coast. During the entire period when its economic development was both remarkable and comparatively exceptional in the region, the Ivory Coast was a major destination for immigrants. Millions of other African nationals (and particularly Burkinabes) were welcomed in the Ivory Coast. When conditions later worsened, immigration slowly stopped.[3]

Economic migrations thus depend on the capacity of countries to create riches, on the variations of the incomes they derive from hydrocarbons (for example, Saudi Arabia or Libya) or on the workforce they need for building works or engineering works. This was the case when Berlin was transformed into the political capital of a reunified Germany.

Beside this first type of economic migration, which results from imbalances between territories, there is a second type, linked to 'technical migrations'. These migrations are the consequence of profound changes in economic structures: they modify the various labour markets of a given territory and can entail population movements. Rural emigration, which arose from the transfer from an essentially agricultural economy to an industrial one, concerned not only internal but also international migrations. Since the mid-1970s, a new type of economic migration has emerged, which I call industrial emigration:[4] It is the result of insufficient anticipation and of the rapid change from an industrial economy to an information society. Like the previous type, this emigration is mainly internal but can also be international – for instance, former workers in obsolete Polish industries went to Germany to find means of subsistence.

In addition to political and economic factors, demographic factors need to be taken into account.

Demographic Factors

The third type of classic factor arises from the demographic differentials that generate migrations. Thus, the low level of previous settlements facilitated immigration to North America, which became particularly intense in the nineteenth century.

Beside density differentials, mortality differentials can also have an influence. The best, though particularly tragic, example is Ireland in 1842, when terrifying mortality figures led to a considerable number of Irish people emigrating to America. This is why, as the introduction mentioned, Ireland still has fewer inhabitants than in 1840.

Age differentials can also be an attractive factor for younger workers who would expect their income to improve more quickly if they provided services to older and richer customers.

Mixed Factors

The borderlines between these three types of migration (political, economic and demographic) are not always clear. It is thus important to stress a fourth type: mixed factor migrations, which is to say migrations which are both political and economic, or economic and demographic, or political and demographic, or even political, demographic and economic all at once.

The American phrase which refers to migrations resulting from political and economic factors is 'migration for bread and freedom'. The case of Algeria is a perfect illustration of this: Algerian emigration, particularly since the oil backlash of the mid-1980s, is based on both.[5]

The second type of mixed migration is economic-demographic migration. Fairly old examples – the Germanic populations who emigrated to Eastern Europe, or the Polish workers who supported France, when that country was recovering from the First World War – and more recent ones – millions of immigrants working in the small Gulf States – can illustrate this.

A complete type of mixed migration results from three interconnecting causes: economic, demographic and political. One particularly telling example, because it occurred over a short period is that of the Moroccans who emigrated to the ex-Spanish Sahara after the Green March of 1975. These

Moroccans emigrated towards a land where their government offered them more profitable economic conditions. They settled on a vast[6] but sparsely populated territory where their arrival was not terribly disturbing, despite geopolitical opposition with Algeria. Politically, this migration gave Morocco a means of reinforcing its sovereignty over the ex-Spanish Sahara.

The four classic types of migration causes will inevitably still be at work in the future. Other factors, which I consider together under the new logic of migration and which result from our new era, also have an influence.[7]

Following from the last decades of the twentieth century, the twenty-first century is characterized by three processes: political globalization, internationalization and economic globalization. The definition that I suggest for each term highlights the usual mishmash use of the term 'globalization' and draws clear distinctions between them. It is thus a useful tool to understand their impact on migration.[8]

Political Globalization and Politically Encouraged Migrations

While the use of the term 'globalization' is generally wide and imprecise, there is a normative dimension to the term *political* globalization. I define political globalization as follows: *all the political processes that aim at setting up regional market organizations and/or a single world-wide market organization. Political decisions mean that markets are less segmented or heterogeneous because of national or regional borders.* Globalization is thus all the political decisions that aim to abolish political borders. It does not simply involve international processes – through decisions taken at the GATT (General Agreement on Tariffs and Trade), or today, at the WTO (World Trade Organization) – but also regional ones – the European Union, NAFTA (North American Free Trade Agreement), or the South American trade bloc Mercosur. Political decisions that aim to reduce the influence of borders – whether on goods, capitals, or individuals – inevitably result in new possibilities for the movement of populations.

First, the free movement of goods accelerates migrations because it puts populations at the heart of exchange networks that turn human mobility into an economic necessity. Similarly, countries that are members of the WTO must be involved in an open economy and abandon autarchic systems. When China joined the WTO, it opened its doors to the exchange of goods but it also opened up to international migrations, with westerners going to China to set up commercial activities.

Secondly, globalization is also financial and thus encourages migrations. Indeed, the prime concern of many migrants from the South is to send money to their families who have remained in their country. Financial globalization, which means for instance the end of exchange controls, makes it easier for immigrants to transfer money.[9]

Third, some decisions on globalization have a direct positive impact on the freedom of movement for populations, as in the European Union (for all residents, whether they are European nationals or not) where it has become a right. When the former Communist countries like Poland joined the European Union after the implosion of the Soviet Union, they linked their future to the logic of European globalization.

Practical modalities for the freedom of movement are likely to increase flows of population. For instance, the end of border controls following the Schengen Agreement makes migration easier.

Other political decisions that reduce the importance of borders result from national migration policies, in keeping with the logic of globalization. For instance, the reforms that several European countries undertook from the 1970s (1976 for France) gave legal immigrants (and immigrants whose papers had just been put in order) the right to bring their families over, which is another way of lowering borders and encouraging migrations.

As a consequence, globalization makes some migrations easier because the barriers that existed previously have been lowered or abolished. Its effects are linked to the impact of transports, made more versatile by internationalization.

Internationalization and Reticular Migrations

Another new logic of migration is linked to internationalization, which has become unexpectedly rapid since the 1980s. According to my definition, internationalization – the technical side of the too general term of globalization – is *the use of techniques and processes that reduce the space/time effect in the material, information and human exchanges between the territories on the planet.*

We can date the acceleration of internationalization fairly precisely, along two phases: the turn of the 1980s and the end of the 1990s. During the first phase, flights become shorter: the record for the fastest flight around the world on a commercial airline was broken in 1980 (37,124 km in 44 hours and 6 minutes), while in 1981, the Caravelle flew for the last time, the Airbus flew for the first time, and the French high-speed train (TGV) link between Paris and Lyon was opened. Contrast this with the technical difficulties of trading with, say, Vietnam in 1933: the flight between Paris and Saigon, with the new airline Air France which took over from the Far-East airline (inaugurated by Maurice Noguès in 1931), would leave on a Thursday to arrive on the Friday of the following week, after no fewer than sixteen stopovers. In 2003, the flight between Paris and Ho Chi Minh City lasts 12 hours and 35 minutes: it is fifteen times shorter. Such shorter trips obviously make trade, investments and thus entrepreneurial migrations easier.[10]

Internationalization makes migrations easier because it reduces the space/time effect. First, no one could have imagined a century ago that Sri Lankans who felt oppressed would be asking for political asylum in Switzerland twenty-four hours later. Today, thanks to the aeronautic revolution, all of this sounds commonplace, all the more so as the reduction of the space/time effect combines with a considerable reduction in transport fares. Internationalization, along with the globalization decisions that abolished some of the air monopolies, also generates migrations closer to home. For example, since the beginning of the twenty-first century, more and more British people have moved to France because low-cost airlines operate between London and Bergerac in the Perigord, Rodez in the Aveyron, or Limoges.[11]

The second phase of internationalization became possible with the extension of computers, which the forecasters of the early 1970s had not taken into account. In the 1990s with the Internet, emails and mobile phones, migration has become easier because of the growth of information, which is readily available. Everyone, whether they are thinking of emigrating or not, can instantly or very quickly benefit from information that enables them to make choices.

Cyberspace means instantaneous contacts: immigrants are not necessarily cut off from their original family whom they can contact at all times via email or mobiles, which are much faster means of communication than rather unreliable posts or telephonic communications via the rather outdated exchanges in the countries of the South.

More generally, the changes that internationalization brings about allow for the development of what I call 'reticulate migrations' – migrations based on the development of networks that partially abolish the notion of frontier and provide the context for a more flexible mobility.

Economic Globalization and Entrepreneurial Migrations

The third cause of the new logic of migration derives from economic globalization, a term which one should use only for the *praxis* of economic agents: I thus define economic globalization, in the strict sense, *as the actions undertaken by companies in order to respond to specific demands everywhere and without any discrimination based on time or price; in order to perform these actions, companies must implement world strategies which are adapted to the evolving context of globalization and internationalization.* Economic globalization comes from the fact that companies have been forced to implement world-wide strategies to satisfy their imperatives and their needs for results. Because of globalization and internationalization, it has become essential for companies to implement world-wide strategies, which give way to migrations on two levels. On the one hand, some migrations are linked to training – both initial and life-long – as distance learning does not entirely exclude real meetings during part of the course.

On the other hand, what I call 'entrepreneurial migrations' come from the fact that companies necessarily must think 'world-wide', even if they simultaneously need to respond to the local specific demands of their customers. Thus, companies organize the international migrations of some of their employees in order to create commercial subsidiaries, production companies, or joint ventures.

Besides the new logic of migration due to the processes of political globalization, internationalization and economic globalization, one should survey the intensity of the migrations that the twenty-first century might well have to face, given the factor of climate change.

Climate Change and Migrations

Though climatic migrations would not be a new event in the history of humanity, they could become a significant phenomenon in the twenty-first century, with an intensity that has not been witnessed for several thousands of years. If the average rise in temperature and sea level, predicted and alread being witnessed in some parts of the globe,[12] were to modify the living conditions in many territories, several types of migration could appear.

The first one that comes to mind concerns forced migrations, which would be linked to the rising sea level or its consequences. True, these consequences could be controlled – as is already the case in many countries such as Argentina, Bangladesh, the United States,[13] France, Japan and the Netherlands – but the costs involved in investments and in the maintenance of protective equipment would go on increasing. Existing dykes could be reinforced, new protective dykes could be built wherever necessary, or habitations could be designed which would be adapted to the new sea level. But not everything could be set up in all the areas involved. People who would want to live on dry land would have to migrate and these migrations could become international in many territories around the world.

A second type of climatic migration would happen in some territories where climate change would lead to temperature levels that would cease to correspond to the idea that some inhabitants would have of a good quality of life: it would thus be more of a voluntary migration. During seasonal hot periods, some people would migrate to other territories with less sun. This process would thus be a sort of negative heliotropism,[14] the reverse of the positive heliotropism that has been observed in various countries over the last decades.

Finally, climate change could lead to economic migrations towards territories that would become exploitable and habitable thanks to a major thaw and to the land or sea routes that would have been created by this thaw. Indeed, many territories in the northern regions of the Northern Hemisphere today are little exploited and inhabited, given the current climate. Their situation could change and generate climatic migrations.

Classic factors and the new logic of migrations combine and multiply the types of migration. Reticulate trends appear, on top of the radial migration trends.[15] This also engenders more and more complex trajectories, for example, migrations from Central Africa to Europe, which take people through several African countries and spaces of transit like Morocco,[16] or Libya.

In the future, all migrations would ideally follow from a deliberate decision. But, tomorrow as yesterday, it is very unlikely that it will be the case: many power-mad and prevaricating leaders very often disregard the principle that priority should be given to peace and development. It is likely that wars, civil conflicts and the existence of 'incapacitating states'[17] that compromise development will generate forced migrations in the twenty-first century, as was the case during the previous centuries.

However, migrations in the twenty-first century, whether voluntary or forced, will stand out because of their specific context, due to the processes of political globalization, internationalization and economic globalization, to which the effects of climate change might be added. Countries, regional organizations like the European Union, and international organizations must take these facts into account to prevent forced migrations and to allow voluntary migrations to take place within the logic of exchange and partnership that would serve development.

Gérard-François Dumont is Rector of the University Paris-IV (Sorbonne).

NOTES

1 Gérard-François Dumont, *Les migrations internationales* (Paris: Editions Sedes, 1995); Gérard-François Dumont, 'Les nouvelles logiques migratoires', Université de tous les savoirs', in Yves Michaud (ed.), *Qu'est-ce que la Globalisation?* (Paris: Editions Odile Jacob, 2004).
2 And not to return, because it was their ancestors who had left Germany.
3 Gérard-François Dumont, 'Les migrations internationales en Afrique', in Gabriel Wackermann, *L'Afrique* (Paris: Ellipses, 2003).
4 In France, for example, the only major urban centres that have lost population through emigration in the 1980s and 1990s are located in the former industrial basins like Lens, Béthune or Saint-

Étienne. Cf. Gérard-François Dumont, *La population de la France, des régions et des DOM-TOM* (Paris: Editions Ellipses, 2000).
5   When the President of the French Republic, Jacques Chirac, went to Algiers in the first half of 2003, young Algerians welcomed him with cries of joy: 'Visas! Visas!'
6   252,000 square km.
7   Dumont, Gérard-François, *Les populations du monde* (Paris, Editions Armand Colin, 2004).
8   Cf. also Gabriel Wackermann (ed.), *La mondialisation* (Paris, Ellipses, 2006).
9   Estimates for 2002 give the following proportions of sums that immigrants transferred to their families: 39 per cent of transfers from the US, 21 per cent from Saudi Arabia, 5 per cent from France.
10 Some, particularly from Europe to eastern Asia, follow from the collapse of the Soviet Union, as the Russian Federation decided to open (and charge for) more air corridors above its territory.
11 'La campagne française prend de plus en plus l'accent british', *Le Monde*, 11 July 2003, p. 10.
12 Cf. the Intergovernmental Panel on Climate Change created in 1986 by the World Meteorological Organisation (WMO) and the United Nations Environment Programme (UNEP), whose headquarters is in Geneva: <www.ipcc.ch>.
13 For example, the events in New Orleans in 2005; cf. Jean-Marc Zaninetti, 'Catastrophes naturelle et pauvreté: le cas de La Nouvelle-Orléans', *Population et Avenir*, 679 (September–October 2006).
14 Gabriel Wackermann (ed.), *Dictionnaire de Géographie* (Paris: Ellipses, 2005).
15 Gérard-François Dumont, 'Les grands courants migratoires dans le monde au début du XXIe siècle', in Jacques Dupâquier and Yves-Marie Laulan (eds), *Ces migrants qui changent la face de l'Europe* (Paris: L'Harmattan, 2004).
16 Mehdi Lahlou, 'Le Maroc et les migrations subsahariennes', *Population & Avenir*, 659 (September–October 2002).
17 To borrow the expression of the United Nations Development Programme; cf. PNUD, *Rapport mondial sur le développement humain 1995* (Paris: Economica, 1995), p. 128.

The Catholic Church and the 'Others': A Universalist Strategy

MANLIO GRAZIANO

'To him who knows not to what port he is bound, no wind is fair', wrote Seneca in his *Epistulae Morales*.[1] In spite of the confusions and questions raised by the state of international relations in the context of globalization, this maxim does not seem to apply to the Catholic Church.

General De Gaulle was convinced that the Catholic Church was 'half a century ahead of the greatest officials in the world'.[2] Even today, to understand the evolution of the process of cultural integration on the Old Continent, there are certainly more lessons to be learned from the tendencies set out by the Catholic Church than from those of world leaders. For the Catholic Church is half a century ahead not only as far as Europe is concerned, but also on the question of migration as a whole.

The Church's Long-Term View

Half a century has not yet elapsed since March 1963, when Pope John XXIII instituted in the Holy See a committee to study demography. Five years later, John's successor, Paul VI, drew on the work of this committee when he published his encyclical *Humanae Vitae,* devoted, as mentioned in its opening lines (*incipit*), to what is referred to in the original Latin as the 'very serious duty of transmitting human life'.[3] This encyclical has essentially been read as a doctrinal text against contraception and against abortion. At the time, it was considered by the secular world as the last remnant of a reactionary and sexist mindset. In reality, it was the first step in a firm and, in a way, perceptive, attitude toward the inevitable consequences of development.

One of the mistakes which is most currently made when dealing with the positions of the Church is to take them literally, to consider them as a circumstantial answer to circumstantial problems, as do politicians whose strategies and policies never go beyond the next elections.

Such an attitude ignores the fact that the Catholic Church does not have to bear the burden of democracy; that its time is neither the time of individuals, not that of other organizations in society. The Catholic Church has as much patience as a spider. The results of its work are to be measured in centuries, not in years or decades.

Let us give an example directly related to our subject: the strategic plan which is encompassed in the phrase 'Asia's Millennium'. In January 1995, in a public speech delivered to the faithful in Manila, Pope John Paul II declared, among other things, 'Just as in the first millennium the Cross was planted on the soil of Europe, and in the second on that of the Americas and Africa, we can pray that in the Third Christian Millennium a great harvest of faith will be reaped in this vast and vital continent [Asia].'[4]

When reading an 'agenda' with this kind of timescale, it should not be forgotten that during the first millennium, Christianity spread in Europe between the fourth and the eleventh centuries, and that in the second, Africa and America were not reached until the sixteenth century. When Karol Wojtyla talked about 'Asia's Millennium', he did not necessarily mean that this result could or had to be reached during the twenty-first century.

Prophetic Opposition

In this article, we will try to answer five questions. How is it that the Catholic Church has managed to be 'half a century ahead' of all the other officials in the world? Why is it the only international body to have a structural position on the migration issue, that is, a position which does not change according to circumstances/What is this position, in which attitudes is it expressed and what are its consequences?

First question: the Church is the only international institution to have a structural position on this question because it is the only international body which knows to what port it is bound, to use Seneca's words. It has what is called in political and military jargon 'a strategy': to plant the Cross in Asia in the third millennium. To implement a strategy, people must carry out tactics. The *sine qua non* for conquering Asia is then to strengthen its bases there; Europe was first (which largely explains the Church's unfailing Europeanism), the Americas, and then Africa. The new evangelization advocated by Karol Wojtyla consisted essentially in this priority of strengthening the Catholic Church's bases.

It is difficult to understand the positions of the Vatican if they are not put in this perspective: to carry out every possible means to strengthen the Catholic – and more generally Christian – identity on the three continents where the Cross has already been raised; and to carry out every possible means to thwart a weakening of Catholic – or Christian – citadels in Asia, which are, moving from west to east: Lebanon, Iraq, Israel, Syria, India, Korea, and even

China (the Philippines being, in this context, a special case).[5]

However, it is well known that in order to know where you want to go, a strategy alone is not enough. One also needs to know the wind, to be able in reality to implement one's strategy and to attain one's goals. The war in Iraq in 2003 corresponded to a well-defined strategy of the US's ruling classes. However, the way this strategy was implemented brought about results which were the opposite of those set out by the Bush administration.

For a strategy to be successful, it must be based upon the most accurate knowledge of the conditions in which it will be implemented. It is precisely in this field that the Catholic Church is ahead of any other world leaders. Indeed, not only does the Church have a two-thousand-year-old experience of human affairs – an advantage which is far from insignificant – but it also has two additional assets which almost all other social organizations lack: two centuries of experience in the struggle against capitalism, which allows it to know the strongest and weakest points of the capitalist societies; and its ability never to 'bend the knee to the world', to use Jacques Maritain's famous phrase. Before he became Pope, Joseph Ratzinger called this the 'duty' of 'prophetic opposition'.[6]

At Regensburg, on 12 September 2006, the Byzantine Emperor, Manuel II Paleologus was not the only issue on the agenda. Among other observations, Benedict XVI declared: 'The positive aspects of modernity are to be acknowledged unreservedly: we are all grateful for the marvellous possibilities that it has opened up for mankind and for the progress in humanity that has been granted to us.' But, he added, 'While we rejoice in the new possibilities open to humanity, we also see the dangers arising from these possibilities and we must ask ourselves how we can overcome them.'[7]

In his book *The Salt of the Earth*, published in 1997, Ratzinger was already considering this contradiction between the 'possibilities' and 'threats' of development: 'The Catholic Church can only be modern if it is anti-modern,' he wrote[8] – which is an form of dialectic of which Marx would have been proud. The Catholic Church's anti-modernity is evidenced in its ability to point out the flaws of modernity before anyone else (for esxample, 'the greatest officials of the world') even becomes aware of them.

This ability allows the Catholic Church to be one of the protagonists of modern times, a protagonist who can even afford to take the lead – because of its 'prophetic opposition' – of a disoriented modernity, like a scout who points out dangers and signals wrong tracks.

'Half a Century Ahead'

The Church's positions on abortion or contraception are simultaneously anti-modern and modern because they are answers to a crucial threat: the demographic crisis of developed countries. They are an integral part of the policy of supporting a high birth-rate which the Church had adopted even when there was huge enthusiasm for Malthus' theories about the danger of overpopulation: this was the case, for instance, when *Humanae vitae* was issued.

It does not matter if these positions are unpopular: Ratzinger wrote in 1997 that 'statistics are not one of God's criteria.'[9] He said, 'We are not a trading company which can use figures and opinions as units of measure: our policy has been having good results and the sales have increased.'[10] These battles have sometimes ended in bitter defeats – for instance when abortion was legalized in Italy in 1981 – but they encouraged the 'convinced minorities', as Ratzinger called them in 1997, to rally around a set of basic principles which clearly define a collective identity.

The battle against demographic decline must be considered as the foundation stone of the supporting wall of a structure built in order to save, and, if possible, strengthen Christian identity on those continents where the Cross has already been planted.

Only recently have 'the greatest officials in the world' begun to adopt policies supporting a rising birth rate. They came to the same conclusions as the Catholic Church had already had half a century earlier. *Quod erat demonstrandum.*

Migrations: Above all an Issue Concerning Catholics Themselves

Catholic hierarchies know perfectly well that the Church's struggles over principles are fundamental in order emphasize its positions and to consolidate its moral authority. But they also know that these principles have little influence on actual conditions. In this case, the hierarchies know perfectly well that opposition to abortion and contraception are not sufficient to reverse the demographic decline in Europe.

According to specialists on the Catholic Church, the declining birth-rate among the more prosperous classes in most countries is not the only consequence of modernity. There is also the phenomenon of migration which principally concerns those countries most affected by modernity, namely prosperous industrialized states.

All the Church's documents on migration – first and foremost of which is the Pope's 'pastoral instruction' *Erga migrantes caritas Christi* ('The love of Christ towards migrants') of 3 May 2004[11] – consider the migratory phenomenon as structural, as a 'clear indication of a social, economic and demographic imbalance on a regional or worldwide level'. But what is the Catholic Church's real attitude to this phenomenon?

Official documents deliver little information on the subject. The above-mentioned directive says that Catholics 'are struggling for the rights of migrants, forced or voluntary, and for their defence', that they involve themselves in favour of the 'establishment of educational and pastoral systems that educate people in a 'global dimension', that is, 'a new vision of the world community, considered as a family of peoples for whom the goods of the earth are ultimately destined when things are seen from the perspective of the universal common good'.

It is clear that these are very general propositions, with no real practical consequences by themselves. But what emerges from these official documents is the fact that the Catholic Church's priority is to look after Christian migrants. The document quoted above, written by the Pontifical Council for the Pastoral Care of Migrants and Itinerant People, is divided into 104 points. Of these, one-third (thirty-four points) deal with internal organizational matters; eleven describe the phenomenon of migration; twenty-two (one-fifth) deal with doctrine and canonical standards on the subject; twenty-two deal with Catholic migrants of all rites; three deal with non-Catholic Christians; eleven (10.5 per cent of the document) deal with non-Christian migrants, among which four deal with Muslims (3.5 per cent of the document); and more than a dozen points address obscure legal and pastoral matters which are incomprehensible to those who are not part of the ecclesiastical structures.

Here are the Catholic Church's official priorities on the subject, which are not shared (it must be repeated) either by the 'other greatest officials in the world', or by many members of the flock who are unable to see the structural coexistence of the possibilities and the threats generated by modernity, and who would like to enjoy the former while being protected from the latter.

At every level, when the Catholic Church expresses itself, people in Western countries are often only interested in the Church's attitude towards Islam and the Muslim world. But this tendency to ignore everything else tends to give a partial and biased idea of the Catholic Church's position. From the lecture at Regensburg, only the quotation of the Byzantine emperor was emphasized; from the Pope's trip to Turkey, only the relationships between Christians and Muslims and the attitude of the Catholic Church towards the negotiations between Brussels and Ankara held people's attention. People think that the Church is interested only in the problems posed by the integration of Muslims into even secular societies which, in this case, are rediscovering their Christian roots.

The Dialectic of Witness and Respect

Of course the problem of the relationships with the Muslim world is also an issue for the Catholic Church, but it is not its main problem.

The directive *Erga migrantes caritas Christi* tackles the issue but leaves it to the national bishops' conferences to interpret the general lines it sets out. These broad lines (§59) are calling for dialogue: 'The Church is called upon to open a dialogue with them [non Christian immigrants]'; but quoting the encyclical *Redemptoris Missio* of December 1990 (§55) and the post-synodal apostolic exhortation *Pastores Regis* of October 2003 (§68), both written by Pope John Paul II, it is also specified that this dialogue 'should be conducted and implemented in the conviction that the Church is the ordinary means of salvation and that she alone possesses the fullness of the means of salvation'.[12]

In a document of 29 June 2001,[13] the Italian Bishops' Conference defined far more clearly the position which Catholics should adopt in 'our more and more multi-ethnic and multi-religious society' The Italian Bishops advocated an 'original missionary task' of 'evangelising the people settled among us by current migrations'. The document continued: 'In a way, we are called to accomplish the mission *ad gentes* here, on our planet. With all the respect and attention we owe to their culture, we must be able to bear witness to the Gospel and, if it is God's own Will and if they wish, tell them the Word of God.'

Here is the Church's policy on the subject: to welcome the immigrants with respect, which does not mean only keeping Catholic identity intact, but also being proud of it, bearing open witness to it. It calls on Catholic not to renounce their duty to convert, while of course accepting that it will be difficult to convert all immigrants.

Such instructions can be followed both by those who emphasize the Christian's duty to bear witness and by those who lay the emphasis on respect for other cultures. The Church's cleverness and strength lies in its ability to have a single firm line yet to be flexible enough to offer it in a great variety of forms. After all, this is a Church which has remained faithful to her true self while keeping within its ranks military chaplains and pacifist vicars, aristocratic bishops and worker-priests, Opus Dei and Caritas.

Reciprocity

The last issue which helps to understand the attitude of the Catholic Church towards the presence of non-Christian immigrants on Christian continents is that of 'reciprocity'. This theme has been raised many times by hierarchies (for example, the meeting of the Italian Episcopal Conference in February 1989), when the difficulties encountered by Christians in non-Christian countries are raised. In particular, the problem lies with Muslim countries where religious proselytism is either illegal or frowned upon. Reciprocity then implies that if in Christian countries, Muslims are free to practice their own religion and even to proselytize, then Christians should have the same rights in countries with a Muslim majority.

In May 2006, on the 150th anniversary of the foundation of 'Oeuvre d'Orient', a French organization which assists Eastern Catholic Churches, bitter notice was taken of the increasing numbers of Christians leaving the Middle-East, all the more so because of conflicts in Lebanon, in Palestine, and now Iraq. Other elements explain this exodus, including a lower birth-rate among the Christians who usually belong to the better-off classes of the population, and an Islamization of society that some non-religious governments authorize in order to cut the ground from under the feet of radical religious movements[14]. For example, in Bethlehem, in which all these circumstances are combined, the percentage of Christians in the population has decreased from 75 per cent to 25 per cent in the last fifty years.

But the issue of 'reciprocity' is a double-edged sword. On the one hand, as we have said, it is a call for more freedom for Christians who are living in Muslim countries; on the other, if the demand for reciprocity is not met, then it could be exploited by some for retaliation, or the threat of it, against Muslims living in countries with a Christian majority.

The Catholic Church refuses to assent to the latter interpretation but the choice of the word itself, and the positions of some groups and trends of opinion supporting the idea that Islam has to be 'contained', leaves this last hypothesis open.

Christian Europe

In conclusion, the position of the Catholic Church's hierarchy on the issue of non-Christian immigration is based on three main themes which cannot be separated:

> 1. an uncompromising defence of Christian identity and its specificity against every other religious or secular identity;
> 2. the firm belief that migration is an inevitable consequence of unequal economic development;
> 3. the need to avert at all costs a conflict between Christian Europe and the Muslims settled on its soil, if only to avoid retaliation against Christian minorities in Muslim countries.

On these positions, the Catholic Church can hope to be the cement of a European identity. It is precisely what Benedict XVI said in the homily delivered on 10 September 2006 at the Neue Messe esplanade in Munich:

> People in Africa and Asia admire the scientific and technical prowess of the West, but they are frightened by a form of rationality which totally excludes God from man's vision, as if this were the highest form of reason, and one to be taught to their cultures too. They do not see the real threat to their identity in the Christian faith, but in the contempt for God and the cynicism that considers mockery of the sacred to be an exercise of freedom and that holds up utility as the supreme criterion for the future of scientific research.[15]

Put in a nutshell, according to the Pope, the 'clash of civilizations' can be avoided 'only if faith in God is reborn, if God becomes once more present to us and in us'.[16]

Those who are worried about the social and cultural consequences of immigration must be prepared, whether they like it or not, to put themselves under the direction of the Catholic Church.

(Translated from the French by Virginie Roiron)

Manlio Graziano is a lecturer at the École Supérieure des Relations Internationales, Catholic University, in Lyon.

NOTES

1. 'Ignoranti quem portum petat nullus ventus suus est', *Epistulae Morales*, VIII, 71, 3.
2. Interview with the former Archbishop of Paris, Jean-Marie Lustiger, *Le Monde*, 23 April 2005.
3. The English version of the encyclical translates 'munus' (duty) as 'role'.
4. John Paul II, Address to the Sixth Plenary Assembly of the Federation of Asian Bishops' Conferences (FABC), Manila (15 January 1995), 11: *Insegnamenti* XVIII, 1 (1995), 159, in <http://www.vatican.va/holy_father/john_paul_ii/apost_exhortations/documents/hf_jp-ii_exh_06111999_ecclesia-in-asia_en. -html>.
5. It should not be forgotten that, after the Philippines, India is the Asian country with the largest Catholic community (thriteen million) among a total Christian population of forty million. Although such statistics are not available in the case of China, it is generally agreed that the total number of Catholics of both Churches amounts to ten million there (the total number of Christians amounting to about one hundred million according to some sources). In South Korea, there are four million Catholics: this is the largest community as a proportion of the whole population (about forty-eight million), after that of Lebanon.
6. Joseph Cardinal Ratzinger, *Salt of the Earth: The Church at the End of the Millennium: An Interview With Peter Seewald* (San Francisco, CA: Ignatius Press, 1997). The references in this article are to the Italian edition: Joseph Ratzinger, *Il sale della terra. Cristianesimo e Chiesa cattolica nella svolta del millennio* (Cinisello Balsamo: Edizione San Paolo, 1997).
7. Meeting with the Representatives of Science, Lecture of the Holy Father, Aula Magna of the University of Regensburg, Tuesday, 12 September 2006 <www.vatican.va>.
8. Ratzinger, *Il sale della terra*, p. 271.
9. Ibid., p. 17.

10 Ibid.
11 This document was written by the Pontifical Council for the Pastoral Care of Migrants and Itinerant People <http://www.vatican.va/roman_curia/pontifical_councils/migrants/documents/rc_pc_migrants_doc_20040514_erga-migrantes-caritas-christi_en.html>.
12 Paragraph 61 also indicates that 'it is not considered opportune' to allow the members of non-Christian religions to use Catholic places of worship for their own religious practices, and 'still less ... to obtain recognition of demands made on the public authorities'.
13 Comunicare il Vangelo in un mondo che cambia. Orientamenti pastorali dell'Episcopato italiano per il primo decennio del 2000.
14 Cf. Pierre Prier, 'Le grand exode des Chrétiens d'Orient', *Le Figaro*, 13 May 2006.
15 Homily, Outdoor site of the Neue Messe, Munich, Sunday, 10 September 2006.
16 Ibid.

Farewell Italy: Demography and Immigration

GIUSEPPE SACCO

> Si chiamava
> Moamed Sceab
> discendente
> di emiri di nomadi
> suicida
> perché non aveva più
> Patria
>
> Amò la Francia
> e mutò nome
>
> Fu Marcel
> ma non era Francese
> e non sapeva più
> vivere
> nella tenda dei suoi
> dove si ascolta la cantilena
> del Corano
> gustando un caffè.
>
> Giuseppe Ungaretti, *In memoria*

('He was called Mohammed Sceab, the descendant of emirs and nomads, who committed suicide because he no longer had a country to call his own. He loved France and changed his name. He was Marcel but he was not French and he no longer knew how to live in his people's tent, listening to songs from the Koran and drinking coffee.')

"Es wurden Arbeitskräfte gerufen, aber es kamen Menschen."

Max Frisch
('A workforce was called in but instead it was people who came.')

There was, in the 1960s, much talk of 'revolution', in Italy no less than in France and the US. All in all, seen with the benefit of hindsight, most of it can be considered more a case of collective infantilism that an explosion of extremism. Still, in Italy the turmoil dragged on from 1968 until 1978, producing a wave of still unexplained terrorism and reaching its apex and conclusion with the killing of Aldo Moro.

In political terms, such a 'revolution' failed completely, but some of the consequences of that attempt are unexpectedly ravaging Italy in the new century. For something revolutionary did happen, and it took the form of a radical change in the behaviour – in particular in the sexual behaviour – of subsequent generations, above all among the women of Italy. Although this gave rise to long-overdue advances in the professional and social equality of women which are now irrevocable, it also brought about a change in reproductive habits that – along with the invention of the 'pill' – has rapidly turned out to be incompatible with a satisfactory population renewal cycle.

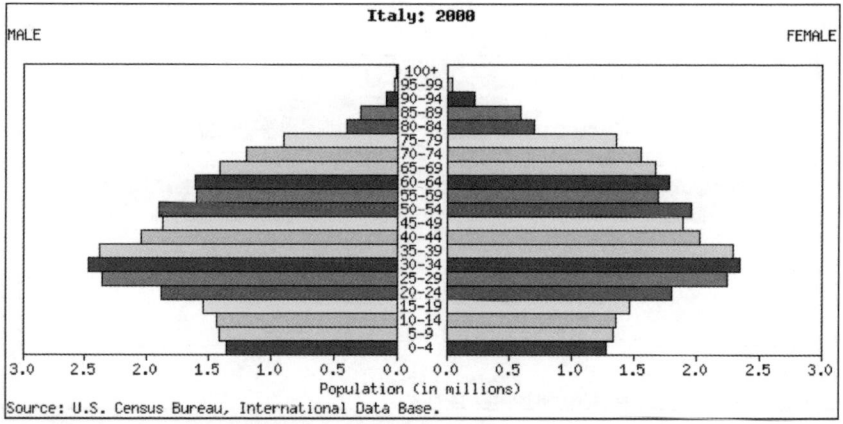

Figure 1

In fact, for the peninsula's demography, the birth-rate peaked in the middle of 1960 and then went into uncontrolled decline, which appears bound to continue in the coming decades as well. Today, fertility is around 1.3 children per woman: a rate which is not compatible with the maintenance of a minimum balance between age groups and thus between the classes of the population that are of the age to work and create wealth, and those classes of an advanced age which need the support of society.

Italy has been one of the first developed countries in which the number of births and deaths have balanced each other, resulting in zero population growth. Consequently, since the 1970s, another phenomenon which is fully under way, that is, the progressive increase in length of life expectancy, has weakened natural dynamics still further. The results can be seen in the following graphs.

From Figure 1 it can be seen that in the year 2000 the adult sectors of the population in the Italian peninsula were greatly in the majority, compared with the younger sectors. It is thus easy to understand why such a decrease in the birth-rate is the phenomenon that has characterized society since the second half of the fatal 1970s. Indeed, demographic decline has been so strong that the outline of this graph, usually called an 'age pyramid' no longer looks like a pyramid, but rather a rough rhomboid. Moreover, the future prospects seems even more negative. In Figure 2, which illustrates the forecasts for 2025, the base of the pyramid has now shrunk and appears to be an inherently unstable construction.

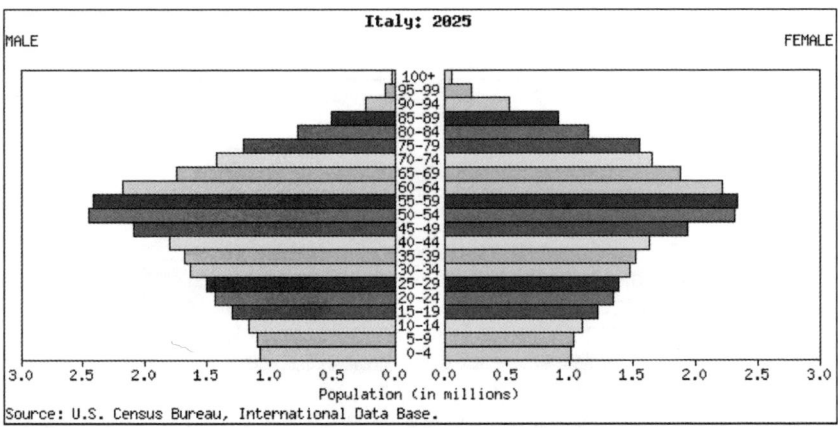

Figure 2

In addition to the many groups over 65 – the traditional age for retirement, and which, because they no longer generate earnings, are a burden upon the younger generations – there can also be seen the formation of a vast stratum of men and women over 85. Above all should be noted the appearance of a category which for the first time has statistical significance: women who are over 100 years of age. As well as being unable to contribute in any way to the collective good, these people are as a rule not self-sufficient and need personal care and assistance.

Projecting the picture further forward until 2050, in Figure 3, it then seems clear that Italy must expect a further decline in births and a growing increase in the number of people over 85. The situation promises to become completely unsustainable.

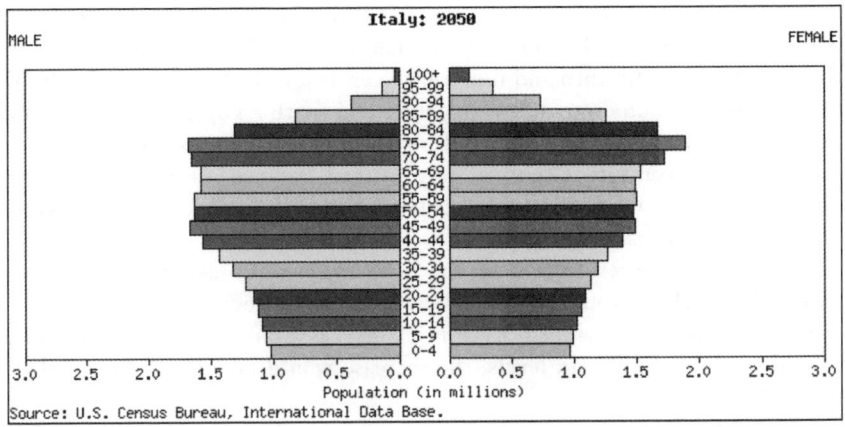

Figure 3

Looking at these three graphs in succession in order to obtain a dynamic picture, the impression is that the very plentiful post-war generations, the so-called 'baby boom' generations, have moved upwards through the whole demographic structure, like an embolism, until they have reached its head. What should be a pyramid has at this point taken on the form of a moneybox. But, on the contrary, the outline of Figure 3 effectively demonstrates that in the next few years there will not be enough resources to deal with the needs of broad sections of an extremely old population. In practice, this had already started to become evident in 2006, when the most numerous of all the generations alive, that born in 1946, reached 60 years of age.

A Rise in the Birth-Rate?

For two or three years there has frequently been talk in Italy of a rise in the birth rate. It is said that the Italians have been having more children for several years now. This is not entirely untrue. A certain change in outlook can be seen, namely in more educated and progressive women; and the female fertility rate went from 1.26 children per woman in 2003 to 1.33 in 2004.

Even if still a long way from the 2.1 rate necessary to safeguard reproduction of the species, this is undeniably an encouraging sign.

However, it would be an illusion to take this slight rise as evidence of a reversal in the trend which might generate its own answer to the problem. This would be an even more serious delusion than the well-known one Napoleon I wished to spread around war-decimated France, that is, that the French would be capable, in one night, to fill the gaps left by his battles. Our delusion would indeed be more serious than the French emperor's, because at the cusp of the eighteenth and nineteenth centuries, the female population of France was very numerous and very young and (with a population of more or less half that of today) the number of births in France was double its present-day figure. In twenty-first-century Italy, instead, because of the overall ageing of the population, the percentage of potentially fertile women –those between 18 and 35 – has drastically reduced.

Thus, a possible resumption in the birth-rate would take place on a very narrow base. There would be a very few women to produce more children if in reality this were to happen. Even if fertility, as a pure working hypothesis, should reach Third World levels, the increase would take much longer than the decline has.

The unlikelihood of a possible resumption is made greater by the fact that the rhythm of demographic events in Italy has been greatly slowed down by the propensity of women – who are by now established in the working world – increasingly to delay the age of maternity, to the point where the span of one Italian generation is today double what it is, for example, in Colombia.

Finally, it is necessary to show clearly how the fact that in 2004, for the first time in many years, the number of births was not much lower than the number of deaths does not indicate any reversal of the trend. In 2004, there were not many more births than usual, but only far fewer deaths. And this abnormally low mortality resulted from the fact that the terrible heat wave in August 2003 had killed many old people who in normal climatic conditions would probably have lived some months longer and died in 2004 instead.

Immigration as a Corrective

Given that one ought not to count on a resumption in the birth-rate as a corrective to the demographic situation in Italy, or at least not expect anything of it in terms that might be significant enough to avoid a complete collapse of the socio-economic system, the alternative that is generally considered is immigration. But this raises other questions which are not easy to answer.

The flows of immigration to Italy from abroad have caused a very real upheaval in a quite short space of time. In the 1970s, Italian emigration to other countries dried up, and was replaced by a reverse population flow. Movement into the country – previously limited only to an economic and social élite – became a mass phenomenon.

In fact it was a very different form of immigration from the one observed before, when enriched English and German politicians and public intellectuals followed the fashion of buying sumptuously restored farmhouses amid the vineyards of Chianti, so they could play 'gentleman farmer'. The phenomenon was so widespread in the SPD in Germany that there could be found there a parliamentary group ironically baptized 'the Tuscan Faction', just as the area where the London political elite likes to transfer its social life in the summer months has been branded 'Chiantishire'.

In the last two decades of the twentieth century, immigration into Italy became a very different phenomenon, as it was above all derived from economic factors. And indeed its effects are to be seen in the economic field rather in Italy's demography. In practice, it consisted of a spontaneous flow of labourers attracted by job possibilities in the peninsula's economy and – as was to be expected – it ended up by filling those scarcities that existed in the workforce, but not the gaps in the population which resulted from Italy's insufficient birth-rate. And as there was no sign of any policy aimed at attracting an inflow of men and women who might rescue Italy from demographic catastrophe, economic factors proved unable to trigger any substantial change in population trends.

Contrary to general belief, migrations are not a simple phenomenon made up of interconnected pots from which the flow tends to balance differences in demographic density or wealth. They are a complex sociological phenomenon where cultural and political factors come into play. The immigrant is neither a thief who seeks to steal jobs from Europeans, as is believed by the most fearful and extremist of the latter, nor – as portrayed in the gutter press – is the immigrant an invader who wants to replay in the reverse the past phenomenon of European colonization of the Third World.

In fact, from an economic point of view, the immigrant today no longer belongs to the most destitute sections of the Third World population. Rather, he or she is someone who has already absorbed a certain number of western skills, and is motivated by their own aspiration to a better economic situation, and by the aspiration of their enlarged family, who have contributed to finding the necessary resources to pay for the immigrant's passage. Furthermore, from the political and cultural perspective, the immigrant is a person who has to some extent rejected the ideals of the society in which they were born. This attitude is not to be confused with that of the second and third generation, that is, of the immigrant's children and grandchildren who, if they grow up in a western society, and face all the difficulties related to the social status of their parents, end up idealizing their culture of origin, and risk becoming suburban thugs or members of some extreme Islamist group.

The immigrant himself, the man – and, more and more frequently, the woman – who takes the decision to abandon their archaic society in order to come and live in a very different and much more secular one, is a semi-westernized person, already seduced by one of the myths of the 'globalized' world, the myth according to which cultural differences have been greatly reduced or are even in the process of disappearing altogether, and according to which national identities have been replaced by the homogenized tastes of people who are no longer citizens (or subjects) of various countries, but simply 'global' consumers. In short, the phenomenon of modern migration is very different from the barbarian invasions of the past, or the English colonization of North America, which resulted in the acquisition of a territory being taken away from a demographically weak population by one which was more numerous and prolific.

Thus immigration into Italy, as a spontaneous, market-driven phenomenon, seems unable to meet the demographic needs of the country. For this to happen it would be necessary for the government to have a policy on population; but this does not exist. The authorities limit their activity in this field – crucial as it clearly is – to assessing shortages (in quality as in quantity) in the labour market, in order to set annual immigration quotas – quotas which are always greatly exceeded anyway.

The most serious workforce shortages – as becomes evident year after year – are in a quite specific sector of the labour market: those jobs that Italians refuse to do. This is clearly indicated by the fact that about 2.7 million immigrants of working age find work in Italy while almost 1.5 million Italians who are 'in the prime of life' are officially unemployed.

There are, however, some more indisputable facts which must be taken into account if one wishes to draw a more realistic picture. First, to any official figures for the Italian economy one must always add 15 per cent to cover the 'black' market. Secondly, the sectors in which 'irregular workers' are most frequently found are often made up of very labour-intensive activities. Finally, the number of Italians who have two or three jobs is far more than the number of the truly unemployed. One thus arrives at the conclusion that unemployment in Italy could be expressed by a negative number. Which is to say that – in spite of the weight of some 'disaster zones' such Naples, Palermo, or Calabria – the overall national level of unemployment is below zero. Within this framework, immigrants have as their function (and their aim) only that of taking up work that Italians refuse to do. Indeed, indispensable as they are, they cannot take the place of Italians in the economic domain, and even less so in the demographic one.

Until now the labour market's need for immigrants has been quantitatively more important than the deficit in the population's birth-rate. That is, in absolute terms, the positive balance in migration is greater than the negative balance in natural dynamics. The total population of Italy has continued to grow in the recent years because the immigrants who have come to work in the peninsula have been more numerous than the 'unborn' Italian children. It is possible that this will last for some more time, since the potential for influx is still considerable, as is evidenced by the constant pressure from the entrepreneurial sector for a relaxation of the bureaucratic restrictions on entry to the country. But there is no guarantee that immigration can continue to compensate for the decline in population. Indeed, the two phenomena simply coincide in time with one another, while remaining fundamentally different and only marginally related.

The Limits of Immigration

According to projections based upon UN data, any strategy aimed at having recourse to immigration in order to fill in the gaps left by the decline of Italian population could bring about a complete collapse of the situation, and, in reality, the end of Italy. Everything revolves around the level of quantitative relations between the non-productive age groups. It is indeed estimated that the ratio between the population of working-age people and the over-65s could increase from 3.94:1 in 1999 to 1.73:1 in 2049. In other words, while there are today about four people of working age for every person over 65, in 2049 this ratio will have been more than halved. In order to compensate for the falling birth-rate, the population of Italy would in effect have to reach the levels indicated in the following table.

As can be seen, these forecasts involve extremely large flows. Merely to maintain the Italian population at the quantitative level equal to the last year of the twentieth century – and without paying heed to its internal composition, average age and the ratio between active and inactive members – a yearly influx of a total of 235,000 immigrants would be required for fifty years in succession.

When one thinks only of one year, this is not a huge figure. In the end, it corresponds more or less to the number of foreigners who have entered Italy annually over the last few years whether within the regulations or not.

However, if one looks at the internal composition of the Italian population, rather than at absolute numbers, and if the conditions which at present allow the welfare state to function are to be left unchanged – that is, a ratio between active and inactive members of approximately 4:1 – it would be necessary to have an influx of immigrants of around 2.2 million per year for almost half a century without interruption. And it would be necessary, of course, to create

Table 1

	Without immigration	Flows needed to maintain constant		
		population	working-age population	active:non-active ratio
Immigrants 1995/2050		13,000,000	19,160,000	119,000,000
Total Population 2050	40,700,000	57,300,000	66,400,000	193,000,000
Population of working age.	21,600,000	33,000,000	39,000,000	126,000,000
Immigrants as a percentage of the total population		29%	39%	79%
Ratio between active and inactive members of the population	1.52:1	2.3:1	2.25:1	4.00:1

Source: UN Population Division, *World Population Prospects* (New York: United Nations, 1999).

the number of jobs which meet the needs of this new component of the Italian population. By the year 2050, the population of Italy would amount to 190 million as opposed to the 60 million today. Of these 190 million, only about 40 million would be of Italian 'stock' while 150 million would be of immigrant origin. And about 100 million new jobs would have to be created because the active population should increase from about 20 million to 120 million approximately.

These data speak for themselves, and do not need much comment. They demonstrate clearly enough that in the present state of things, immigration is able to maintain the peninsula's total population, looked at as crude data, at a little more than constant, but it cannot rectify its rapid and growing ageing. In fact, one should not overestimate the contribution that immigrants may be able to make to the birth-rate. One should not count on female immigrants maintaining the characteristic birth-rates of their country of origin once they are settled in Italy. On the contrary, the example of the local population, and the conditions of life in the host country, cause them rapidly to adopt local customs in this respect. The fertility rate of foreign-born women is indeed around double that of Italian women, and slightly over that of foreign-born women in France, but still too low to compensate for the demographic decay of the country.

Such a substantially equivalent propensity to have children is apparently illogical, as foreign-born women in Italy have been exposed to Italian low-fertility culture for a rather short period, and are on the average younger than foreign-born women in France, where the great immigration waves have taken place around twenty years before migrations into Italy even began. But it can be explained with the fact that a great number of female immigrants to the Italian peninsula originate from eastern Europe, where fertility rates are very low. Moreover, a non-negligible share of Albanian and Romanian immigrant women are trapped before they leave the home country into rings that exploit them as prostitutes, while women from Muslim countries – more numerous in France than in Italy – tend to marry and have children.

It is clear that in reality an influx of immigrants such as that required by the UN forecast to maintain the 4:1 ratio between the active and inactive population has no chance of happening. Probably, long before the level of immigration reached the consistency necessary to bring the peninsula's population to the level of the estimates, or even to half of the latter, rejection of immigration would reach crisis proportions; the present favourable attitudes would be replaced by a policy of closing the doors, or at least very selective choice of immigrants and a very rigorous admission into the country.

Moreover, it is easy to foresee that such an attitude of rejection would affect not only the population of Italian stock – probably supported by the governments and public opinion of neighbouring countries, which would regard Italy as a bomb ready to explode in their backyard – but also by those strata of the population originating in the first waves of immigration. In fact, it is quite clear that if it continued for years, a phenomenon as massive as that forecast by the UN would end up jeopardizing whatever the first immigrants had gained in terms of affluence and quality of life. This would be the result not only of the 'political' fear that the rejection crisis would also eventually affect those immigrants who at the beginning had been accepted without hostility, but also of the 'economic' fear of strong demographic pressure on Italian resources. Notwithstanding the 'clearing-out' effect of the quantitative decline in the population of Italian 'stock', these resources are definitely limited, whether in terms of housing, infrastructure or simply space.

Public Catastrophe, Private Tragedy

Thus, according to the most realistic predictions possible, one may conclude that there is no 'demographic' solution to the Italian crisis, but only a 'political' solution, in particular through reforms to the pension laws which would prolong the working age until 75. This would postpone the problem until about 2020. But beyond that date, one can very probably foresee a situation in which it will no longer be possible to maintain a decent level of pensions and an acceptable quality of standard 'welfare state' provisions. In other words, in a short time from now, Italian society will be incapable of feeding its numerous old people and of guaranteeing the medical care necessary to maintain them in good health.

In this, however, Italy will be a situation not very dissimilar to that in all European countries, including France. In fact, even if, in France, births stay slightly above of two children per woman, and thus correspond more or less to the reproductive rate needed to maintain the generations, such a rate is still completely insufficient to offset the destabilizing effects of longer life expectancy, a recent phenomenon which is of concern to France as much as to Italy. If, leaving aside immigration, it is predictable that the ratio between the active and inactive in Italy will go – by mid-century – from 4:1 to a little more than 1.5:1, in France it can be predicted this will be about 2:1.

Thus there is an imminent danger that in Italy the burden of taking care of the old will become too heavy for the government, and will largely come to fall upon the family and upon religious institutions, as used to happen before the introduction of the welfare state. The consequences could be dramatic since, once there has been a drastic reduction in expenditure on pensions and medical care, the problem of insufficient numbers in the generations of working-age relative to those who do not work will no longer be a collective problem for Italian society. It will be a problem for the family and individuals. In 2025, couples who today have one child – not to mention those who have none – will find themselves in an unsustainable situation, and the problem will inevitably be made worse by the fact that only 50 per cent of present-day couples, legal or *de facto*, will reach old age without separating, that is, without aggravating the complex problem of aid for old people living alone.

Consequently only a very small number of Italians who are today nearing old age will have enough offspring to be able to rely on the help of their own children. The severe planning put into practice by the 'baby-boomers' who launched the failed revolution of 1968 now risks turning against them, leaving them bare, alone, devoid of strength and help in the face of a society which appears to find it impossible to guarantee the goods and services which in the welfare state all Italians have grown accustomed to considering their personal right.

Without any doubt, it is possible to consider the hypothetical situation in which the excessive number of old people is brutally solved by a kind of 'revolt of the slaves' and a redistribution of power and resources from the 'old' to the 'new' Italians. Within a few decades the 'old' Italians – those who make up the Italian population in the strict sense – will be too weak and exhausted, too dependent on the support, including physical support, of their nurses and immigrant domestics, in any way to oppose the rebalancing of the economic and social security system to their disadvantage. This might mean an acceleration of the death-rate, and a drastic reduction of the 'burden' represented by decrepit 'old Italians'. The process might not be smooth, of course, as there will surely be one – or numerous – communities of 'new Italians' who will become active politically and claim their right not only to get rid of their defenceless hosts, but also to prevent new arrivals from abroad: a phenomenon that might be very similar to American 'nativism' at the end of the nineteenth century, when those who had become 'natives ' of the North American continent opposed new waves of immigrants. In the New World as well, those who had become 'natives' were not of true 'American stock'. They were British, Dutch and Scandinavian colonists who derived their 'right' to exclude new arrivals from that part of the world from the fact that they had been the first to exterminate the indigenous population and take control of their territory.

The presence of various immigrant communities in Italy – and not of one group in the absolute majority like the North Africans in France and the Turks in Germany – will certainly influence this process, even if it is not yet possible to predict in what way. It will perhaps delay the 'power takeover' by the 'new Italians', or it might lead to this being accompanied by disputes and fights between groups of different origins.

The Islam Factor

One of the main groups to play an important role in the Italy of the future will assuredly be the Muslims. This is made likely by a variety of factors, first of all, by the sheer number of Muslims in Italy. Already in March 2006, indeed, they numbered approximately one million out of a present total of about three million immigrants in Italy.

During the last two decades there has formed in Italy – without any significant socio-cultural or political reaction – a religious minority which would have been unthinkable even yesterday in a nation whose history is strongly bound to the history of Christianity. For a long time this did not create any serious problems; in fact, it went almost unnoticed, or was seen as a further contribution to the richness, variety and vitality of the Italian scene. Only in the last five years – in practice, since September 2001 – have there begun to emerge in Italy fears about the less positive aspects of the immigration phenomenon and the problem of the coexistence of very different habits and customs, along with occasional hostility as well.

Acknowledgement that immigrants practising the Muslim religion pose a problem beyond those surrounding immigrants of other cultures – European, African, Asian and, of course, South American – is a very recent phenomenon.

This is why it is only very recently and incompletely that the idea – still not examined in depth – has started to emerge that it is a problem which would be better confronted at the moment of entry to the country, rather than when its most serious negative consequences make themselves felt.

This is especially true in Italy. In fact the liberal conviction has always prevailed in Italian public opinion that coexistence between people of different nationalities, traditions and religion does not pose an insoluble problem in itself, and that it is really possible to produce a 'multicultural' society without the human and social costs being too high. Not much importance has been attached to 'the nation' as a factor because the Italians are not conscious of it themselves and find it difficult to imagine that it could exist for other peoples. It is the same for religion, because the majority of Italians are convinced they live in a 'secularized' society. And above all, there has been no desire to attach too much importance to the fact that the two religious traditions, Islamic and Christian, are different enough to impose moral codes which are sometimes antagonistic, but near enough not to be indifferent to each other and also to offer more than one field for quarrels and disputes.

Italian society – with a few exceptions in the North, in some Alpine valleys on the border with Switzerland – has been very open-minded in its relations with immigrants. This was very evident in the first stages of the inflow, and still is, especially south of the Appenines, where the impact of the hate campaign conducted by the Northern League has not been felt.

Undoubtedly, this welcoming attitude has been motivated by fresh memories of the time when Italians looked to Germany in more or less the same way as Moroccans look to Italy today. In addition, as a factor in reducing Italians' feeling of the immigrant's insuperable 'foreignness', there is their own weak consciousness of their cultural identity – on the contrary, almost an absence of the very concept of collective identity, as well as their lack of national pride and weak capacity to act in concert. These characteristics of Italian identity, which have always been a negative factor in their relations with other countries, whether in time of peace or war, have turned out to be positive when confronted with the new phenomenon of immigration, in particular Muslim immigration. They have been a moderating element in anti-Islamic propaganda, and thus a very useful factor in establishing good relations with the new arrivals.

In this respect, Italy can be seen as having, from a certain viewpoint, an 'arm's length advantage' over other European countries, which in contrast are societies with a deep consciousness of themselves and their own identities, often to the point of being ridiculous and arrogant, and which have created invisible barriers to the integration of foreigners. But – and in order not to attribute to Italian society more merit than it deserves – it could be argued that the Italians' watered-down 'rejection' of immigrants comes from the fact that immigration into the country has characteristics which make it less of a problem than in countries like France, Germany, or Britain. In particular as we have already pointed out, the immigrant community in Italy is characterized by the great diversity of national, ethnic and religious origin.

This is very clearly different from France, where immigration in the last forty years has created a relatively homogeneous block of several million people with a common North African origin, their own single language and a generalized but fairly homogeneous link with a great culture, that of the Arabs, as well as their own, Islamic, religious tradition. Thus, the immigrants in France have all the characteristics needed to form a separate nation.

Hardly less serious is Germany's problem with several million Turks, who however are neither culturally nor in terms of religion and ethnic grouping as homogeneous or as strong as the Arabs. They have, however, been transformed into an foreign bloc, by being parked for decades in a kind of legal 'no man's land' under the pretence that they are '*Gastarbeiter*' or guest workers, even when they have been born and have grown up on the edges of large German cities. The difference with – and the advantage over – Britain is even more distinct. In that country, there is a significant minority of Pakistanis and Arabs who are very well integrated into economic life, but who are far less culturally assimilated than their French equivalents, and who are determined to assert their difference and identity to the point of giving rise to incidents like the 7 July 2005 bomb attacks in the London Underground and the *fatwa* pronounced against Salman Rushdie in 1988.

The results of differences in the composition of immigrant communities are easy to observe in public transport and at the workplace. In France, immigrants often speak Arabic among themselves, in Germany, Turkish, and in Britain, Urdu. In contrast, in Italy the immigrants speak most of the time a kind of crude and bizarre Italian among themselves. Certainly it is a much impoverished Italian, deformed, debased and simplified, a language from which the conditional and gerundive forms have disappeared, and in which everyone uses the familiar '*ciao*', and addresses other people with '*tu*' (for 'you'), rather the more formal '*lei*'. It sounds a funny 'pidgin Italian' to an educated ear, but it is still Italian. Mutual understanding will doubtless improve with time, as both sides make an effort towards communicating with the other. The Italians do not indeed perceive their language like the French, as a monument to be polished and preserved unchanged generation after generation, but more the way the English-speaking peoples do. The Italian language is – one might say – like a garden made of living and permanently growing plants, where new species can easily put root, and grafts happen all the time. A new Italian language is thus being born, together with the New Italians.

However, it may not be forever thus. In the long term, Italian policy towards Muslim immigration cannot avoid being influenced by the fears that this arouses in other countries with which the society on the Italian peninsula maintains a permanent cultural osmosis. As we have seen, the ageing of the population and the rapid decline in numbers to which Italian society is condemned makes it indispensable to introduce an increasing number of adult workers from abroad into the labour market. And some of the latter come from, and will inevitably continue to come from, countries which are geographically close and which have a population surplus. These are principally the Arab Muslim countries of North Africa.

Some rumours suggesting a 'sigh of relief' appeared at the margin of Italian politics, for example, among the 'populists' of the Northern League, when it became known that in the last few years the percentage of Christians among immigrants had exceeded 50 per cent, if only by a little. This has been due to the influx from eastern European countries that had joined or were on the point of joining the EU. But it can only be a temporary phenomenon. The former communist Europe from which the great majority of Christian immigrants originated does not in fact have the demographic resources to cope with Italy's — and in general western Europe's — workforce needs. Actually, in Romania and Ukraine, the very two countries that have been providing the main influx into Italy, the population is subject to a sharp decline and ageing process.

Coexistence with Immigration?

By now, Italians have been coexisting with immigration and its consequences for about twenty-five years. These issues have become the ones of which Italians are most aware, although it is a subject which is relatively little discussed as it is known to generate contradictory feelings, if not disputes. On the one hand, there undoubtedly exists in Italy a sense of solidarity with the immigrants, a sense of solidarity which is quite widespread, also for the good reason that there are not many other causes to stimulate moral involvement. On the other hand, a sense of anxiety is appearing in daily life, the beginning of a certain 'Angst', a feeling which until now has been least characteristic of the Italian people. Finally, and most seriously, the Italians have been facing the necessity of coexisting with immigration without any real guidance, as their intellectuals have not undertaken a true analysis of the problem and their political leadership has not adopted a policy to tackle and control it.

In everyday matters, successive governments, and the state authorities in general, have gone about their dealings with immigrants in an uncertain fashion, passing theoretically stringent laws and then permitting mass exceptions, for 'humanitarian' reasons which are partly a pretext and which partly correspond to a significant body of public opinion, especially among the young, that is sympathetic to the immigrants. In effect, extra-legal situations have been created and tolerated, while the habit has spread of 'making rules on the spot', with successive laws and repeated amendments, which seems bound to cause more problems in the future than they solve today.

Neither the government authorities nor, if the truth be told, public opinion have really ever succeeded in finding the right balance between the different objectives to be followed in dealing with the influx of immigrants: satisfying the demand of humanitarian behaviour; putting this new labour force to productive use according to both their needs and the economic requirements of the host country; taking into account the political needs of public order. In no case has this been seen so clearly as when tens of thousands destitute Albanians arrived *en masse* in Apulia, the heel of the Italian 'boot', aboard two incredibly over-crowded ships. The first wave was accepted with excessive benevolence while, a short while later, the second was rejected with equally excessive harshness.

If the laws and policies concerning immigrants have been improvised and patchy, it is not only because Italy had no previous experience in dealing with immigration, but also because there has not been a calm and realistic debate on the subject. Furthermore, public opinion has only recently appeared really to be worried enough by these problems for the question of immigration to have become a new cause of political debate and discussion about the economic problems which such a phenomenon is bound to bring with it in the long term. Skirmishes among politicians and interest groups have instead prevailed with, at least so far, a clear majority of the political and economic forces of society in favour of the ever-growing inflow of outsiders.

The Pro-Immigration Lobby

Confindustria – the Confederation of Italian manufacturing industry – is normally cited as one of the organizations which appear to favour an 'open door' policy. And indeed, in a 'normal' year, it requires the admission of some hundred thousand new immigrants – which is nearly double the government's forecast – especially to work in sectors refused by Italians. Plainly, this is a requirement for people able to work, a requirement for labourers to be employed full time in factories, and does not take into account families. On the top of that, it comes as no surprise that there are more active organizations such as the Farmers' Confederation. At the end of every winter, they apply vociferous and strong pressure for the admission into Italy of seasonal workers without whom, they repeat every year, the early fruit and vegetable harvest would collapse. Employers in the tourist sector join in with reverse seasonal requirements.

The position of *Confindustria* and of the farmers' and hoteliers' organizations is very understandable. They represent the economic interests which have always gained from the presence of immigrants and which made up the initial hard core of the pro-immigration lobby. But later, the impact of immigrants as consumers of goods and services, which did not seem to be very relevant in the last years of the past century, has become more significant. From an initial stage, in which the only sectors positively affected were the used-car business, plus some illegal activities, and the rent markets in areas of deteriorating housing,[1] the most recent developments have shown the positive economic consequences of immigrants becoming settled.

Thus, in 2005 and 2006, around 13 per cent of all residential units were sold to immigrants, an amount that looks disproportionate if one thinks that they represent only 3 per cent of the total population, but a percentage more or less in line with the share of their contribution to the total number of births in Italy. In short, the number is the result of the fact that immigrants represent a young and fertile component of Italian society which tends to create families, have children, settle in the country and invest in real estate that they have largely contributed to build (construction obviously being one of the sectors that has most recourse to immigrant semi-skilled labour).

Being usually young – some immigrants who arrive alone are even in their teens – they obviously are a healthy component of the Italian population. A rough assessment of the relationship to the national health service has come to the conclusion that an immigrant contributes to it about €1,000 a year, getting back – in services and drugs – around €600 (one-third of this in assistance to pregnant women and newborn babies). Thus only the fact that hospitals are a public service prevents them from being part of the pro-immigration lobby.

On the other hand, there are no sectors threatened by immigration. If one excludes the hawkers and souvenir sellers in Florence who, several years ago, rioted against immigrant peddlers at the cost of the lives of three unfortunate Moroccans, it seems clear that the immigrants move into sectors abandoned by Italians or sectors which are in such strong expansion, like the catering trade, that the explosion is absorbed without shocks (for example, the Chinese restaurant business).[2]

The Politics of Immigration

However it must be said that the employers' organizations are only one of the components in the line-up seeking the opening of Italian society to migratory movements, and they are not even the most vocal and active. On the contrary, the attitude most favourable to making it easier to cross frontiers comes from the political left and above all from Catholic organizations. In the case of the latter, it is easy to understand the ethical imperative that is well known for its frequent expression in well-intentioned small actions, while remaining incapable of seeing what are the long-term complexities of immigration. This has been a problem of historical importance for western society.[3]

It remains however to be explained why it is the left-wing political forces – the very forces that should represent the social classes most directly economically threatened and whose lifestyle is daily damaged by immigration – that have created a real 'pro-immigration lobby'. Their political stance is indeed one of 'opening the door wide', which claims the indiscriminate granting of rights to the 'new Italians'. Above all, the progressive parties are at the origin of the repeated amendments of regulations on behalf of illegal immigrants, amendments that have spread around the whole Third World the idea that Italy is the soft underbelly of Europe, the promised land for all destitute people, that the important thing is to get into the country, legally or otherwise and, once in, 'things will be fixed' by joining in one of the periodic 'regularizations' of semi-illegal immigrants.

In a crude attempt to exploit the backlash from the immigration phenomenon, that is, the growing irritation and nascent xenophobia among workers in the north-east, some groups on the extreme right have no hesitation in advancing the explanation that the left, having lost the support of the Italian electorate, aims to let millions of immigrants into the country so that they will form a reservoir of voters once they have become citizens.

This clearly is an absurd explanation. It is true that in the dying years of the last century, Livia Turco, the Minister for Social Affairs, proposed giving the vote to immigrants quickly. But this proposal was limited to communal, regional and provincial elections only.[4] And Turco was evidently inspired by a political commitment to treating immigrants as people, not just numbers; an inspiration which undoubtedly corresponded, and corresponds, to an ethical and egalitarian demand which is traditional in socialist political groups, and which the recent transformation of world politics has not been able to suffocate. In short, Turco's point of view revived, in a positive sense, Max Frisch's famous conclusion about Switzerland's experience with immigration.

The explanation of the commitment to immigration of left-leaning political forces must be sought closer to hand. It must be sought first of all in the fact that the political left is the natural electoral representative of those strata of public opinion who most painfully feel the lack of values in contemporary Italian society and generally in western society. And this part of the public, which consists widely of young people, sees solidarity with the immigrants as one of very few opportunities for moral engagement, even if most young people are often ill-equipped to understand the problems which the influx of immigrants brings to the host society, or the true requirements and aspirations of people who come from such distant and different countries, and whose life has suddenly been so traumatically torn apart. These strata of the population and the electorate are larger than the cynicism of the media would have us believe, and they form the better part of a pro-immigration lobby which the political class feels must be taken into account.

It is only too obvious that – to balance the benefits of an influx of immigrants – there are, in the Italian economy and society, non-negligible costs as well. These costs are typically 'externalities', that is, they fall to be paid outside of the enterprise to which the working immigrant is providing their labour. It is all about the cost of the social welfare problems caused by the 'men' who moved from one country to another in order to bring the 'force of his right arm', which is necessary and very useful to the enterprise. These men, when outside the factory where they work, must sleep, eat, look after themselves. They must fill their spare time and satisfy their religious, cultural and moral needs. And thus begin the problems and the costs to the host country.

In the cases in which the family has come along with, or has joined the worker, a part of these needs will be satisfied within it. But the family is the cause of more problems, as its presence compounds the number of men and women to be accommodated in a society very different from that of origin. And while the worker himself learns a lot about the host society and its habits while at the workplace, the same does not apply for wives and children. Their presence imposes instead new problems in new fields, education first of all, but also in the areas of leisure activities, drugs and youth crime.

The sum of all these problems is the physical and economic 'disintegration' of the areas in which immigrants are concentrated, a degeneration the price of which is mainly paid by the people who lived in these areas before the immigrants arrived. And it certainly will not be the central or 'good', upmarket areas of the cities which will feel the biggest impact, but above all the areas where the mass of people live. Thus, to physical and economic and social degeneration – typically represented by formation of bands of child thieves – can also be added the 'political degeneration' deriving from the reactions of these social classes which will rapidly give birth to extreme and xenophobic political positions. An opinion poll – conducted by a foundation operating in the north-east of Italy, where the Northern League is active – found that 43 per cent of Italians believe that immigrants are 'a threat to public order and personal security', and that as many as 60 per cent have concluded that 'The country is no longer able to accept immigrants, even legal ones.'

This conflict of interests affects all of Italian society and is the determining factor in the approach of all political parties. The impetus from the employers and above all the pro-immigration lobby had in fact brought the centre-left government to an immigration policy which had or should have had implications for three areas of government action in this field: programmed entry with fixed annual quotas (based upon objectives relevant to the structure of the population and the composition of the workforce), a more efficient fight against clandestine immigration and criminal exploitation of immigrants, and support for immigrants' integration into Italian society.

It is easy to confirm that, with the introduction of another immigration law (Bossi-Fini) by the centre-right government which had followed the centre-left, neither the second of these intentions, much less the first, was translated into reality, while the initiatives for integration of immigrants and the related expense proliferated like a cancer.

As can be seen, the political attitudes of the Italian about immigration are very different from that of the country's European partners. This is primarily the result of the already mentioned lack of a strong sense of national identity in post-fascist Italy: a feature that makes it impossible in Italy for there to be the sort of extremely violent reactions to immigration that, especially since the Madrid attacks, have been occurring periodically in Spain, a country where national feeling is in contrast quite strong.

The comparison with Spain is interesting because, like Italy, it is a country which has only recently been subject to a wave of immigration. The differences with the other European countries come from the fact that the social fabric in North-West Europe has by now been so profoundly modified by decade after decade of immigration that these countries are bound to political attitudes which are not yet prevalent in Italy.

In France, for example, the country which, from legal/administrative and political perspectives, is the most like Italy, it is clear that the number of voters of recent migrant origin is so high as to make it difficult to have a simple approach to the problem. There is a large bloc of voters with feelings of solidarity and definite fears when it comes to immigration policy, and thus there are those who shun excessively severe positions, while others go to extremes. Part of the Front National's electoral support comes from people of purely French stock, who are driven by feelings which are to a greater or lesser degree racist. But another firm section of its support is to be found among recently naturalized French, who exhibit extremist feelings in an effort to appear more French than the French. In so doing, they seek legitimacy in their own eyes and believe they gain legitimacy in the eyes of others.

This basic difference in its situation makes it difficult for Italy to adjust to any immigration policy agreed with its community partners in Brussels. Italy's weaknesses (such as undeveloped national feeling) and Italy's strong points (such as the recent immigration phenomenon) clearly make it expedient that immigration policy should remain a national policy for all of the predictable future.

Men or Labour?

The cluster of questions revolving around immigration is strongly political in its nature, but not just in the squalid sense of creating a framework of dependents for small-time politicians. It is political in the sense that for Italian society, as for all societies which have an influx of immigrants, the arrival of a mass of labourers, mostly men, from the Third World brings with it the need for some hard choices. The first choice – which is the most important but also the one of which public opinion is least aware – concerns the strategy of the very aims of immigration policy. Another choice, tactical in nature, concerns the tools and methods of this policy. And, as it lends itself to the facile electoral demagogue, this choice is the one which causes the most fuss.

So far as the policy aims are concerned, firm distinctions must be made between two major problems tied to the immigration phenomenon, and which can already be glimpsed in existing legislation: the problems of our country's demographic decline and that of the availability of the workforce.

To compensate for demographic collapse, that is, for the lack of men and women, brought about by the drop in fertility levels to half of that required to re-create the population, it is first necessary to define a long-term 'population policy', which should take account of the fact that a reduction in the total number of inhabitants would not perforce be a reduction in Italy's 'power' but, on the contrary, would eventually improve the quality of life. On the other hand, in order to deal with the quantitative ratio between active and inactive members of the population, that is, to deal with the lack of labour, a different and separate policy is needed, a 'workforce policy', aimed at bringing about immediate results.

It is easy to see that the two policies, for population and workforce, have different technical requirements as well as different objectives. And even if the lack of labour force is partly also a result of Italy's demographic collapse (but only partly, given that Italians will not do certain jobs), it would be impossible even with a big demographic programme over many years to expand the workforce by increasing the number of births per woman. In short, dealing with the demands of the system of production means that recourse to immigrants, 'ready to work' adults, is inevitable, and not only to fill the jobs refused by Italians but also for many new types of work.[5]

Immigration will, however, not suffice. In fact, the more rapid their entry into Italian society, the more the behaviour of the immigrants will come to resemble that of the local population. Thus the immigrants' availability to take on jobs refused by Italians, and their willingness purely and simply to be exploited in terms of working conditions and schedules, will decrease rapidly.

The Arm, not the Soul

Italy's economy needs labour and immigration provides it. Labour, however, never comes all by itself. With it – the 'working right arm' – come inevitably men and women. And it is then that the problems we have been describing come together, and the facts of being a poor man exposed to all the charm and temptations of an affluent society, of being a stranger in a potentially hostile environment, of being an immigrant competing for work and shelter with the poorest of the natives, of being a Muslim living among the infidels, of being a lonely man away from home, begin to form a knot difficult to untie.

As long as one is just a worker, 'selling the strength of his right arm', integration is very easy. Generally speaking, the company and the working environment immediately create homogeneity and coordination in gesture and deed, timing, activities, needs and behaviour. In addition, since the immigrant has come with the sole aim of making a bit of money, he tends to devote as many hours of the day to work as possible. It is outside the workplace that the problems begin, for it is there that the needs and behaviour of men take over from the uses and functionality of labour.

At work, a Muslim immigrant can hammer, cut, sew, fix parts together, or cast steel just like a non-Muslim. But when he is away from work he must eat meals prepared according to the rules of Islam, resist the temptation to drink like westerners or avoid drinks the effects of which he is not used to controlling. In his non-working hours, he ends up 'entertaining himself' in ways to which he is unaccustomed, and has social relations – especially with the opposite sex – which follow far more liberal rules than those of his original society. For the Muslim immigrant, the shock of European-style social relations is both extremely violent and tempting, if only for their great freedom and the absence of state control. He has the feeling that, by coming to work among the infidels, he is trading for money not only the strength of his right arm, but the whole of his personality, including his soul. The Muslim immigrant can avoid such problems and complications only if he comes away from work totally exhausted, and sleeps the sleep of the dead until it is time to return to work.

This, of course, applies to the immigrant who is single. For, as we have already mentioned, when there is a woman and a family behind every pair of hands, the problems increase exponentially and take on complex moral aspects that finally interfere with the work itself. How can the immigrant worker ever concentrate fully on his job when he knows that away from his workplace the women of his family are exposed to all the corrupting temptations of a society that seems to him to be profoundly immoral from the sexual and family perspective?

And how can he work long hours, when he has reason to fear that his children are absorbing the values of a world in which there is no more respect for parents and old people? And if, on the contrary, he knows that his children react negatively to their new environment how can this labourer devote himself totally to his job knowing that there is a risk of his son's becoming an anti-western fanatic prepared to destroy himself and others? Were the London Underground suicide bombers maybe not the children of immigrants, and not first-generation ones? And was it not the children and descendants of 'new Frenchmen', who seemed to be more or less successfully integrated culturally, the ones who brought turmoil to the life of the French city-suburbs for three weeks?

Although few Europeans realize it, even when the Muslim immigrant is economically integrated, even when he enjoys a decent income and some social services inconceivable in his home country, he does not live in tranquillity. On the contrary, the Muslim immigrant in Europe lives in a situation of fear and anxiety over 'cultural contamination' and its consequences very similar to, if not greater than, the anxiety felt among the most ill-informed and susceptible sections of the European population. And his fears are much more firmly based than those which are spread in Italian public opinion by the most virulent critics of immigration, namely by journalists imported from the Middle East, and in general by the most ill-bred parts of the right-wing press.

It is fear that pushes the Muslim, or other, immigrants to request the authorities in the host country to recognize his own cultural identity. What he seeks is a sort of endorsement of an irremovable quantum of the very 'diversity' that endangers him from moment to moment. It is again fear that is behind the tendency for the Muslim immigrant living *in partibus infidelium* that pushes him to prefer living inside a ghetto, that in some vague way reminds him of his cultural community. There are understandable reasons why he is inclined to defend his own identity.

Actually, what is pushing him to assert his difference, and to ask the host country to recognize his right to be 'different', are the same reasons as those which lead more than one journalist and more than one European politician to ask him to integrate, to deny himself, to abdicate, or at least conceal his faith, the way the Jewish 'Marrans' and the Muslim Moriscos had to do under Torquemada. In short, the Europeans ask the immigrants to behave in such a way as not to frighten them. Alas, no one – with the exception, it seems, of the Abu Ghraib torturers who exploited it in depth – has realized that these men are timid and chaste, and for them the mere presence of women causes profound embarrassment. The Muslim immigrant in our cities feels lost and uneasy, and he is as afraid of us as we are afraid of him. This negative symmetry does not make relations easy. On the contrary, it can create a very dangerous situation, in which all irrational and unforeseeable behaviour becomes possible.

We Europeans too often forget this aspect of immigration, the fact that human beings are involved as well as labour. We do not imagine, indeed it does not even enter our heads, that immigrants have their own morality, their own sense of personal honour and dignity which is expressed in a very different way from ours. We do not realize that Muslims, even when they come from countries which proclaim themselves 'revolutionary', in reality are the trustees of a mentality which is primitive in some things (namely, concerning relations between the sexes), but very delicate and fragile under other respects, and above all extremely conservative, much more conservative than that of any European. And that they give an extremely severe judgement of our society, of our being de-Christianized, of the laxity of our customs, of our treatment of our elders.

Our negligence and incomprehension is what sometimes provokes hostility in our relations with them, most forcefully from western conservative political forces. And this is all the more paradoxical as, if his political vision stretched further than the end of his nose, a true western conservative ought to see them as natural allies. A true Italian conservative, rather than calling for help, should probably ask himself whether the spread of new conservative trends in Italian society which go hand in hand with its ageing, do not make it possible to come up with an immigration policy that meets the demands of both sides. Such a policy would need to create the tools to allow the immigrant to profit fully from the work potential which Italy offers, while discouraging everything that produces insoluble problems for social integration.

In partibus infidelium

It is easy to predict that the distinction between the needs of the labour market and those of population structure, the difference between 'labour' and 'men', will provoke many negative reactions and these will be paraded in every egalitarian and supposedly humanistic argument. It will be said, with reason, that the two things cannot be separated and that to see immigrants only as a labour force is a negation of their humanity, a violation of the ethical principles that demand that a man should never be considered as a means but always as an end in himself.

But it is easy to reply by emphasizing how, given all the understanding and readiness which people claim to show towards immigrants, no one has ever taken the trouble to try and see how they themselves, the people most directly affected, perceive the phenomenon of immigration, and whether or not they are interested in integration with Italian culture. Everyone in our society can see the presence of the incomer, but it occurs to almost no one that in order to come among us he first of all has to have emigrated, and is someone who has been torn away from his own country and culture by economic exigencies. Almost no one allows for the fact that when the immigrant arrives he is already profoundly marked by this severance. Not by chance do the Muslims of North Africa who make up the majority of immigrants to Europe call emigration *'elghorba'*, 'the exile'.

If anyone had been bothered to understand how the immigrants see their own circumstances, their feeling of belonging to the society they have left and the feeling of diversity towards that in which they have arrived, it would have been simple to ascertain that immigrants themselves make a clear distinction between 'labour' and 'men', between their readiness (and ambition) to become part of the workforce in the country to which they travel, and the much more painful process of ceasing to be that which they are, in order to become Italian (or French, or Spanish). In this context, it is of very great interest to see the distinction which Moroccan immigrants in France make between the 'green passport' (the Moroccan passport which is seen as a kind of permission to leave the country and go to work abroad), the 'blue passport' (which French citizenship used to confer, and which was seen as the sentencing to the loss of one's own identity), and the most recent 'brown passport', indicating the European Community, which is seen as a true guarantee of freedom because it permits the holder to live and work in a free-market, free-movement area which is not defined in cultural and national terms, where everyone can be himself and it is possible to earn one's living without having to pretend to be what one is not.

Anyone who had taken the trouble to understand the thinking of the immigrants which they claimed to be concerned about, could also have easily found out what are immigrant's real priorities. They are a lot less interested in their families joining them than the Italian authorities believe, and especially a lot less than many Italians who see immigrants as future Italians, and are convinced that one day there will be erected on a beach in the island of Lampedusa (the point of Italian territory nearest to the African coast) a monument to the disembarkation of the 'barefoot armada' that came to rescue Italy's population from decline. In fact, the majority of immigrants are single men who set out in the belief they will stay abroad only for a more or less short period of time, and whose most ardent ambition is to return home with some savings, to live according to their own customs. They are certainly not longing to be assimilated in western society, and even less to bring their own families to be assimilated also.

In reality, the majority of those who face 'exile' alone, thinking they will remain abroad only for a short period, nurture as their most keen desire that of returning to their village not only to enjoy their family life, but most of all in order display the tiny affluence they have gained as the price of the sacrifice of emigration to those who saw them suffering hunger before they left. They certainly do not leave with the intention of assimilating into western society, and even less of bringing their own family to be assimilated also. This applies most of all to immigrants from North Africa – whose morality is especially rigid in family and sexual matters – but not only to them.

For the Senegalese, who are also Muslims, but from a society where sexual morality is less strict, the lack of interest in putting down roots in Italy is – if possible – even stronger. Senegalese immigrants, who can be seen on every street corner in Italian cities, are not individuals who are adrift and who seek to make a place in someone else's home. On the contrary, they are part of a very efficient and well-structured network of hawkers who belong to a Muslim brotherhood, the Muridya, which manages and helps young people enter the world of work in Senegal, usually the commercial sector, after they put together a little savings on the international hawkers' circuit. The elders who give spiritual direction to the brotherhood from the mosques in Touba in Senegal certainly do not want to see them integrated with Italian society, a society with which their relationship is moreover excellent, perhaps the best among all the immigrant communities, but which is nevertheless still a different and foreign one.

The provincialism of Europeans, not to mention that of Americans, is such that they see the Muslim immigrant as someone who wants to steal their jobs, their houses and their country. The average American or European believes that no human being could wish for anything other than to become like them. They cannot imagine that the majority of Muslims despise the West as well as fearing it, as even the new Pope Benedict XVI has noticed. For Muslims, the freedom they enjoy, in all senses of the word, in Italy and in Europe is very close to moral ruin, and is thus 'excessive'. And they see us less as a lay society than as a society which has dramatically lost contact with the common God.

There are therefore also moral reasons, and not only economic ones, for the Muslim immigrant to prefer to leave his family behind, in his home country. To be sure the economic reasons are extremely important. Indeed, with the small sum of money even a precarious job has allowed him to save after a few months in Europe, an immigrant can secure for his family (and for himself when he returns at harvest time) a style of life which, in his home country, appears affluent, with all the advantages it gives him, a better social standing for his children and the respect which his society accords to the image of the person who has achieved a certain success abroad. His meagre savings are a fortune there, while in Italy they would be small change, too little to feed his family, let alone securing somewhere to live other than a shack in some quarter peopled with drug dealers, transvestites and prostitutes.

In this context, German policy towards the *Gastarbeiter* is instructive, especially for its failings. It started from the premise that it was more or less impossible to become German, and that the foreign worker should maintain a link with his country of origin so that one day he could return there. This policy however rapidly contradicted itself. Germany has in fact accepted that it should play 'host' to the workers' families as well and that their children should be born and should grow up in Germany. However, in its handling of these *de facto* roots, the German Federal Republic for a long time stubbornly took a legal position which treated both the worker and all his family as *Wandervogel*, 'birds of passage'.

The temporizing and wavering typical of the 'host country' were thus set in theory but denied in practice, and became a fiction, creating a problem which became increasingly worse with time and turned out to be extremely difficult to solve.

The Creak of the Closing Gate

If all this were true, one could object, how do you then explain certain traits of behaviour among the Muslims who live in Italy? If immigrants from the Maghreb countries, the nearest geographically and the furthest culturally from which the influx originates, truly preferred to come and sell their labour for quite short periods, when they were forced to survive in highly straitened circumstances, and returning home at intervals with trifling savings, why is it that a lot of them end up with their families rejoining them, as encouraged by Italian law?

The answer is clear and immediate; because in Italy and to a lesser degree in the rest of Europe, too many immigrants live in a situation in which their rights are very uncertain. So far as the future is concerned, they are in a permanently uncertain state, because in the space of twenty years Italy has seen four immigration laws and something like six mass regularizations. Actually, the majority of immigrants entered the country irregularly and are working without secure rights, but on the basis that the important thing is to enter Italy and wait for a law to regularize any irregularity.

The widespread feeling in the Third World that being in Italy physically counts for more than the legal position reinforces this state of insecurity. In fact, from the moment he returns home, the immigrant is always in fear that he might be prevented from re-entering Italy. He knows that this is an improbable scenario, but the stake is too serious for him to take the risk. He cannot live decently as a 'man' in Morocco or Tunisia and look after his family unless he can be sure of selling his 'hands' in Italy where their value on the labour market is incomparably higher than in the Maghreb countries.

It is the alternative between staying 'shut inside' with no possibility of being able to return home or remaining outside of Italy that drives the immigrant to consider an intrinsically aberrant idea, moving his family. He knows that this decision means definitive uprooting, that once the women in his family will have tasted the European way of life they will oppose with all their might any idea of returning home. He knows that, when he will be at the age of retirement, of returning home to enjoy the fruits of his sacrifices, his children will be at a stage in their life when the opportunities offered by the western society will appear within their reach. He knows that his sons and daughters will be exposed much more than at home to the risk of drugs, criminal behaviour, prostitution, and so on.

He knows that moving his family away from his country of origin to the country in which he has a job will mark the beginning of a series of terrible economic difficulties for himself and his family. But he still prefers this alternative to the risks of not moving; and – given the way in which the situation appears to him – his behaviour is rational. And for Italian society his decision marks the beginning of all kinds of problems, in suburban neighbourhoods and on the sidewalks of major cities, in schools and in mosques, in juvenile courts and in reform institutions: all problems which at first look small and unlikely, but then grow more and more formidable the more immigrants there are, and the more uprooted and non-integrated they are.

This fear of being shut out of the country where the immigrant has found work is naturally aggravated each time some common-or-garden little politician from the Lega del Nord seeking popular favour starts bawling out 'Enough immigration!' And this fear provokes a decision taken reluctantly, not only because of the moral and material costs which follow, but also because of their loss of status. It is indeed well known that, when they periodically return home, immigrants enjoy the pleasure of showing off and boasting to their families and acquaintances about their successes, much inflated for an audience who only knows by hearsay the marvels of Europe. Moving the family therefore means suddenly finding oneself forced to reveal all the misery of the true situation.

The threat of a ban on immigration, campaigns by the gutter press to stir up religious and racial hatred, stupid rumours put about by those seeking popular acclaim by opposing immigration, the creaking of gates which threaten to close, all encourage a race to enter before it is too late, even if the immigrant is in no way prepared to provide decent conditions for his family. Thus, one should not be surprised that such a hasty and badly prepared transfer means that the family ends up living in disintegrating outer city edges among all sorts of marginal people. One should not be surprised that the children, male and female, in an environment where their own moral code does not apply, and in which they are lacking any cultural endowment to help them resist contamination, finally take to working the pavement and the drugs traffic.

Rights, not Fear

Contrariwise, were there reliable legislation and a framework of secure, irrevocable rights, it would be possible for the immigrant and especially the Muslim to be fully satisfied to offer himself to Italian society simply as a worker. It would be enough for him to be certain that once he had passed through the inevitable filter needed to select those who can be integrated into the Italian workforce he would have gained the right to enter and leave Italy when he wished, on the condition of respecting the criminal code.

We have seen that this solution would also serve the interests of the host country. Obviously, the important immigrant community which has already settled in the peninsula should enjoy all the rights Italians have, and be fully protected by the law. But new work permits allowing entry and exit would be given exclusively to those seeking professional activity and having the real capacity to pursue it, thus creating an incentive to look for an alternation of periods of work in Italy and regular trips home.

Such an approach would indeed offer the guest worker not just the chance to display the fruits of his labour to those who have seen him poor and humble but also a chance to be politically active: a right that the European are ready to concede to the immigrant only in the framework of the political choices offered by the western political spectrum. Retaining a role in the country of origin would instead empower the migrant to promote his political views in his own cultural environment, not in Italy. And one may bet that in his own country he will be less conservative, less extreme and more westernized. As often happens when someone lives across two worlds, the immigrant will turn into a spokesman for both sides and their values and advantages. He will probably be a spokesman for European values in the Islamic country of origin as much as he tends today to protect his Islamic values in the country in which he is obliged to work. Would it not perhaps be an intelligent strategy to make the immigrants ambassadors for our civil and religious values in the Islamic world rather than homesick champions of the word of Islam in our own house?

On the basis of this convergence of interests between immigrants and the host society it is possible to construct an immigration policy which is liberal while being also protective. Liberal in the sense that it would effectively open the doors to a large number of foreigners with the will and capacity to work. Protective in the sense that it would avoid the socio-cultural upheaval – and thus xenophobic and fascist-leaning reactions – which has occurred wherever the need for labour has been met by allowing in a mass of immigrants with profoundly different cultural characteristics, thus giving rise to an interminable series of problems and reciprocal rejection.

Within – perhaps – the limits of a fixed annual quota, the Italian Republic could issue a permit for unlimited entry and exit to immigrants of a working age and with the required physical and mental qualifications who were looking for a job in the country. This document would have no expiry date, be impossible to counterfeit and could at all times serve as identification. It would be a document issued exclusively to those able to work or, if one wished it to be more restrictive, to those who had already found work in Italy. But once in possession of such a document, crossing the frontier should be a right acquired forever, and with no possible restrictions, which could only be lost as a result of criminal behaviour, ascertained through regular judicial procedure.

Giving the immigrants well-defined rights to work in Italy would create a situation in which they would have a clear interest in respecting the laws of the host country. And this would be much more effective then trying to enforce the same laws by menacing the immigrant with any form of punishment. As the situation now stands, most of them are indeed too desperate to fear punishment, they have literally nothing to lose. Giving them rights would reverse their condition: they would have same real privileges to protect by abiding to the law.

Such an immigration policy would help create a mechanism with which the Italian system of production would have available all the labour it needed, and in which the inward and outward flow of workers would tend to reach a balance in the medium term. In short, one would see fewer Muslim workers in the situation they are in today, which greatly resembles that illustrated in the 'prisoner's dilemma', one in which uncertainty forces him to prefer the second-best choice of emigration and family transfer rather than the optimum choice of keeping separate his place of work and the society to which he belongs. Of course, those who wished to transfer their whole family and be assimilated into the host country, would always have the possibility to do so: not – as it has happened in the last quarter-century – in a chaotic system of retroactive regularization, but according to procedures laid down by the law in the framework of 'population policy', that is, the set of legislative measures that will be necessary to enact if the decline of the total population of Italy – that cannot be compensated by immigration alone – will have to be dealt with.[6]

There is, however, one condition necessary for this theory to become possible and translate into reality the concept of a truly libertarian immigration policy, one which opens the door to immigration at the same time as conserving the host country's culture. It would be indeed necessary for the Italian political class to free itself from the prison of ideas and ideologies that belong to the nineteenth and twentieth centuries, and no longer have recourse after the clean break that has marked the last decade of the latter. Public opinion, in Italy as in many other countries that immigrants cherish, must get rid of the platitudes and demagogy in which even the best-intentioned part of that opinion has been confined for the last few years.

Both the public and its political leaders would have to come up with a rational and 'political' vision to deal with the immigration problem, which has too often been put under the distorting lens of our partisan prejudices.

On the right, it has too often been seen more – if not solely – in terms of public order or national identity, and sometimes through a darkly xenophobic hostility. In a majority of cases, insufficient attention has been given by conservative political forces to the fact that migrations are just one aspect of an open world system, in which borders tend to weaken, if not disappear; in which market forces and the pursuit of economic opportunities play a growing role in the destiny of each man and woman, and in which more and more people are learning to live multifaceted lives, moving back and forth from one culture to another.

On the left, the immigration issue has brought about a generous attempt to invent an policy inspired by a political commitment of solidarity to the weakest and most oppressed among humanity. Not enough attention has been paid to the limits that such an attitude is bound to encounter, and fairly fast: the limits set by the reaction of the lower middle classes in the destination countries, whose feeling have been well expressed by a French socialist leader when he happened to say *'nous ne pouvons pas accoeuilir toute la misère du monde'*: we cannot play host to all the world's misery. Too often, their aim gets lost in a generous wave of solidarity with fellow humans and exhausts itself in an effort at helping the man or the woman who is today a poor emigrant, just as so many Italians were barely fifty years ago. A much stronger intellectual effort is needed in order to invent a progressive immigration policy.

Today's old and rich Italians cannot indeed forget what Max Frisch wrote in the introduction to a book entitled *We are Italians* and devoted to their compatriots who – barely half a century ago – had to migrate to Switzerland in order to find work. In a single phrase he summed up the paradox and tragedy of immigration: *'Es wurden Arbeitskräfte gerufen, aber es kamen Menschen'*: We asked for labour, but it was people who came!

Professor Giuseppe Sacco is the Head of the Chair of International Relations and World Economic Systems, Department of Political Science, at the University of Rome.

NOTES

1   Nevertheless, it should be noted that while the demand for housing of recently arrived destitute immigrants makes for quite a lucrative rental market in suburban areas, letting to what are, frequently, single males, leads very rapidly to deterioration and dilapidation of buildings, and economic depreciation of entire neighbourhoods. The problem is made more serious by the fact that, together with immigrants who come to work, there is a flow from eastern Europe (mainly

Romania) into Italy of nomadic groups. These refuse all type of occupation or integration, do not send their children to school (in violation of Italian legislation on compulsory education), live by begging and illegal activities, and bring about the rapid devaluation of real estate near their encampments. Starting on 1 January 2007, the admission of Romania in the EU, with the ensuing abolition of all limitation to travel into Italy seems likely to compound the problem, as around 60,000 gypsies are expected to enter Italy.

2   It must be noted, however, that this is a market which is distorted by the licensing mechanism.
3   And that is leaving aside the ignorance, pure and simple, of those of whom we are talking. One example covers all, that of the bishop who in order to encourage Italians not to see all immigrants as criminals (a most praiseworthy intention) could find nothing better than to ask the question: 'What would have happened if Italian immigrants to America had all been considered potential criminals?' These makeshift opinion leaders would benefit from a little reading on the subject, for example, the infamous Sacco and Vanzetti case. They might then realize that in the United States, even in the last decade of the twentieth century, Italians were still considered to be just that, a natural and incurably criminal race, a human material immune to the effects of 'Americanisation', which could turn ordinary humans beings into disciplined citizens of the Stars and Stripes republic. In order to find out about American hostility against Italians – considered to be a race apart, non-white 'dagos', for the simple reason that they treated the American blacks like human beings – it would have been enough to ask the late founder of the most important pro-immigrant organization, *Caritas*, Don Di Liego, whose father, a fisherman with ten children to look after, made four attempts to emigrate to the US and was four times refused.
4   Turco also proposed lowering the voting age to 16 probably with the idea of giving responsibility to those sections of the population which can most easily fall prey to the attitudes of xenophobic organizations.
5   See Giuseppe Sacco, *L'invasione scalza* (Milan: Franco Angeli, 1992).
6   In this article, only immigration policy is taken into consideration. Population policy will have to be considered a different matter.

Humanitarian NGOs and the Migration Policies of States: A Financial and Strategic Analysis

MARC-ANTOINE PÉROUSE DE MONTCLOS

This article addresses the role of NGOs in the management of migrations in countries to the south of Europe, and more specifically with forced displacements due to armed conflicts or natural disasters. The question is whether humanitarian organizations are involved in a containment policy, which consists in slowing migrations to the North and containing potential asylum seekers in refugee camps. I will show, through a financial and strategic analysis, that NGOs are in fact very dependent on institutional donors. More often than not, their budget leaves them very little room for manoeuvre and they are very susceptible to the pressures or requests that determine their geographic zone of intervention. The field of humanitarian action is thus extremely complex and reveals a great diversity of associations, including NGOs that simply do what governments tell them to.

***

When one looks into the migration policies of states, it is clear that refugee flows are allocated a specific place and are linked to both asylum rights in the North and containment policies in the South. Simultaneously, western industrialized countries take a global view of forced displacements. Their specificity is due to the fact that in the South, the management of refugees is mostly in the hands of intergovernmental agencies and international solidarity associations (ISA) rather than in those of the immigration services of the Interior or Justice Ministries of the host countries. NGOs (non-governmental organizations) are thus agents in their own right in migration policies, whether by denouncing the restrictions on asylum imposed by western countries in the North or by helping the populations in the South who have been displaced by conflicts or natural disasters and who have found themselves in camps under the legal protection of the United Nations High Commission for Refugees (UNHCR).

A brief economic analysis confirms the importance of international solidarity associations in the management of humanitarian crises. Depending on data and years, NGOs now deal with 6–15 per cent of public development aid, whereas this figures was less than 0.2 per cent in 1980.[1] In 2004, they received $4 billion from the various budget allowances which the Organisation for Economic Cooperation and Development (OECD) classifies as governmental cooperation, on top of their other funding sources. The proportion is even greater if one considers only emergency operations and not longer-term development programmes. The absolute and relative value of the funds devoted to emergency humanitarian actions has thus increased tremendously: from 3 to 10 per cent of all public development aid between 1970 and 1999. The funds that donor states allocate to emergency operations are divided into roughly equal shares between the Red Cross, the UN and NGOs. NGOs actually get the biggest slice of the 'cake', as they also receive between one-third and one-half of the funds that governments give to UN or regional agencies – for instance, the European Commission Humanitarian Office (ECHO) gives NGOs up to three-fifths of its budget.[2]

The current situation is most telling. At the international level, NGOs that specialize in emergency operations manage around half of humanitarian assistance and deal with around $1.5 billion a year, and double this amount when their own private financial sources are taken into account.[3] All in all, the yearly budget of international solidarity associations is probably around $100 billion, one-tenth of which goes to countries in crisis. This trend also concerns NGOs which work only in the country where their headquarters is located. If one includes all the social, educational, cultural, sports, or humanitarian private organizations, the so-called 'non-profit' sector accounted for around $1,600 billion in 2002![4] In a country like France, for example, one inhabitant in four is a member of one or more associations and 170,000 NGOs have employees, as opposed to 120,000 ten years ago. The impact on the national economy is far from minor. With 960,000 paid posts in 1996, the non-profit sector – religious congregations excepted – provides 4.9 per cent of non-agricultural jobs and 3.7 per cent of the GDP; this increases to 6.3 per cent when the value of voluntary work is included.

The Reasons for the Economic Success of NGOs

Several factors explain this craze for NGOs. First of all, with the end of the Cold War and the triumph of liberalism, private initiative became pivotal and the state was rolled back, along with its social policies and its ambitions for intergovernmental cooperation in the Third World. It used to be the opposite: subsidies to international solidarity associations barely increased in the 1970s and 1980s. During this period, they represented an ever-smaller share of public development aid.[5] The collapse of the USSR was actually a turning-point: in a world that had become multipolar, as opposed to bi-polar, new fields of action for international NGOs opened up, humanitarian intervention became more legitimate and the pioneering work of 'without borders' associations was applauded. Donor states have thus focused on emergency services: they had the advantage of costing far less and of being far more visible and spectacular than the development programmes whose efficacy always seemed dubious and whose budget had been decreasing since the 1960s.

To some extent, NGOs have also been a success because states were willing to organize their diplomacy by proxy, and because they preferred more discrete channels of influence than the official context of bilateral government-to-government cooperation could provide. As everyone knows, the funds devoted to humanitarian services and the zones where they operate are determined to a large extent by political, security and economic considerations.[6] During the Cold War, the main objective was to help the United States' allies to contain Communist penetration; today, the aim is more to promote and strengthen so-called democratic regimes, as opposed to dictatorships likely to threaten world peace.[7] To paraphrase George Orwell, some refugees are thus more 'equal' than others, and the attention of institutional donors is more focused on the more equal ones. Very often, crises in the Third World are assessed on their strategic relevance and the risk involved rather than on the intensity of human suffering. In 2003, for instance, half of the UN's humanitarian aid went to Iraq where the overthrow of Saddam Hussein's dictatorship caused very few forced displacements. In Washington, the war on terror has also modified the geopolitical map of regions that have priority. Since 2001, Afghanistan, Iraq, Pakistan, Turkey and Jordan have thus benefited from a spectacular increase in US aid, even if the geographic repartition of the other recipient countries has not changed fundamentally, essentially because of the administrative inertia of the cooperation agency, USAID (United States Agency for International Development).[8]

In this context, the generosity of donor states is all the more problematic as most humanitarian aid (60 per cent) is bilateral, which helps turn it into a foreign policy instrument. The workings of the system implicitly show a division of labour that assigns an executive role to NGOs that are subsidized by their own government while international agencies and states give the orders. The most recent trends even show a form of nationalist withdrawal from governmental cooperation. At the beginning of the 1980s, it was considered good form to finance a few 'foreign' NGOs. Today, public development aid is increasingly in the hands of associations whose headquarters are located in the donor country.[9]

All the same, it would be too restrictive to argue that NGOs working with displaced people in the Southern Hemisphere are systematically the subcontractors of western states and their migration policies. The extreme diversity of the humanitarian movement makes it more complex and one should beware of simplistic analyses, whether they exaggerate or minimize the independence of associations and their room for manoeuvre. According to the idealistic vision of the representatives of this sector, NGOs are groups of citizens interacting with states.[10] International solidarity associations are thus neither the agents of a parallel diplomacy nor instruments for the privatisation of international relations. None the less, more detailed and individualized analyses show the extreme complexity of a very heterogeneous field. Nicknamed 'GONGOs' (Government NGOs), some 'para-governmental' NGOs are in effect foreign policy instruments. Eager to preserve their independence, others try to distance themselves from states. To avoid amalgams and excessive generalizations, it is thus necessary to analyse NGOs on a case-by-case basis.[11]

Some Elements for a Financial and Strategic Analysis

The nature and intensity of the relations between international solidarity associations and the public authorities can be decoded by means of a financial analysis, among many strategic indicators. Three main sources of data make the budget analysis of NGOs easier:

- the percentage of private and public resources,
- the diversification of income and
- the proportion of earmarked funds.

At first sight, the importance of government subsidies in the budget of an international solidarity association demonstrates its political dependence on institutional donors. As a matter of fact, most humanitarian NGOs are mostly financed by states.[12] In most industrialized countries, this is actually true for the entire non-profit sector, where public subsidies represent around 40 per cent of the budget on average.[13] The funds given by private firms or foundations are indeed negligible and instead, NGOs must rely on their ability to finance themselves, essentially through commercial activities (charity sales, fair trade, etc.). The great wave of generosity for the victims of the Asian tsunami at Christmas 2004 was rather exceptional. Contrary to preconceived ideas, donations from individuals actually account for a very small proportion (around 10 per cent) of the income of the non-profit sector.[14] Compared to the Anglo-Saxon world, these levels are particularly low in countries like France, despite a slight increase during the last decade. Many French people remain very attached to the public service system: they argue that they already pay a lot of taxes and they are reluctant to finance private charities. Historically, the development of the Republican state has actually led to a correlative decrease in donations, legacies and philanthropic involvement since the early twentieth century.[15] Moreover, the omnipresence of the public authorities may have impeded civil society initiatives, in contrast to the principles of subsidiarity in Germany or free enterprise in Britain.

NGOs are not generally believed to have close relations with states. Many state-funded associations have actually no desire to reveal their dependence on institutional donors: they either try to hide the reality of the structure of their accounts or they justify it with various strategic arguments. The low level of private funds contradicts the official discourse of organizations which claim to be 'non governmental'. It also makes it more difficult for associations to receive state subsidies because often they must show that they have provided a minimum of funds themselves in order to qualify for grants. A recent report from the French Senate has thus highlighted the various (well-known) ways to increase the proportion of private funds in the budget of an NGO, artificially and completely legally.[16] One method consists in counting only subsidies from the Ministry of Foreign Affairs, leaving out those from local government, para-state firms or European agencies, as well as the civil servants or buildings given free of charge by the public sector. Another method is to use government allowances to build reserves that are then transformed into private funds over the following years. A similar process can be applied to the funds given by agencies that are financed by public authorities but are registered as NGOs. Similarly, some associations campaign to have their contributions counted at their 'true value'. In particular, they want to include the free manpower of their volunteers in order to raise the percentage of their private resources. In France for instance, the contribution of public subsidies to the non-profit sector falls from 57.8 to 33.4 per cent if the value of voluntary work is counted.[17] In the same fashion, the proportion of private resources in the budget of an NGO can easily increase when donations in kind are valued according to their retail rather than wholesale price, even if it means overestimating the contribution of individuals.[18]

Besides, other associations justify state backing by arguing that public subsidies lower the fundraising cost and increase the ratio of social mission, that is, the percentage of expenditures effectively devoted to humanitarian actions. As a matter of fact, those NGOs whose work depends on the generosity of individuals have to use a fairly substantial part of their budget for marketing and advertising campaigns. The reactions of the general public are rather contradictory in this respect. On the one hand, 56 per cent of French people find it normal that NGOs should spend money on advertising. On the other hand, 50 per cent do not accept that their donations may be used to cover the publicity costs of the associations, and 59 per cent do not trust organizations who send them mail adverts.[19] In an attempt to minimize overheads, some NGOs change perspective and pay more attention to the ratios of social missions than to the proportions of public funding. In this way, the percentage of the budget spent on the headquarters of an organization becomes a decisive criterion for assessing the efficacy of actions in the field. German NGOs that want to get subsidies from their government thus must show that the ratio of their social mission is at least 80 per cent net, once management and funds collection costs have been deducted.[20] Thus follows a system that is tacitly sustained by both sides. Despite large management costs, governmental aid agencies actually refuse to bear the administrative cost of the humanitarian programmes set up by NGOs. The ratios officially given represent around 5 per cent of the total budget but are not realistic: they force associations to minimize, or even falsify, the amount of money they spend on the running costs of their headquarters.

Tales and Legends About the Synergies Between the State and NGOs

Nevertheless, some people will still believe that the subsidies of western states leave NGOs room for manoeuvre. According to this school of thought, one should not be deceived by the freedom of action given by private funding. After all, non-governmental 'donactors' do not give without any strings attached. Local and multinational firms impose geographic or thematic constraints when they finance emergency operations in countries in crisis where they have interests at stake. Individuals also have their priorities and preferences: this is demonstrated by the controversy over the international solidarity associations that wanted to transfer to other areas the financial surplus they received during the Asian tsunami of 2004. Conversely, state subsidies are not necessarily synonymous with restrictions. Researchers have shown that they do not stop NGOs from criticizing the authorities.[21] Given the great variety of humanitarian financing, international solidarity associations can also diversify the sources of their public income to play one off against the other and escape from the political constraints of aid programmes. It is even a question of economic survival: in France, NGOs like EquiLibre in 1998 or Hôpital sans frontières in 2002 have thus gone bankrupt because they were too dependent on a single donor who suddenly decided to stop funding them.[22]

If one believes the promoters of a partnership between states and international solidarity organizations, the autonomy of humanitarian operators would be guaranteed by the diversification of their income sources. This indicator would be more relevant than the simple proportion of public funding in a budget. However, a case-by-case analysis shows that such a claim must be qualified. In today's Iraq, an NGO financed by the United States, Britain and Australia would in reality have very little room for manoeuvre since these three states are involved in the conflict. In itself, diversifying public subsidies does not guarantee anything if it merely reflects the consensus in the foreign and migration policies of the states, in particular of European states.

In the financial structure of an NGO, it is rather the proportion of unrestricted funds that shows the strategic independence of an association. Earmarking determines which areas will be chosen for action and its percentage is so crucial that humanitarian operators disclose it rarely – the International Committee of the Red Cross and ActionAid excepted. Large NGOs thus consider that it is not necessary to indicate what percentage of governmental contributions are earmarked: the British at Oxfam argue that public funds are necessarily earmarked, while the Americans at CARE argue that state subsidies are free from all political strings.[23]

Along with these very mixed statistics, which make comparison difficult, the relative opacity of humanitarian operators emphasizes the analytical limits of a purely financial vision of NGOs and their donors. In fact, more subjective and qualitative elements interfere, because the nature of partnerships between governments and associations depends above all on the institutional culture of each side. As far as NGOs are concerned, the dynamics of organizational development and programme profitability can well lead them to favour economic survival over a critical appraisal of their humanitarian involvement. In fact, very few organizations will risk losing a source of income and break a public contract if an aid programme is counterproductive and serves only to promote political interests.[24] It sometimes happens that relations between international solidarity associations and the authorities of their host country display a certain osmosis. A director from Oxfam-USA, an organization known for its independence from (even for its insolence to) public authorities, explained that NGOs who work as subcontractors for the US government and who obtain most of their budget from the US have 'the mentality of a civil servant. They think like the government. They do not need to be told what to do to satisfy Washington's expectations.'[25]

Everything depends on the attitude of the authorities. Individual cases are very different in this area. On the one hand, the United States generally uses international solidarity associations as a 'soft power' to relay its foreign policy. On the other hand, relations between the Jacobin and centralized French state and associations have long been tense, on both sides. Until recently, the French state has not, or has rarely, tried to use international solidarity associations for its own ends through subsidies. Despite attempts at partnership and decentralized cooperation from the twenty-two regions, ninety-five departments and 36,000 'communes' of France, the proportion of development aid that public authorities give to NGOs remains one of the lowest in Europe.[26] Between these two extremes of the US and France exists a variety of situations. Scandinavian states in particular are very generous with NGOs and are known for respecting the independence of their partners.

In the end, it seems difficult to generalize. At a quantitative level, earmarked funds (more than the proportion of public funding) reveal several degrees of dependence. At a qualitative level, subsidized NGOs are undeniably linked to public authorities through overlapping interests: NGOs depend on the funds of institutional donors and states need NGOs to access areas in crisis, to obtain specific information and to implement humanitarian programmes, even if this means containing displaced populations in camps to stop them from coming to the West for asylum. From such an assessment, relations between the two types of actors appear more or less interconnected. Case-by-case analyses show a great variety of situations, from mere subcontractors to NGOs that have been manipulated against their will to those that have been able to reject subsidies in order not to compromise their freedom of action. A more in-depth study would also consider the lack of public funding for some crises, 'forgotten' because they are not very strategically relevant, or deliberately 'left aside' because they do not correspond to the political criteria of donors. It is here that the limits to cooperation between NGOs and states appear most clearly, as do the structural contradictions between the pursuit of national interest and the moral imperatives of humanitarian aid which is supposed to be universal and impartial, even to the extent of helping out the 'enemy' of the moment.

(Translated by Mélanie Torrent)

Marc-Antoine Pérouse de Montclos is a researcher at the Institut de recherche pour le développement, Paris.

NOTES

1   Catherine Agg, *Trends in Government Support for NGOs* (Geneva: UNRISD, 2006), p. 14.
2   Ian Smillie and Larry Minear, *The Charity of Nations: Humanitarian Action in a Calculating Wrld* (Bloomfield, CT: Kumarian Press, 2004), p.184.
3   Judith Randel and Tony German (eds), *Global Humanitarian Assistance 2003* (London: Development Initiatives, 2003), p. 8.
4   The following figures are taken from the updated data of the impressive study carried out at Johns Hopkins University on the non-profit sector in the world: Lester M. Salamon, Helmut K. Anheier

et al. (eds), *Global Civil Society: Dimensions of the Nonprofit Sector* (Baltimore, MD: Johns Hopkins Center for Civil Society Studies, 1999).
5   Claude Emerson Welch, *Protecting Human Rights in Africa: Roles and Strategies of Non-Governmental Organizations* (Philadelphia: University of Pennsylvania Press, 1995), pp. 278–9.
6   Alberto Alesina and David Dollar, 'Who gives aid to whom and why?', *Journal of Economic Growth*, 5, 1 (2000), pp. 33–63.
7   James Meernik, Eric Krueger and Steven Poe, 'Testing model of US foreign policy: Foreign aid during and after the Cold War', *Journal of Politics*, 60, 1 (1998), pp. 63–85.
8   Todd Moss, David Roodman and Scott Standley, *The Global War on Terror and US Development Assistance: USAID Allocation by Country, 1998–2005* (Washington, DC: Center for Global Development, 2005), p. 19.
9   Agg, *Trends in Government Support for NGOs*, p. 23.
10 Henri Rouillé d'Orfeuil, *La Diplomatie Non Gouvernementale: Les ONG Peuvent-Elles Changer le Monde?* (Paris: Ed. de l'Atelier, 2006), p. 25.
11 On this, see the data of the Observatoire de l'Action Humanitaire: <www.observatoire-humanitaire.org>.
12 Peter Uvin, 'From local organizations to global governance: The role of NGOs in international relations', in Kendall Stiles (ed.), *Global Institutions and Local Empowerment: Competing Theoretical Perspectives* (Basingstoke: Macmillan, 2000), pp. 9–29; Joelle Tanguy, 'The sinews of humanitarian assistance: Funding policies, practices, and pitfalls', in Kevin Cahill (ed.), *Basics of International Humanitarian Missions* (New York: Fordham University Press, 2003), pp. 200–40.
13 Salamon et al. (eds), *Global Civil Society*, p. 24.
14 Helmut K. Anheier and Lester M. Salamon, *The Emerging Nonprofit Sector: An Overview* (New York: Manchester University Press, 1996), p. xix.
15 Jean-Luc Marais, *Histoire du Don en France de 1800 à 1939: Dons et Legs Charitables, Pieux et Philanthropiques* (Rennes, Presses universitaires de Rennes, 1999).
16 Michel Charasse, *Les Fonds Octroyés aux Organisations Non Gouvernementales (ONG) Françaises par le Ministère des Affaires Etrangères* (Paris: Sénat, Rapport d'information no. 46, 2005).
17 Salamon et al. (eds), *Global Civil Society*, p. 87.
18 Ian Smillie, 'NGOs and development assistance: a change in mind set?', in Thomas Weiss (ed.), *Beyond UN Subcontracting Task-Sharing with Regional Security Arrangements and Service-Providing NGOs* (Basingstoke: Macmillan, 1998), p. 189.
19 From a survey on the donations given to associations, carried out in September 2005 by the Institut CSA (Conseil Analyse Sondage), available on <http://www.csa-fr.com/dataset/data2005/opi20050915e.htm>.
20 Judith Randel and Tony German, 'Germany', in Ian Smillie and Henny Helmich (eds), *Stakeholders: Government-NGO Partnerships for International Development* (London: Earthscan, 1999), p. 118.
21 Samuel Lucas McMillan, *Fueling Funding Dependency? Northern Governments, NGOs and Food Aid* (San Diego, CA: Annual Meeting of the International Studies Association, polycop., 2006).
22 In 2002 for instance, HSF (Hôpital sans frontières) was put into official receivership. Founded in 1976 by Tony de Graaff with the support of the Rotary International French clubs and in partnership with Médecins du Monde, HSF built rural hospital, very often relying on French Air Force Transaal planes for transport. Typically, as the NGO grew, it took its distance from its initial donor and witnessed a rapid rise in its pay budget thanks to EU subsidies. In 2000, HSF's own funds represented a mere 4 per cent in a €1.8 million budget. Today, nothing is left of HSF but a small structure that purely relies on voluntary work – it is actually a Belgian branch, which was created in Namur in 1992 and remained linked to Rotary International. The case of Medicus Mundi, of which only a Spanish branch remains, is a little different: this NGO disappeared because it was unable to adapt and rely on media professionals to gain the generosity of the public.
23 Agg, *Trends in Government Support for NGOs*, p. 14.
24 Marc-Antoine Pérouse de Montclos, *L'aide Humanitaire, Aide à la Guerre?* (Bruxelles: Complexe, 2001).

HUMANITARIAN NGOs AND THE MIGRATION POLICIES OF STATES

25 Joseph Short, quoted in Brian Smith, 'US and Canadian PVOs as transnational development institutions', in Robert Gorman (ed.), *Private Voluntary Organizations as Agents of Development* (Boulder, CO: Westview Press, 1984), p. 164.
26 Religious NGOs are very telling in this regard. In France, the 1905 Act on the separation between Church and State has led Catholic missions abroad to be self-financing, more so than their counterparts in Germany where the contribution to parish churches is collected and redistributed by the government. Today, many Christian international solidarity associations still depend less on public subsidies than lay NGOs do; in terms of private resources, they demonstrate a better financial balance: lent collections and donations from parishioners actually provide them with an annuity.

PART II
Fortress Europe – The Solution?

PART II
Eastern Europe & the Wilderness

What Migration Policy for the EU?

EBERHARD RHEIN

The Global Scene: Past and Present

Human history has been an eternal sequence of migrations, more frequent in some parts of the world than in others, in some periods more intense than in others, often peaceful but often accompanied by military conquests and extermination of indigenous populations. Most migration takes place at short distances; say of less than 1,000 km, within a given country, from rural to urban areas, from regions short of employment opportunities to those with wider job markets and from poor to wealthy regions.

From 1850 to 1930, humanity saw huge waves of intercontinental migration, from overpopulated, repressive, hungry, poor Europe to the wide spaces of the Americas, Australia and South Africa. More than fifty million Europeans emigrated during those eighty years, up to 5 per cent of a world population of about one billion.

Since the turn of the twenty-first century, humanity is once again on the move, with even larger numbers involved. Today we are faced with economic migration, political migration and – increasingly – environmental migration. Most migrants move from poverty, misery and repression to the wealthier parts of the world. But we also see increasing numbers of people fleeing from political and military conflicts. More than two million Iraqis have migrated to tiny and poor Jordan and Syria since the US invasion in 2003! More than half a million Sudanese from Darfur have migrated to poor Central Africa and Chad in order to escape from brutal repression at home.

Globally, the number of migrants has increased to some 200 million. Yet as a percentage of the planet's population of 6.5 billion, this number is smaller than a hundred years ago. This is a big difference compared to a hundred or a hundred and fifty years earlier: the wide stretches of largely uninhabited land that European emigrants found in the Americas are no longer available. Illegal and 'chaotic' migration has replaced the well-organized ships full of migrants in the late nineteenth and early twentieth centuries. Migrants have become *personae non gratae* practically everywhere on earth, from the Gulf States to Japan, Europe or North America. They must secretly enter the countries of their preference and often spend many years before being admitted as legally residing citizens. There is not a single major country that still welcomes large numbers of immigrants. The standard policy paradigm has become 'migrants unwanted'.

The reasons for this rejection are simple. Most immigrants are illiterate and poor. They have difficulty integrating into what have become selfish societies and demanding labour markets. The risks of immigrants failing and becoming either a burden for social welfare (including unemployment benefits) or turning to petty crime are much higher than a hundred years ago. The gap between the incomes and lifestyles of potential immigrants and the citizens in reception countries has greatly widened.

Migratory pressure has become stronger than ever. Living conditions in many parts of what remains of the Third World – Africa in particular – are abominable: miserable housing and sanitary standards, health care and education, unemployment exceeding 20 per cent of the labour force. No prospects for improvement are in view. A grim life awaits, especially for youngsters who constitute more than half of the population, as a consequence of high birth-rates and low life expectancy.

We should not expect migratory pressure from the South to subside in the foreseeable future. Africa, parts of South Asia and Latin America house a huge proletariat of roughly a billion people who are ready to face the hardships and risks of illegality and suffering in any of the rich countries, if they could only get inside the supposed paradise. Every day they can see on television how rich people live elsewhere in the world. It would be against human nature if they were not dreaming of a similar life for themselves, whether in Manchester, Chicago, or Singapore. However, only the few dynamic and educated migrants, however, find it easy to integrate into rich countries. But their migration will unfortunately have a negative impact for the countries of origin: the so-called 'brain drain'.

Migratory pressure will run totally out of control after 2050 if humanity proves unable to tackle global warming successfully. Hundred of millions of 'environmental refugees' may lose their livelihood because of droughts or floods. Europe will not be spared: it will need to relocate several million people living on its coastline into its hinterland, in addition to tens of millions of environmental refugees from other parts of the globe.

Migration will therefore become a major concern for humanity in the course of the twenty-first century. Every country will be faced with this concern, in one way or another.

Intra-European Migration

For European societies, global migration is compounded with intra-European movements of workers and citizens. We have had such movements since the late nineteenth century: from Italy to France and Belgium, from Poland to Germany. With the beginning of European integration in the 1960s, the scope of these movements has increased: the affluent North and West have had to cope with successive waves of migrants from the South – that is, Italy, Spain, Portugal, Greece, Turkey, Yugoslavia – and since 2004, from the East – Poland, Romania, Bulgaria, Moldova, Ukraine, and so on.

Though most of these migrants came from similar cultural backgrounds – Christian, mostly literate, often with some professional training and work discipline – they still posed problems for citizens in the receiving countries. These include problems associated with language, cultural difference and the black labour market. However, the overall economic and social benefits from these migrations for European integration is undoubted. Intra-European migration has become part and parcel of European integration. It is now an integral part of the 'European constitutional order' – that is, the four freedoms – which no one contests.

Intra-European movements of people are here to stay and set to become even more intensive. More and more Europeans will spend part of their lives in other EU countries. (Almost 150,000 students spend at least half a year in a university of another EU country.) This will become an integral part of the future European identity, though it will take one or two decades before Europeans will have fully digested this new social reality: Europe will become a European multicultural society and a 'demographic melting pot' in ways never experienced before.

Migration from Turkey remains a case apart. The fear of another wave of Turkish migration, after the first not very successful one in the late 1960s, has induced the EU to exclude free circulation in the case of Turkish membership. According to the 2004 negotiation directives, Turkey will only join if it renounces free movement of workers, one of the four basic freedoms of any EU citizen. And it might take another decade or more after membership before Turkey is authorized to join the Schengen area and obtain the freedom to travel without visa requirements.

Migration from Outside Europe

For the EU, migration is a new policy area. Before the 1990s, migration from third countries was hardly a problem for any member state. The eastern borders were 'safe' thanks to the Iron Curtain; and from overseas very few people tried to enter Europe.

Formally, EU migration policy derives from the principle of free movement within the EU. In 1997, in the Amsterdam Treaty, the EU extended free movement also to third-country citizens living within EU borders. This in turn obliged member states to harmonize the rules on refugee/asylum status and visas. The Constitutional Treaty (Article III-267) goes a step further by stipulating that 'The Union shall develop a common immigration policy aimed at assuring, at all stages, an efficient management of migration flows and preventing illegal immigration.' In 2005, the EU has stepped up the control of its external borders by creating FRONTEX, a European Agency for Border Control, situated in Warsaw and with a small budget and a staff of fewer than a hundred officials.

Member states have aligned their visa regimes with third countries. They apply a very restrictive common visa regime which sets out the countries whose citizens need a visa to enter the EU, the type of visa (Schengen or national), the fees (€60 for a single-entry visa), and so on. Despite the restrictive visa regime, it is estimated that between half a million and a million migrants have annually entered the EU during the last few years, mainly from Latin America and eastern Europe (Ecuador, Moldova, Romania, Ukraine, Russia, Morocco and the former Yugoslavia). Most of these have gone to France, Spain, Italy and Germany, where they found jobs in agriculture, construction, tourist services, and so on.

This influx has led to successive waves of legalization, the most recent being in France and Spain. Indeed, member countries consider legalization the lesser evil compared to either tolerating hundreds of thousands of illegal workers not paying social security contributions, or repatriating them to their countries of origin, which is becoming increasingly difficult.

For the time being, the implicit ground rule for the EU remains that only small numbers of well-trained people are welcome, but no large-scale immigration from third countries. All governments seem scared at the social costs of illegal migration: tensions between migrants and citizens, criminality, unemployment benefits, housing ghettos, and so on. There is therefore at present no question of opening the EU fully to migrants from outside Europe. This basic position is unlikely to change before 2015 at the earliest, when Romania and Bulgaria will fully enjoy free movement of workers and when EU citizens should have become more used to a 'Europe without internal borders', which will include Switzerland, Norway and Iceland.

Under the impact of increasing anxieties about Muslim and Romanian migrants the few more open countries like Spain, the UK, Italy and Ireland have recently also taken more restrictive views on additional immigration. In the foreseeable future the EU is likely to reinforce border protection, especially at its border with Africa, and to put a cap on the swelling stream of Africans trying to establish a foothold in Europe. The year 2006 saw new routes being tested, for example, the Canary Islands, leading to a tripling of the numbers of illegal boat people. About 40,000 have landed in Spain and Italy. These numbers are not by themselves a cause for concern: instead it is the trend that they demonstrate. Millions are waiting south of the Sahara and in North Africa to follow the example of those courageous enough to risk their lives at sea. Those EU member states most directly affected are likely to deploy more effective naval interception in 2007, with increased financial and logistical support (night-vision equipment, speedboats, and so on) from the European Border Agency.

For the EU Ministers of the Interior, border control and the prevention of trafficking in human beings have become a priority. They have formed a Group of Six (Germany, France, the UK, Italy, Spain and Poland) in order to improve cooperation among national police and to ensure more effective border protection. However, border control will never be fully effective without cooperation from neighbouring countries. The EU and individual member states have therefore engaged in a close dialogue with Morocco, Tunisia, Senegal, Mauritania and other countries on how to come to grips with migrants from the South.

The EU has offered the Maghreb countries financial assistance and equipment to enable them better to control their own borders in the South. It is in the process of negotiating readmission agreements with its Mediterranean neighbours and some Sub-Saharan countries. It has also expressed a willingness to consider 'managed immigration' (from Sub-Saharan Africa and the Maghreb) in areas where Europe faces a labour shortage. Finally, the EU has promised some extra funding (€40 million) for job creation in the Sahel countries on the border between the Sahara and the more fertile South. These gestures of good-will are , however, meant to detract from the main thrust of EU migration policy, which is preventing illegal immigration.

Guidelines for the Future

What migration policy should the EU pursue in the future, whatever its constitutional rules? Should it allow substantially more temporary or permanent immigration? From what countries? For what jobs? According to what procedures? Why should it do so? There is no consensus on these sensitive issues, neither within specific European countries nor between countries.

The most common argument advanced to justify a more forthcoming immigration policy is the projected ageing and the decline of European society

Based on past trends, the EU's population might decline by some 75 million people (15 per cent of the present population of 500 million) between now and 2050. The percentage of people over 65 years of age will rise by 30–35 per cent over the same period. By 2050 one active citizen will have to finance one inactive citizen, compared to a ratio of 3:1 in 2005. It will therefore become increasingly difficult to cover future labour demand from domestic sources and to finance the rising burden of pensions. Those who cry most loudly for additional immigration are employers, especially those used to hiring cheap illegal immigrants whom they can exploit at will. The trade unions have a more cautious position, for good reason.

There is no doubt that the greying of Europe will radically improve the situation in the European labour market and make it substantially more difficult to finance the mounting burden of pensions. But this should not seduce governments into taking wrong decisions. Governments should carefully weigh the advantages and cost of immigration from outside Europe.

First, Europeans are far from fully exploiting their existing labour potential:

- There are some twenty million registered unemployed people in the twenty-seven member states of the EU. Our efforts should be directed at these as a matter of priority. The easy availability of young immigrants will distract attention from the huge army of unemployed people and keep the costs of unemployment and welfare benefits unduly high.

- The twelve new member states will have to make redundant another twenty million people in agriculture and industry in order to reach West-European productivity and income standards. This process will last for the coming twenty years.

Intra-European migration and full mobilization of this huge potential within each member country should accelerate the process of putting these people to work:

- West European citizens retire between the ages of 55 and 65, far too early compared to a rising life expectation that will soon exceed eighty years. In order to cope with the rising pension charges, all countries will progressively need to raise the statutory retirement age and to squeeze the level of pensions. This process, already under way, will progressively release some 10 per cent of the labour force. That is roughly the equivalent of another twenty million employed, which constitute the third element of our hidden labour reserve.

- Europe should massively invest in preventative medicine, especially against pandemics like smoking, drinking, high blood pressure and diabetes. In doing so it will enable people to remain fit beyond the age of 60 and minimize early retirement because of sickness or invalidity. Better health policy will mobilize millions of people for active life and employment beyond their fifties and sixties.

- The EU should massively invest in life-long learning. This is a precondition for absorbing the high unemployment, especially of older people (above 50 years of age).
- Last not least, Europe is better off with labour scarcity than with excess, because this will push for higher productivity and higher retirement ages (65–67).

Secondly, the immigration of active people will, of course, lower the average age of society and relieve the pension burden. But even young immigrants will retire, at the latest forty years after their immigration. European society will therefore only buy time. Immigration is a matter of convenience more than of necessity.

Third, immigrants represent an extra cost to society. They require special educational and integration efforts (language, technical skills). They may cause intra-societal tensions, and sometimes higher unemployment charges because their limited skills may expose them more easily to unemployment. These tensions are likely to increase with substantially higher share, say above 15 per cent, of immigrants.

Fourth, with appropriate family policies and incentives it should be possible to reverse the trend of the one-child family. France is an excellent example of the success of such policies. Its fertility rates are higher than in any other European country, except Ireland.

Fifth, there is no reason for Europe to keep up its population at the present level of some 500 million. What matters is not the GDP but the GDP per capita! A declining population does not require economic growth; zero growth will mean in fact higher per capita incomes. It seems neither desirable nor necessary to fill the emerging 'demographic gap' through immigration. After all, Europe is the most densely populated continent on earth. Countries like the Netherlands, Belgium, Malta and parts of Germany would enjoy a higher well-being with lower population levels.

In conclusion, there is no urgency for either the EU or individual member countries to decide on substantial immigration programmes. Europe can quietly observe demographic trends and the evolution of its labour market and study more thoroughly different options to confront demographic changes.

Another argument advanced, though less frequently, in favour of stepping up immigration is to relieve developing countries from their demographic and social pressures, in particular, Africa

The prosperity disparities between Europe and Africa are the primary motivation for migratory pressure. Rapid population increase exacerbates that pressure. Sub-Saharan Africa has the highest fertility rate (over five children per woman) and the highest population growth (over 2 per cent) world-wide. Its labour markets must absorb almost twenty million new entrants every year. This is an impossible mission for any society. But emigration should not be the answer. In the global world, every society must assume the responsibility for its offspring. African societies have failed to curb excessive population growth. The West in general and Europe in particular shares some responsibility for this failure.

It is therefore more urgent than ever that Europe tells African leaders that they must invest much more effort in sexual education and the distribution of contraceptives. And Europe must be more ready than in the past to support those efforts, instead of closing its eyes and shunning common responsibility. Population, health and educational programmes must become the top priorities for development assistance. Africa should learn from China's experience and promote the 'one-child family' – for the coming thirty years.

Europe should therefore take a dual approach:

- Make it crystal clear to African heads of government that it is neither willing nor able to absorb their demographic surplus.

- Direct its huge development aid into channels that will lower population growth, keep the population in the countryside, focus on agricultural development and, more generally job creation.

Conclusions

In conclusion, the EU should

- Be able to meet the challenge of its ageing population without large numbers of immigrants. Immigration would only offer a temporary respite. All societies will age and therefore have to put up with the phenomenon. Europe should offer an example of how to cope with ageing without resorting to immigration.

- Cover its labour demand by using the hidden European labour reserves, that is, some twenty million unemployed, some twenty million from new member countries, migrating from agriculture and industry, and some twenty million from a much longer active life.

- Strengthen its border controls to stem the flood of illegal immigration. It should provide the necessary means to FRONTEX to that end. The EU should establish zero tolerance as an objective.

- Refrain from using the 'brain drain' that stymies the efforts of development in the Third World. It should therefore minimize immigration of qualified labour – doctors, nurses, engineers, scientists, and so on – from developing countries, especially Africa. If it does, it should, when there is immigration, compensate the countries of origin for the educational expenses they have incurred. It should take the initiative to propose an international convention – for example, in the framework of the OECD – concerning immigration of highly qualified personnel from poor countries. Qualified personnel should only be authorized for a limited period, say up to three years.

- Make clear to all concerned – its own businesses, potential emigration and transit countries, interested international organizations – that it has no intention of increasing immigration from outside Europe.

- Enhance its development assistance, especially to Africa, and target it more to population, education and employment programmes.

- Mobilize the US, China, Japan and Russia for an effective campaign against global warming in order to prevent chaotic and large-scale environmental migration across the globe in the course of the coming fifty years.

The time has come for the EU to get its act together on migration/immigration policy. The European Council should invite the Commission to submit, before the end of 2008, a comprehensive study and recommendations on its future migration EU strategy. Europe needs a critical and comprehensive debate on a subject of great complexity and far-reaching consequences for its future.

Eberhard Rhein is a senior policy adviser at the European Policy Centre, Brussels.

Immigration: A French Taboo

IVAN RIOUFOL

In France, for the first time, immigration is no longer a completely taboo subject. The issue was largely absent from the presidential campaign in 2002. However, with the social and economic crisis, it is likely to be at the heart of the campaign in 2007. The official position has broken down under the burden of proof: through a desire to adhere to a strict anti-racist and universalist ideology, this position denied all problems posed by the constant and huge-scale arrival of immigrants from North Africa and Sub-Saharan Africa. The riots in the suburbs in November 2005, which were called 'the first intifada of the French banlieues' by some commentators, caused people to wake up to the threat posed to national cohesion and identity by mass unregulated immigration and by a refusal to impose a duty on immigrants to integrate. This phenomenon of immigration comes on top of an unprecedented emigration: more than 2 million French people have left the country, which has led some to say that France is exporting graduates and importing illiterates.

I would like to examine the mechanisms by which reality is denied. The right-wing candidate at the 2007 presidential election, Nicolas Sarkozy, has said (on France-Inter on 9 October 2006), 'We have not had the courage to speak of immigration. For thirty years, we have been passively experiencing the phenomenon without taking control of it.' I would add that we have not even been able to measure the phenomenon of migration or to evaluate it. There are no reliable statistics on the arrival of new immigrants. The number of persons who are either immigrants themselves or who have an immigrant parent or grandparent is said to be around 15 million, a figure given by the Minister for the Equality of Opportunity, Azouz Begag. Concerning Muslims alone, demographers agree that the figure is between four and six million. They believe that by 2023 18 per cent of births in France will be of African or Turkish origin.

In reality, for more than thirty years, a political and media coalition of silence has kept the French people ignorant of the enormous ethnic change under way in their country. Officially, multiculturalism was not recognized until 11 December 2003, on the occasion of the submission to Jacques Chirac of a report by the Committee for the Application of Secularism in the Republic, a report better known as the Stasi Report. This report eventually led to the law forbidding the wearing of the Muslim veil in schools. That document said, 'France today is one of the most diverse European countries ... This major break in its history gives it the chance to enrich itself through a free dialogue between its diverse component parts.' On this occasion, Jacques Chirac invited the French people 'to rediscover a new political community' (*communauté de destin*).

The problem is that no political leader had mentioned this 'major break' before or said that 'a new political community' needed to be invented. On the contrary, everything had been done to make people think that this extra-European immigration, which started in the 1960s, was a non-issue and that the French were wrong to be worried about living side by side with new cultures and especially with Islam, which was presented as a religion of peace and tolerance, easily compatible with the secular state and the modern world. 'The richness of a nation, the richness of a people, is in large measure a result of multiculturalism, that is to say of the presence of different cultures,' said President Chirac, adopting for himself the rhetoric of the left wing and extreme-left pro-immigration lobby and, with it, the praise of multiculturalism of which we are only now beginning to appreciate the damage it has caused to national cohesion and identity.

It is because people's minds have been lulled into a daze, a daze made worse by the weight of political correctness which forbids anyone from making any critical judgement about these new minorities, that many French people are now shocked when they discover, little by little, the extent to which their country and its culture have been overturned without them having been asked or even informed by their representative and the media. People themselves make the previously forbidden link between immigration and crime when they see that urban violence has become a daily occurrence. According to recent figures provided by the Minister of the Interior, every day fifteen policemen are wounded in such incidents and no fewer than 115 cars burned. There is now a genuine risk of separatism between communities, especially since the prefect of Seine-Saint-Denis, the *département* which borders Paris, revealed in September 2006 the state of affairs in his area, where no less than two-thirds of its 1.5 million inhabitants are immigrants or of immigrant origin. Youths are often attracted to Muslim extremism.

It is essential to briefly mention the National Front. From the 1980s, Jean-Marie Le Pen's party warned against such unregulated immigration from former colonies. From this point of view, the French admit that the FN was right before the other parties, which explains why it continues to enjoy strong electoral support. However, by putting himself beyond the pale politically, with too many racist, xenophobic and anti-Semitic outbursts, Le Pen and his party has also put beyond the pale all the political issues it addresses. The *Front national* has made it impossible to have a calm debate about immigration, and has thereby caused us to lose precious time.

I would like to look a little more closely at how the law of silence has been imposed on the question of immigration, by means of two truly Orwellian slogans, 'Immigration is an opportunity for France' and 'France is not a country in which there is mass immigration.'

Before it became a slogan, the first of these propositions was the title of a book, written in 1984 (appropriately enough!) by Bernard Stasi, a politician whose father was an Italian immigrant. The book surmised that the new black African and North African immigrants would integrate into French society without difficulties, as previous immigrants had done in the past. Thus, for twenty years, the party line was maintained without anyone looking too closely at the social disorder caused by such large-scale immigration. During all these years, the French model of integration was accepted as gospel truth, on the basis that it had worked in the past.

It is true that immigration has long been good for France. While Louis XIV was a child, France was governed by Anna of Austria and the Italian Cardinal Mazarin. In more recent history, immigrants from Belgium, Russia, Poland, the Balkans, Armenia, Italy, Spain and Portugal in the twentieth century mixed in remarkably well with the indigenous population. This is all the more surprising since the majority of the new arrivals suffered hostility from parts of the population. There was a massacre of Italians in Aigues Mortes in 1893, for example. Immigrants also suffered discriminatory treatment by the French administration which is unimaginable today.

In fact, everything which might have contradicted this image of peaceable immigration has been simply suppressed for the last thirty years. Left and right followed their orders: don't mention the *banlieues*! The slightest question or criticism of mass immigration was forbidden. Although today politicians have started to talk about immigration, they carefully avoided it in the past because even to mention it was to render oneself suspect of racism in the eyes of those who spoke only of universal man, the respect of the other and the abolition of differences.

It was not until 2000 that the media finally admitted, under pressure from what was obvious, that the French model of integration had failed and that there was a link between immigration and rising crime on housing estates, in schools, on public transport and in hospitals. Today there are 630 problem housing estates in which there is a sort of apartheid. The refusal to acknowledge this situation, to actually visit these areas, has led the press and antiracist movements to underestimate during this very same period the rise of anti-Semitism in Muslim suburbs; this culminated in 2006 in pure barbarism when the young Ilian Halimi was tortured to death because he was

Jewish. Anti-white racism is also largely underestimated by the media and politicians.

This same desire not to contradict the dogma that immigration is a good thing for France has also caused the media to ignore the fact that the Muslim community is become increasingly fundamentalist. It was not until 2003 that the Stasi Report described how society was fragmenting. It was not until 2004 that the Obin Report revealed Islamist pressure on state schools. It was not until 2005 that the Dénécé Report expressed concern about the rise of Muslim fundamentalism in the car construction and large-scale retail sectors. Finally, it was not until 2006 that the Muslim fundamentalist networks operating at Paris's main airport were revealed by the conservative politician, Philippe de Villiers. The press played no role whatever in any of these revelations.

The second slogan – 'France is not a country of large-scale immigration' – has enabled any claims to the contrary to be denounced as extremist fantasies. Precisely this dogma was taught by the Director of the National Institute for the Study of Demographics, INED, François Héran, who published a report in January 2004 entitled 'Five illusions about immigration'. In this document, which was hailed with near unanimity by the press, Héran wrote:

> France has certainly been a country of immigration for a long time but it has not been a country of large-scale immigration for at least the last twenty-five years. It has become, on the contrary, the European country in which demographic growth depends the least on immigration. The current influx of immigrants remains well below the levels of thirty years ago. The French authorities are managing the flow much better than people say.

INED estimated that there was net immigration of 75,000 people a year with 13,000 illegal immigrants.

Other studies by the national statistical office (INSEE) supported these claims. In a study published in January 2006, for instance, the National Statistical Institute claimed to show that the increase in the population did not stem from immigration but instead from the high birth-rate of people of French origin. According to this official body, immigration accounts for only 25 per cent of the population growth against 80 per cent in the rest of the EU. The author of this report concluded, 'It is true that immigrant women have more children than others. But the proportion of mothers of foreign origin is so low that their contribution to the birth-rate in France is very limited.'

No one questioned these findings, in spite of the fact that the Court of Accounts in 2004 expressed alarm at the fact that there were 300,000 people entering the country every year, including illegal immigrants, and in spite of the fact that the Department of Population and Migrations (DPM) within the

Ministry of Social Affairs revealed, three months after the publication of the report by the National Institute for the Study of Demographics, that the number of legal immigrants settling permanently in the country had jumped by 36 per cent between 1999 and 2002. It is clear that we are dealing here with a desire not to see the truth because this sudden increase was well known to those who listened to the concerns of certain demographers instead of to the invented calculations of INED and INSEE.

In particular, people should have listened to the demographer Jacques Dupâquier, a member of the Institut de France. Writing in *La nouvelle revue d'histoire* (Jan.–Feb. 2006) he said:

> The scandal is that the real figures on immigration have always been hidden from the public. There is a genuine smokescreen hiding the statistics. The total number of legal entries was 125,000 in 1995 but has reached 217,000 in 2003. This figure, however, does not include minors or asylum seekers. Consequently, we can estimate that there are in total about 300,000 immigrants a year. This is a considerable rise, especially since it is now immigration of a different type. This new wave of immigration, principally from Africa and Turkey, is not an influx of labour, as some people continue to pretend, but an influx of population. The worst aspect, in the current context, is that this immigration is concentrated in the Paris region. In my view the situation is very serious.

As far as the supposed high birth-rate among French women is concerned, Dupâquier quickly set the record straight when he pointed out that the French National Statistical Office (INSEE) was including figures from the overseas *départements*, without making this clear, and that the figures did not take account of the birth-rate among women of Turkish and African origin, which is around 3.4 children per woman. He insisted that in reality the birth-rate of the European population was not being calculated and he put it at 1.7 children per woman.

It turns out that it was Dupâquier and his friends who were right. INSEE admitted as much in a report published in August 2006, when it said in an inset that it had 'underestimated the number of immigrants'. According to the corrected figures, immigration from black Africa rose by 45 per cent in the period 1999–2005. The head of ISEE's demographic department, Guy Desplanques, was forced to recognize 'the rise in immigration in the last ten years' thereby refuting the official line which had been peddled up to then. It in interesting to note that the media took no notice of this admission that public authorities had practised disinformation, and that it had been cooking the figures, even though the media is normally hungry for scandals and quick to denounce lies. In keeping silent about this, the media have themselves contributed to this state of denial.

However, the taboo on immigration has not been totally lifted. There is still much reluctance among politicians to tackle the issue head on, out of fear of giving the National Front a helping hand. But the public is increasingly aware of what is going on. It is precisely because of opposition to Muslim immigration that there has been such strong opposition to the accession of Turkey to the European Union.

France is today at a turning-point, like many other western democracies faced with large-scale immigration which creates ghettoization and resentment. Even if France's model of integration has weathered somewhat better than the Anglo-Saxon model (according to a study conducted by the Pew Research Center, 46 per cent of French Muslims define themselves by their religion rather than their nationality, while the figure is 80 per cent for British Muslims) France none the less must ask itself a number of disturbing questions.

Whereas until now tolerance and generosity has been the unthinking response to these new challenges, French people feel that, by continuing to apply humanist principals unilaterally and on their own, the nation, born from a Judeo-Christian and Greco-Latin heritage, is in danger of becoming impoverished or even of disappearing before the end of this century by an influx from another culture which has little in common with its own and which presents itself simultaneously as both victim and master. Already the Ministry of Education is hesitating to teach French culture alone and the school system has had to accept ethnic particularism in numerous cases: some prayer rooms have been opened, teachers decide not to teach Voltaire, schoolchildren demand separate changing room for their gymnastics classes. Town councils and job centres, who are obliged to communicate in French, have started distributing forms in Arabic. Polygamy is tolerated by the administration in spite of the fact that it is illegal, while the Muslim veil is becoming ever more common on housing estates. Ramadan, which was rarely observed ten years ago, is now becoming the norm.

Thirty years of taboo and of a lack of debate on these questions mean that it is impossible to find answers to the following questions. Is democracy compatible with Islam? Should a nation require new arrivals to learn the language and the rules of society? Should there be a code of integration? Should ethnic communities be separated? Is a reverse colonization process underway? What can Europe do to help countries of immigration? Should immigration be stopped? These are all questions which cannot be suppressed by the ideology of anti-racism, of whom the philosopher Alain Finkelkraut, said recently that it was 'the Communism of the twenty-first century'.

Ivan Rioufol is a journalist with *Le Figaro*, Paris.

Why is the Vlaams Belang so Popular?

PAUL BELIEN

The Vlaams Belang (Flemish Interest) is the largest single party in Belgium. It has not lost a single election in twenty-five years, growing slowly but continuously from 3 per cent of the vote in 1987, to 10.3 in 1991, 12.3 in 1995, 15.8 in 1999, 18.2 in 2003 to 24.1 per cent in 2004. It is the most successful of all the so-called 'Euro-nationalist' parties.

Originally founded in 1977 as a radical Flemish secessionist splinter group which broke away from the Flemish nationalist party, its main political goal is the establishment of Flanders, the Dutch-speaking northern half of the federal kingdom of Belgium, as a sovereign state and an independent republic. In the late 1980s, the party managed to transform itself into a mainstream conservative party, aptly filling the gap that the free-market Liberals and the Christian Democrats created when they moved to the left. The Vlaams Belang is also the only Eurosceptic party in Belgium and the only one that rejects the official state ideology of multiculturalism. The Vlaams Belang says it is firmly established on three pillars: Flemish independence, opposition to multiculturalism, and the defence of traditional western values.

The party differs from the other Euro-nationalist parties in three ways. First, it has a collective leadership and does not depend on strong charismatic and often impulsive leaders such as Jean-Marie Le Pen in France, Jörg Haider in Austria, or the late Pim Fortuyn in the Netherlands. The media, who by their nature want to personalize politics, often depict Filip Dewinter as the leader of the Vlaams Belang, but Dewinter is not the party president. Dewinter is the chairman of the Antwerp chapter of the party and the president of its group in the Flemish regional council, but he is not the party's national leader nor the chairman of its group in the federal Belgian House of Representatives. The Vlaams Belang is not a one-man-show.

Secondly, though generally depicted as an Islamophobic anti-immigration party, the Vlaams Belang is not a one-issue party but rather a coalition of people who for various reasons oppose Belgium's traditional parties and their policies.

Third, the party is seen by the Belgian establishment as a mortal enemy of the state, not only because it rejects the multiculturalist state ideology but primarily because it wants to dismantle the state itself. Imagine a Front National in France, with a young, collective leadership, with a broad electoral base enabling it to attract a broad range of conservative voters which other parties no longer cater for, and a party programme that states as its primary goal that it wants to abolish France.

The Vlaams Belang is a unique phenomenon unlikely to occur elsewhere, except perhaps if the European Union should ever develop into a fully fledged federal superstate. Then one could easily imagine a kind of pan-European Vlaams Belang which would oppose multiculturalism and Islamization while at the same time aiming for the abolishment of the European superstate and the re-establishment of the old, former nation-states.

The Vlaams Belang's opposition to Belgium and the European Union are motivated by the same considerations. It considers multinational Belgium, an artificial state constructed by the European powers in 1831, to be the prototype of the federal European state that Eurofederalists want to establish. The Vlaams Belang opposes Brussels and does not really differentiate between Brussels as the capital of Belgium and Brussels as the capital of Europe. It sees both Belgium and Europe – in other words Brussels and Brussels – as enemies of Flanders and of the Flemish people, whose national identity these enemies want to subvert and destroy.

According to the Vlaams Belang – and its success in conveying this message explains its continuous electoral successes – immigrants are being used by the Belgian establishment as a weapon against the Flemings. In other words, immigration has been deliberately promoted with the goal of undermining national loyalties which people adhere to because it gives meaning to fundamental existential questions such as 'Who are we?' and 'What is our identity?'.

Belgium is a country consisting of different peoples: Dutch-speaking Flemings, French-speaking Walloons, and a small number of Germans. Islamic immigrants who settled in Belgium and have acquired (or rather been given) Belgian nationality, tend to be the only group of Belgians whose loyalty is to Belgium, rather than to Flanders or Wallonia. Similarly it is imaginable, and even probable, that if Europe ever becomes an artificial superstate, non-European immigrants would be the only group to identify with this European state and new nationality rather than with that of the old former national entities.

The Belgian authorities have always seen Flemish nationalism as endangering the existence of the multinational Belgian state. They have also explicitly stated that immigrants are to be used by the ruling establishment as a weapon against the Vlaams Belang.

Until 2004, the Vlaams Belang was called the Vlaams Blok. Its party slogan was 'Our own people first'. After the Vlaams Blok had become the largest party in the local elections in Antwerp in 2000, the Belgian establishment decided to extend the right to vote in local elections to non-Belgian residents. It also decided to make it easier for non-Belgians to acquire Belgian nationality, with the intention of creating so-called 'new Belgians' entitled to vote in the general elections. In September 2000, Leona Detiège, the then Socialist mayor of Antwerp, told the press (*Knack Magazine*, 13 September 2000): 'The Vlaams Blok is currently over-represented because the immigrants are not allowed to vote', while Johan Leman, the then director of the Centre for Equal Opportunities and Opposition to Racism (CEOOR), a government agency working for the Belgian prime minister, announced (*De Standaard*, 15 January 2000): 'What will "our own people" still mean fifteen years from now? We will get so many new Belgians that this slogan becomes meaningless. The Vlaams Blok is a thing of the past.'

In the Belgian local elections of October 2006 the immigrant vote tipped the balance in favour of the ruling establishment. In Brussels more than one-fifth (21.8 per cent) of the municipal councillors are now immigrants of non-European origin. Most of them are Muslims, and most of them have been elected as Socialists. The non-European immigrants' vote is overwhelmingly socialist, owing to the fact that many of them migrated to western Europe attracted by the subsidies of its generous welfare states. The immigrants have become the electoral life insurance of European socialism.

In the Brussels borough of Sint-Joost-ten-Node (where the Vlaams Belang party headquarters is located), eleven of the sixteen Socialist municipal councillors are non-European immigrants, as are four of the five Christian Democrats, two of the three Greens and two of the three Liberals. In Antwerp, almost one-third of the Socialist councillors (seven of twenty-two) are Muslims, as are one-third of the Christian Democrat councillors (two of six). In Ghent, one-quarter of the Socialist councillors are Muslims. In Vilvoorde, a Flemish town 20 km north of Brussels, half the Socialist representatives are Muslims. Many of these Muslims have radical Islamist sympathies. It bothers many traditional indigenous Socialists that their party has sold out to radical Islamism. After the elections one of them told the Brussels newspaper *Le Soir* (11 October 2006): 'Whenever one of the Belgo-Belgians [the indigenous Belgians] complained about this, he was considered to be a racist' (*quand l'un des candidats belgo-belges avait le malheur de parler de ça, il était perçu comme raciste*) No wonder that even many French-speaking indigenous former Socialists have begun to vote for the Vlaams Belang.

The Belgian establishment's harassment of the party has only brought it more sympathy from the public. On 9 November 2004, the Supreme Court in Brussels declared the Vlaams Blok, then already the country's largest party, a criminal organization. In the 1990s, in an effort to destroy the party the Belgian Parliament changed the Constitution and voted a series of new laws, including an Anti-Racism Act and an Anti-Discrimination Act, which define 'discrimination' so broadly that anyone and everyone is vulnerable to hate-crime prosecutions. These new laws were used against the Vlaams Blok.

The party was convicted on the basis of an anthology of sixteen articles published by various local Vlaams Blok chapters between 1996 and 2000. Though many of these articles simply quoted official statistics on crime rates and social welfare expenditure, the court posited that they had been published with 'an intention to contribute to a campaign of hatred'. One article, which dealt with the position of women in fundamentalist Muslim societies, was written by Belkiz Sögütlü, a female Turkish-born Vlaams Blok member who had herself been raised in such an environment.

Following the verdict, the Vlaams Blok was forced to disband. Its leaders, however, promptly founded another party, the Vlaams Belang (which means 'Flemish Interest'). It went on to win the next elections. At the time of writing, the Vlaams Belang is again in court. The Council of State, a Belgian administrative court, will rule some time in 2007 whether or not to cut off the party's state funding. This is an attempt to eliminate the party by depriving it of all its sources of income. Ten years ago, the Belgian authorities decided to make it illegal for political parties to accept private donations. Instead they have since been subsidized by the state in accordance with the number of votes gained in the last elections. Parties that are considered to be 'enemies of the state' can, however, be deprived of funding on the grounds that it is illogical for the state to fund its own enemies.

There is little doubt that the party will, indeed, be stripped of funding. The party leadership says that they have set aside funds in a so-called 'war chest' to allow it to participate in the June 2007 general elections. There is little doubt that the Vlaams Belang will also win these elections. If the Belgian authorities take the party's money away, this will cause an outrage and even more indigenous voters will flock to the Vlaams Belang.

The party does realize, however, that its ability to keep winning elections depends on demographics. Its electoral potential will decline when the number of indigenous voters goes down. The Vlaams Belang was the only party in the October 2006 local elections that did not put forward Muslim candidates.

In Antwerp, Flanders' largest city, it won 33.5 per cent of the vote, compared to 33.0 in 2000. This may look like a small gain, but it is significant since the demographic makeup of Belgium's cities is rapidly and dramatically changing. The municipality of Antwerp has half a million inhabitants. Every year on average 4,000 indigenous Flemings move out of Antwerp, while 5,000 immigrants settle there. Most immigrants are

Muslims. These people do not vote for a party that has no Muslims on its list, that opposes multiculturalism and demands that immigrants assimilate and respect the values of their host country.

According to the Marxist sociologist Jan Hertogen, 'The immigrants saved democracy in Belgium' (*Gazet van Antwerpen*, 9 October 2006). Hertogen calculated that if the franchise had not been extended to immigrants the Vlaams Belang would have polled 40.4 per cent in Antwerp instead of 33.5 per cent.

Before the elections, Filip Dewinter, the party's leader in Antwerp, told the Brussels newspaper *Het Laatste Nieuws* (9 September 2006) that his party could not continue to win. 'I am a realist,' he said:

> The number of potential voters for our party is declining year by year ... In ten years' time the number of new Belgians in Antwerp – half of whom are Moroccans – has doubled ... If the number of foreigners in Antwerp continues to grow by 1.5 per cent a year, as it currently does, then in twenty years from now there will be more people of foreign than of indigenous extraction in this city ... Our party has foreign members, but I do not want to be a hypocrite. At present we do not put forward 'alibi Ali' candidates. But I know that it is bound to happen some day. We extend our hands to welcome every foreign-born person who wants to become a Fleming among the Flemings.

It is doubtful, however, that many rent-seeking immigrants, attracted by welfare benefits, would want to join a conservative party that is generally perceived as being opposed to the European welfare system.

This is why I think that the shift to the right in Belgium, but also in western European politics in general, will be over by the end of the decade, when the impact of the immigrant vote will shift politics dramatically to the left. The Belgian and European Social Democrat establishments are slowly but gradually succeeding in what Bertold Brecht advised totalitarian regimes to do: 'dissolve the people and elect another.'

Dr Paul Belien is the editor of *The Brussels Journal* <www.brusselsjournal.com>, and an adjunct fellow of the Hudson Institute.

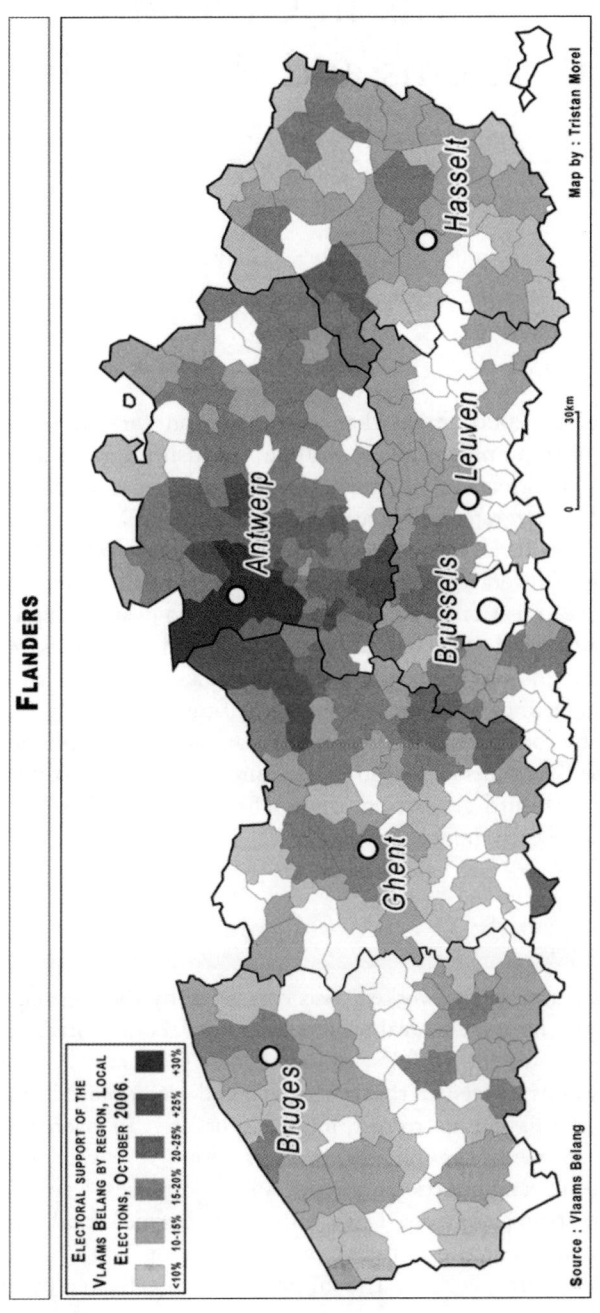

Turks in Germany: Problems and Perspectives

ASIYE ÖZTÜRK

Introduction

At the end of 2006, more than eight million foreign residents live in Germany. With a number of 2.4 million, Turkish migrants are not only the largest national group of foreigners, they are also regarded as the least socio-economically integrated. This article strives to give a short overview of the situation of the Turks in Germany and examines the level of their integration in the light of several factors. It explores aspects of so-called structural integration (the acquisition of rights and the access to positions and status in the core institutions of the German society) as well as aspects of the identificational integration (membership of German society on the subjective level, displayed in feelings of belonging and identification). The level of integration is analysed by comparing Turks with Germans and by differentiating between Turks of the first generation (those who immigrated as 'guest workers') and of the second and third generation (those who were either born or socialized in Germany since early infancy).

History

The enrolment of 'guest workers' was conceived by the German government as a temporary measure to deal with the chronic labour shortage in the second half of the 1950s due to full employment in Germany. As the number of vacant positions exceeded those of registered unemployed, the employers requested permission to recruit foreign workers. To overcome its labour shortage, Germany brought in workers from relatively less developed southern European countries that had labour surpluses – among them Italy (1955), and Spain and Greece (1960). In 1961, Germany and Turkey signed a workforce agreement and in the same year a total of 7,000 Turks migrated to Germany to become workers. The migration was accelerated by the signing of the Association Agreement between the European Community and Turkey in 1963. In 1973, the demand for foreign labour declined when Germany entered a period of economic recession due to the world oil crisis. As a result, the German government declared a recruitment ban for foreign workers. At the time, the number of Turkish migrant workers had reached 910,500. The immigrant employees mainly held low-paid jobs and were for the most part unskilled and less qualified labourers recruited for jobs in the industrial sectors like mining, manufacturing and construction. They were advantageous to the economy because they provided cheap and flexible labour.

Labour migration was seen as a temporary phenomenon as reflected by the term *Gastarbeiter* (guest workers). The term expressed the notion that those immigrant labourers were invited to work in Germany, but they were not expected to become permanent residents. Therefore, German governments and decision makers (as well as Turkish officials) did not take into consideration social and cultural issues in formulating migratory policies concerning the foreign workers from Turkey during the early immigration period. The parameters of political discourse were based on an ethnocentric interpretation of citizenship and nationhood in Germany which led to the political exclusion of ethnic minorities. As a consequence, a significant immigrant minority of 'guest workers' appeared without a real immigration perspective.

Turks in Germany are now no longer *Gastarbeiter* but *immigrants*. As time passed, they delayed their initial plans to earn some money and return home and instead stayed and settled in Germany. The result was an additional migration from Turkey mostly due to family reunification and family formation through 'cross-border marriages', which still continues among the young Turks. The latter is the most common form of migration nowadays. Thus, the social structure of the Turks in Germany has changed too. More than the half of the almost 2.4 million Turks have been living in Germany for more than twenty years. 800,000 of them are under the age of 21 and 445,000 are between the ages of 21 and 30.[1]

Structural Integration: Education and Labour Market

Education and occupational qualifications are the main essentials of human capital. They determine, in large measure, the individual's position in the labour market and the ability for social (and societal) mobility. They are the basic requirements for integration into the job market of a knowledge-based economy and society like Germany's.

Education

Education and language proficiency facilitate the integration of immigrants into the receiving society. It is interesting to examine how far young Turks in particular have succeeded in achieving higher levels of education in comparison to Germans of the same age.

The position of the young Turks is creating problems in access to education, training and employment equal to that of the native young population. The Turks have a worse educational structure than Germans: in 1999, 26 per cent of Germans had an upper school-leaving certificate (*Abitur*) and only 8 per cent left school without any qualifications. In the same year, almost 19 per cent of migrants left school without qualifications and only 7 per cent gained access to a university education. Despite the fact that the second generation of Turkish migrants shows a better educational structure than the first, the share of those without school qualifications was 7 per cent, and only 15 per cent had completed upper secondary school. Whereas 10 per cent of Germans completed university, only 1 per cent of second-generation Turks had a comparable degree. The worse educational background of the migrants seriously compromises their chance to find employment – and very often the attempts to start a self-determined life end up in a blind alley; this is especially true for female immigrants.

The reasons for this disparity are manifold and can be found in the kindergarten and school systems (due to the federal system of Germany every *Bundesland* has its own educational system) which do not take into consideration the social and cultural heterogeneity of the children. Due to the lack of early childhood education and of efforts to promote the educational, vocational and social advancement of immigrants, non-German young people often do not possess the skills they need to undergo and complete a higher education or a vocational training. Another reason for the low rates of immigrants in upper secondary schools and universities is the lack of encouragement on the part of their families. Although most Turkish families desire the success of their children in educational and professional life, at the same time – as they try hard to conserve their Turkish identity and hand it down to their children – they fear that their descendants could alienate themselves from their Turkish origins due to the influence of the educational institutions. In addition, there are more migration-specific problems: financial problems and a low family income put many of the second and third generation under pressure to earn as early as possible money as unskilled workers. The difficulties of finding employment experienced even by skilled Turks reinforce this development.

In addition to school certificates as important parameters of a group's educational structure, more relevant for the integration into the labour market are vocational qualifications. In this context, it is evident that Turks commonly remain about average without formal vocational training.

Unemployment

The rate of unemployment is an important indicator for the integration of migrants into the labour market, for the simple reason that participation in the economic sector determines to a great extent the social status and mobility of immigrants. The unemployment rate among Turks has been constantly above the German average and huge differences between them and other foreigners continue to prevail. Since the beginning of the 1990s, the rate of unemployment among Turks has increased markedly: while at the beginning of the last decade the rate was at 10 per cent, at the end it reached 24 per cent. Even if young Turks have better educational skills than the first generation, more than 50 per cent have no vocational training.

There are several reasons for this development. The main causes of unemployment among the first generation are the decline of jobs and the recession in the manufacturing sector, as well as growing requirements for jobseekers since the beginning of the structural change in Germany at the end of the 1980s. Turks of the first generation were not able to fit in due to their inadequate qualifications. Almost all of them came to Germany as industrial workers and did not succeed in leaving their traditional jobs, either because they did not want to or because or they were not able to do so because of their low education and missing language skills. (One has to bear in mind that they planned to stay only a few years and then return to Turkey.) In consequence, they were among the first to be faced with redundancy, dismissal and unemployment – and among the last to be offered new employment.

The reasons for the high unemployment rate among the Turks of the younger generation are composed of their worse school qualifications and lower vocational training. Even if they are more successful than the generation of their parents in achieving white-collar positions (by the end of the 1990s almost a quarter of the second generation had obtained white-collar positions), they are, in comparison with Germans of the same age, an under-represented group in the public service sector, while their rate in the industrial sector is twice as high as the rate of Germans (about two-thirds to one-third).

Identificational Integration: 'Parallel Societies' or German-Turks?

Contrary to the ruling assumption which persisted until the end of the 1990s, according to which the integration of immigrants would take place automatically the longer the 'guest workers' live in Germany, the development in terms of the tendency towards re-ethnicization, segregation and anti-integration has increased in recent years. The new term of 'parallel societies' was introduced into to political discourse. That discourse focuses on what is defined as a 'cultural' or 'religious' segregation and reflects the notion that the Turks and other, mostly Muslim, foreigners have a 'lack of readiness to integrate' and instead often withdraw from German society and build their own 'society' with their own social, religious and cultural activities as well as their own set of values.

Actually, there do exist formations with almost ethnically homogeneous communities and social structures. Particularly in congested urban areas of big cities like Berlin or Cologne, Turks can manage their everyday life completely within the Turkish community: from daily purchases to visits to the doctor or hairdresser. This kind of 'ethnic self-organization' accelerates the withdrawal to the ethnic group. But it would be short-sighted to draw the conclusion that Turks (or Muslims in general) are isolating themselves against the majority society in Germany. To explain the phenomenon of 'parallel societies,' one has to scrutinize the reasons which favour their emergence – for example, joint leisure-time activities of the immigrants, or similar experiences concerning everyday discrimination – and which purposes this kind of 'ethnic self-organization' serve.

Nevertheless, these communities are not closed clusters of immigrants like the immigrant ghettos in France, Britain, or the Netherlands, nor is there a lack of willingness to integrate among the majority of Turks. It must be taken into consideration that most of the so-called 'parallel societies' emerge especially among those immigrants who are assigned to the underclass and in neighbourhoods which are undoubtedly social hotspots. Political discourse commonly neglects how poverty and discrimination lead to social exclusion and drive the underprivileged classes – among them many guest workers and their descendents – into backwardness. Such discourse also does not mention the high rate of unemployment among Turks, and that the rate of Turks dependent on social security is up to three times higher than for native Germans.

'Typical' migrant labourers of the first generation (currently about a quarter of the Turks in Germany) have low educational and professional credentials, low social status and bad German-language skills. Usually, their chances for social mobility and integration are small. They feel much more connected with Turkey rather than Germany and keep the door open for a possible return. According to this, they show less willingness to apply for German citizenship because they still consider themselves Turks.

In contrast to the first, the second and third generation is characterized by an improved education, a higher vocational training and better language skills. Their chances of participating in German society are therefore better, although the obstacles which are intrinsic to the system still exist. In contrast to the first generation, the second feels connected to Turkey and to Germany. They are trying to define themselves not as being predominantly Turk or German but as being both. The interaction between them and German society is stronger due to the fact that they were either born in Germany or attended a German school. Inter-ethnic or inter-cultural contacts as well as inter-ethnic leisure activities are stronger than within the first generation. Altogether this group among the young Turks is more integrated into, and less marginalized by, German society than the first generation. They also show a more pronounced willingness to apply for German citizenship. However there is another group of young Turks whose chances for participation and integration are still very small and who consequently withdraw from German society – as mentioned above there is still a great percentage of young Turks who have no school-leaving qualifications or vocational education. They more often use self-ethnic structures to cover their insufficiency and to improve their social status. In fact, they have no alternatives but to withdraw to their own ethnic community. Because of the social stigmatization and the ethnic discrimination on the part of the majority society, their own community offers them a social space where they are accepted.

Conclusions – Perspectives

The majority of the young generation of Turks consider themselves as German–Turks and will stay in Germany. For years to come their number will increase and they will demand more representation and participation rights from German society. Even now there are stronger tendencies, especially among young Turks with academic backgrounds and at a higher educational level, to join together and to form pressure groups for articulating their concerns and pursuing their interests more effectively.

Integration understood as bonding in social connections and groups is a process on both sides of the society – by the minority as well as the majority – and it depends on various factors. First of all, there is a need for positive and constructive policies. Improved conditions for entering the labour market, more advancement opportunities through employment and training programmes for immigrants, or additional language courses for young immigrants are only a few measures which can be implemented by the social and political actors in Germany to ensure social and societal mobility. But they require the political will of the decision makers.

It seems that even after nearly five decades, immigrants are only accepted in a very limited way because of the myth that Germany is not a country of immigration – which still dominates especially the minds of conservative circles. This is particularly true for the Turks, as they can only participate in a limited way in civic society due to restrictions stemming from not holding German citizenship, because of its underlying ethno-cultural character. But Germany is a pluralistic society that contains a diversity of cultural forms of life, ethnic groups, religions and worldviews. The only way to meet the challenges and fulfil the requirements of a pluralistic society is to recognize this reality and to formulate new policies that avoid the social, cultural and political exclusion of ethnic minorities. This will allow the different ethnic groups to participate in a political community as full and equal members and help to reduce the tensions between the majority in society and the minorities (as well as between the different ethnic groups).

Asiye Öztürk is a Doctoral student at the University of Bonn.

NOTE

1   Cf. Faruk Sen: Türkische Minderheiten in Deutschland, Informationen zur politischen Bildung, No. 277, 4/2002, <http://www.bpb.de/publikationen/7LG87X,5,0,T%FCrkische_Minderheit_in_Deutschland.html> (30.11.2006). All statistics used in this article are obtained from Faruk Sen if not otherwise indicated.

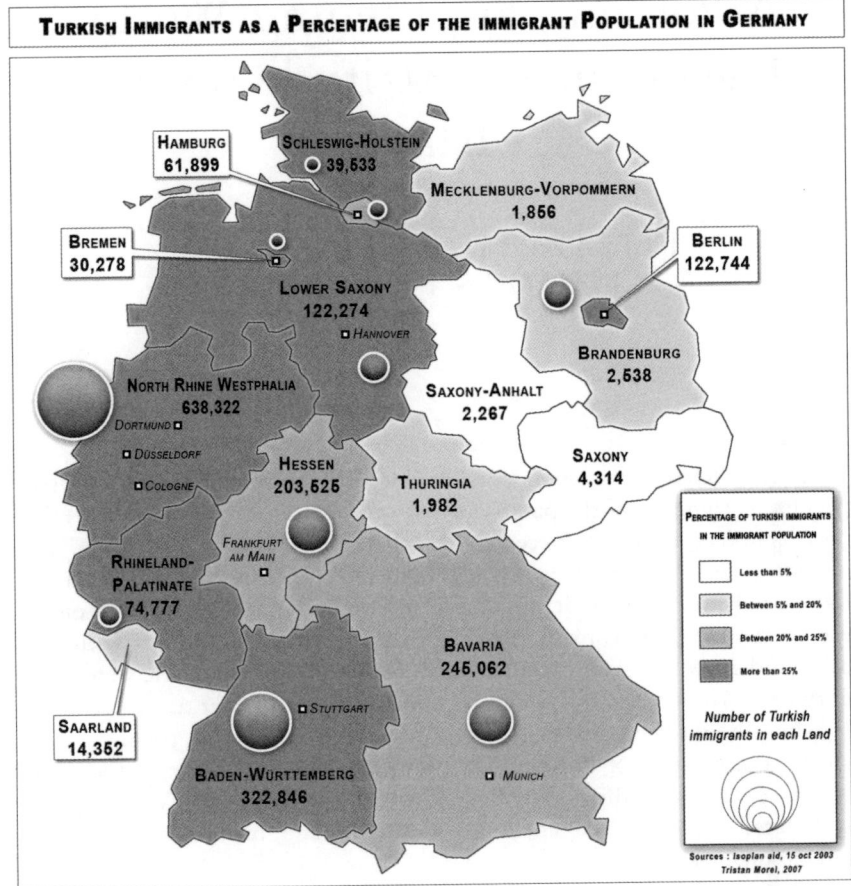

A Foreign, Frightening World: The Germans' Attitude towards Islam

ELISABETH NOELLE AND THOMAS PETERSEN

What is it exactly that causes wars, that leads to conflicts between different peoples or segments of the population? A quick glance at the past shows that open conflicts are often preceded by a lengthier phase of alienation. The opponents determine that their goals are incompatible; people on one side are increasingly unable to understand the concerns and objectives of those on the other side; the parties involved in the conflict speak less and less with each other and grow increasingly mistrustful. Rumours and simplistic stereotypes of those on the other side abound; the language used to refer to the other side is reduced to catchwords and stereotypical phrases. In the end, people come to view the intentions of those on the opposing side as a threat to their own existence — a threat which must be combated by all means. The opposing side is declared to be devoid of any sense of morality and, ultimately, even of any human qualities.

When the American political scientist, Samuel Huntington, published his book, *The Clash of Civilizations,* more than a decade ago, his ideas reaped both considerable attention and criticism. For Huntington's theory that conflicts would erupt in future between the different cultural groups around the world hardly seemed fitting at a time when the Cold War had just ended and the world thus seemed to be heading towards a peaceful future. Historians pointed out that by far the greatest share of conflicts over the course of history had broken out between parties belonging to the same cultural group.

Nevertheless, the findings of a survey on the Germans' attitudes towards Islam, which the Allensbach Institute conducted in May 2006 would seem to indicate that this exact same process of alienation is currently taking place among the German population, with respect both to Germans' relationship to the Islamic world and to the Muslim population in Germany. Viewed pessimistically, the current trend can be interpreted as the first step in the spiralling process leading to an open conflict. For example, the Germans are increasingly convinced that it will not be possible to coexist peacefully with the Islamic world in the long run. In response to the question, 'What would you say: can Christianity and Islam exist peacefully side by side or are these religions too different; will major conflicts therefore keep cropping up again and again?', 61 per cent of respondents today say that serious conflicts will keep arising between Islam and Christianity. If the question is worded somewhat more concretely and respondents are asked whether conflicts will arise in future between the western and the Arabic-Islamic world, the responses hardly change: in this case, 65 per cent anticipate such conflicts in future.

Although German opinions about Islam were already quite negative in past years, they have become markedly worse in most recently. In May 2006, 91 per cent of all respondents associated Islam with discrimination against women, as compared to 85 per cent in 2004. The share of Germans who associate Islam with fanaticism increased from 75 to 83 per cent. Sixty-two per cent say today that Islam is backward, as compared to 49 per cent two years ago, while the percentage of Germans who believe Islam is intolerant grew from 66 to 71 per cent. The opinion that Islam is undemocratic increased over the past two years from 52 to 60 per cent. In contrast, only 8 per cent of Germans associate Islam with peacefulness.

At the same time, the image of Christianity has clearly improved. Eighty per cent of all Germans today believe Christianity is characterized by brotherly love, and 71 per cent associate it with respect for human rights and charitableness. Sixty-five per cent associate the Christian faith with peacefulness, while tolerance is cited by as many as 42 per cent and self-confidence by 36 per cent. All of these qualities are attributed to Christianity considerably more often today than they were in 2004, even though the share of devout Christians in Germany has not increased. The pattern of polarization described above is obvious here: the chasm between the population's own camp and 'the others' is widening.

For a while, it seemed as if the conflict between Islam and the western world was something that was primarily occurring in faraway countries. In the meantime, however, the population is also taking an increasingly wary view of the role of Islam in Germany. This is presumably a reaction to the intense public debate on issues such as the right to acquire German citizenship, so-called 'honour killings', chador-clad girls in Bonn or the problems besetting Berlin's Rütli School, with its extremely high share of foreign students. In the immediate wake of the terrorist attacks in Washington, DC and New York in September 2001, the Allensbach Institute included the following question in a survey of the German population: 'Do you think that tensions will also arise here between the German and Muslim population in the near future, or do you not fear that this will happen?' At the time, the population was divided, with 49 per cent expecting tensions and 43 per cent saying they were not worried about that. In the meantime, opinions have clearly shifted, to the point where 58 per cent of respondents now expect that tensions will arise between the German and Muslim populations in Germany, while only 22 per cent expressly contradict this view. Moreover, 46 per cent of the population expect that there will be terrorist attacks in Germany in the near future and a relative majority (42 per cent) agrees with the statement: 'There are so many Muslims living here in Germany. Sometimes I really worry about whether there might not also be a lot of terrorists among them.'

Contrary to popular assumptions, numerous survey findings indicate that the Germans are not particularly xenophobic and, in fact, display an above-average openness to foreign cultures in comparison with the populations of other countries around the world. Nevertheless, the majority of Germans are becoming increasingly uneasy about the apparent spread of Islam in their own personal surroundings. This is clearly revealed by the results of a field experiment in which the total sample was divided into two representative subgroups of equal size. Half of the respondents were asked the following question, which was developed in light of an ongoing debate in a district of Berlin: 'Assume that there are plans to build a mosque in a district of a large German city. The authorities have granted a permit for building the mosque but the local population is against it. What is your opinion: should the mosque be built even though the local population is against it or should it not be built?' Only 11 per cent thought that the mosque should still be built under these circumstances, whereas almost three-quarters (74 per cent) said the mosque should not be built. The other subgroup of respondents was asked the same question, with the only difference being that the controversy involved not a mosque, but a youth centre. In this case, 59 per cent thought the youth centre should be built despite the local population's objections, whereas only 27 per cent believed it should not be built.

Given their diffuse sense of being threatened by Islam and the perceived intolerance on the part of Islam, the Germans themselves are becoming ever less willing to tolerate the Islamic faith. Fifty-six per cent of all respondents agree with the statement, 'If it is forbidden to build churches in some Islamic countries, it should also be forbidden to build mosques here in Germany.' Less than a third (31 per cent) of the population disagrees. A substantial share of the population even supports the idea of seriously limiting the basic right of religious freedom. As many as 40 per cent of all Germans agree with the statement, 'To ensure that there are not too many radical Muslims in Germany who are ready to engage in violence, we should severely restrict Muslims' right to practice their faith in Germany.'

In the meantime, the Germans' willingness to show understanding for

Islam is largely depleted, as shown by the population's reactions at the start of the year to the controversy over the publication of religious caricatures of Mohammed in a Danish newspaper, which led to enraged protests – some of which appear to have been organized – in many Muslim countries around the world. In February/March 2006, when the public debate on this issue was at full flame, the Allensbach Institute posed the following question: 'Regardless of the violent protests, can you fundamentally understand why many Muslims feel that their religious sensibilities have been offended by the caricatures or can't you understand that?' At the time, 47 per cent of the population said they could understand the reaction of the Muslim world, while 42 per cent could not understand it. Today, now that the debate has subsided – and now that the politicians' and the media's attempts to stem the uproar have faded from people's minds – a majority of 52 per cent cannot understand the Muslims' reactions to the caricatures, while only 35 per cent say their reactions were understandable.

Since the end of the Second World War, Germans have always been particularly averse to conflict, as shown by a number of examples. You could even say that Germans have an especially great need for harmony – a tendency that has remained essentially unchanged until today. With respect to Islam, however, it appears that the line between the fronts is being drawn more strictly. As in the past, Germans are still far from accusing all Muslims of being extremists. In response to the question, 'Do you think Islam as a whole poses a threat, or do only certain radical followers of this religion pose a threat?', two-thirds of all respondents say that it is only some radical adherents of Islam who pose a threat. Nevertheless, the basic tenor of the German feelings towards Islam has changed, as illustrated most clearly by the question: 'People sometimes use the term "clash of cultures", which means a serious conflict between Islam and Christianity. What do you think: Are we currently experiencing such a clash of cultures or wouldn't you say so?' Two years ago, 46 per cent of all respondents thought we were experiencing such a clash of cultures, while 34 per cent disagreed. Today, 56 per cent believe we are already involved in such a conflict, while only 25 per cent say this is not the case. In the minds of many Germans, therefore, the 'clash of cultures' has already begun.

Elisabeth Noelle and **Thomas Petersen** are researchers at the Institut für Demoskopie, Allensbach, Germany.

The End of the Ideology of Multiculturalism in the United Kingdom

JOHN LAUGHLAND

This article deals with a recent and very surprising development in British politics. For many decades, multiculturalism has been a central tenet of political correctness on left and right. Initially a dogma of the Left, the Conservative Party also quickly adopted it. Famously, in 1968, the controversial politician, Enoch Powell, was sacked from the Conservative shadow cabinet by Edward Heath after warning of the dangers of mass immigration in a graphic speech. Ever since then, all major political parties have dealt very gently with the question of immigration, if at all, and they have never questioned either the wisdom of mass immigration or, more particularly, the possibility of integrating such large numbers of new arrivals into British society while continuing to promote the ideology of multiculturalism.

To illustrate the extent to which the ideology of multiculturalism was dominant, it is worth recalling the sad story of Ray Honeyford. A headmaster in Bradford, Honeyford published an article in *The Salisbury Review* in 1984, attacking not immigration as such but instead the ideology of multiculturalism. According to that ideology, new arrivals in Britain should be allowed and even encouraged to retain important parts of their cultural practices while at the same time enjoying the fruits of British citizenship. Ninety-five per cent of the pupils at Honeyford's school were from Asian families, but (unlike Enoch Powell) this was not what he wanted to complain about.

Instead, with a straightforwardly progressive faith in the power and importance of education, Honeyford complained about the fact that Asian pupils were habitually taken out of school by their parents for months at a time, in order to be sent to Pakistan. Their parents wanted these British-born children to be sent 'home' to imbibe Pakistani culture; Honeyford argued that such long interruptions in their schooling were bad for their education, and that in any case it was wrong of the parents to regard British culture as something to be held at bay if they wanted to live in the country.

For making these remarks, and thereby questioning the totem of multiculturalism, Honeyford unleashed a storm of protest. He was vilified in the newspapers and by the then Muslim mayor of Bradford as a racist; he received death threats and had to live for a while under police protection; he was forced to resign, and he never taught again. This was in spite of the fact that he was particularly concerned about the effect of such multiculturalism on Muslim girls who, he felt, were disadvantaged by being prevented from gaining full access to the British educational system by their parents.

Another example to illustrate the same point: in 1990, the former cabinet minister Norman Tebbit made a characteristically humorous allusion to the problems which, he thought, multiculturalism caused. In an interview with the *Los Angeles Times*, he asked ironically whether Britain's Asian felt loyalty to Britain or to the home country of their parents and grandparents. You could tell the problem, he said, when British Asians supported Pakistan at cricket against England: 'A large proportion of Britain's Asian population fail to pass the cricket test. Which side do they cheer for? It's an interesting test. Are you still harking back to where you came from or where you are?' For this innocuous remark, Tebbit was also vilified and treated as a right-wing extremist.

Multiculturalism was therefore the official ideology of both the Labour and Conservative parties for many decades. It stood in marked contrast, at least in terms of rhetoric, to the attitude theoretically adopted towards immigration in France. Members of the liberal establishment in Britain typically reacted with shock and horror whenever steps were taken in France to maintain the integrity of the Republic, most notably when France banned the wearing of Muslim headscarves in schools in 2004. For the liberal British, such an approach smacked of social authoritarianism and was regarded with incomprehension.

All this has recently changed very suddenly. The official abandonment of multiculturalism as a state ideology came on 24 August 2006, in a speech given by Ruth Kelly, the Secretary of State for Communities and Local Government. She was speaking on the occasion of the launch of a new 'Commission on Integration and Cohesion', whose purpose, precisely, was to promote integration at the expense of multiculturalism. Kelly is an arch-Blairite member of the Blair government and was previously Secretary of State for Education.

Kelly began by paying the usual lip service to multiculturalism, claiming that 'Britain's diversity' was 'a huge asset' and that immigration had helped the economy. But there was a 'but'. Kelly claimed that 'the landscape is changing' and that new answers were needed for the 'difficult questions' which were now arising. Among those were the fact that 'global tensions' were now 'reflected on the streets of local communities'. 'Second and third-generation immigrants,' Kelly said, 'face a struggle. Not to adapt to life in the UK – but to reconcile their own values and beliefs with those of their parents and grandparents.' She made the following remarkable statement:

> And for some communities in particular, we need to acknowledge that life in Britain has started to feel markedly different since the attacks on 9/11 in New York and on 7/7 in London – even more so since the events of two weeks ago.

'Two weeks ago' referred to an alleged plot to blow up transatlantic airliners in mid-flight using explosives contained in bottles of Lucozade. The discovery of the alleged plot led to mass cancellations of flights on 10 August 2006. The trials of the suspects are expected to start in January 2008. In other words, Kelly was quite explicitly saying that the new 'war on terror', proclaimed by President George W. Bush after 11 September 2001, had created a new situation inside Britain and that the old shibboleth of multiculturalism had to be reviewed as a result. She then went even further than that and said something which no politician outside the extreme right British National Party would have said at any point in the previous thirty years or more: 'There are white Britons who do not feel comfortable with change. They see the shops and restaurants in their town centres changing. They see their neighbourhoods becoming more diverse.'

This was but a more polite version of the story Enoch Powell had told in his 'rivers of blood' speech about a white landlady who felt marginalized from her own community because of the influx of immigrants. It was considered to the most inflammatory part of his speech, yet here was Kelly expressing essentially the same thoughts. Kelly concluded: 'I believe this is why we have moved from a period of uniform consensus on the value of multiculturalism, to one where we can encourage that debate by questioning whether it is encouraging separateness.'

Multiculturalism, in other words, was the problem not the solution: 'In our attempt to avoid imposing a single British identity and culture, have we ended up with some communities living in isolation of each other, with no common bonds between them?'

It was also, she said, 'not racist to discuss immigration and asylum' – whereas leading lights from the Labour Party typically did precisely vilify as racist people who discussed either issue. Kelly's proposed solution to the problem was to promote 'shared values' and 'a citizenship curriculum for Madrassas'. She said that minority ethnic communities should not receive 'special treatment'.

Kelly referred in her speech to a previous statement by the Chairman of the Commission for Racial Equality, Trevor Phillips, who had questioned whether the nostrums of multiculturalism had done more harm than good. He had argued that it was wrong to present the worst possible interpretation of British history, since this would not foster social cohesion, and he said that certain parts of Britain were becoming ghettoes.

Once the dam had been breached, public complaints about the problems associated with immigrants in general and Muslims in particular turned into a flood. The taboo having been broken, it seemed suddenly as if previously impossible subjects could be aired with impunity. Some of the results of this were astonishing, even shocking. The head of the internal security service, MI5, Dame Eliza Manningham-Buller, said on 9 November 2006 that there was a major terrorist threat in the UK from hundreds of thousands of British Muslim Al-Qaeda sympathizers. This was rapidly picked up by extreme right-wing websites as confirmation of their warnings about 'the enemy within'.

On 22 October 2006, indeed, The *Guardian* had reported new police figures which showed that half of the victims in racial murder cases were white. Whereas the murder of a black teenager, Stephen Lawrence, in London in 1999 had been elevated by campaigners and the media into an event of national importance, thereby giving the impression that racially motivated murders were usually of blacks by whites, Peter Fahy, the Chief Constable of Cheshire and a spokesman on race issues for the Association of Chief Police Officers now said:

> The political correctness and reluctance to discuss these things absolutely does play a factor. A lot of police officers and other professions feel almost the best thing to do is try and avoid it for fear of being criticized. We probably have all got ourselves into a bit of state about this.
>
> The difficulty in the police service is that the whole thing is being closed down because we are all afraid of discussing any of it in case we say the wrong thing – and that is not healthy.
>
> I will be honest with some of this discussion about the alienation of Muslim people. Police officers would tell you there are a lot of young people out there who feel alienated.
>
> There are a lot of young white working-class lads, particularly on the more difficult estates, who are hugely alienated. Yet very little attention is given to that.
>
> Sometimes we forget that ethnic minorities actually make up quite a small percentage of the population.

Earlier on in October, Jack Straw, the Leader of the House of Commons, had launched a storm of protest when he said that he was in the habit of asking veiled Muslim women to remove their veils when they came to see him in his constituency surgery. He said that it was better for face-to-face contact to be established and he criticized the wearing of the veil was 'a visible statement of separation and difference'. The attack on multiculturalism could hardly have been more explicit.

Straw's remarks caused a huge row, carried out in the national media, yet it was clear that there was massive public support for what he said. The traditional British incomprehension for the French position on the veil vanished and suddenly polls found that huge majorities wanted the same ban in Britain. The row crystallized around a Muslim woman foreign-language teacher who was sacked from her school for refusing to remove the veil during classes. The complaint was that it was impossible for her to teach languages without her pupils being able to see her mouth.

The Prime Minister himself intervened in this row and said he supported the woman's dismissal. He, too, said that the veil was 'a mark of separation' which made non-Muslims feel 'uncomfortable'. Blair added oil to the flames by saying not only that he supported the way the school had handled the issue, but also that the row typified a much larger problem, namely 'how Islam comes to terms with and is comfortable with' the modern world.[1]

The argument over the veil coincided with another Blair government initiative at the time (which eventually failed because of strong opposition from the Catholic Church) to force religious state schools to take pupils from other religions. Blair said, 'We would not be having this debate were it not for people's concerns about this question to do with integration and separation of the Muslim community.' In other words, the project to force 'faith schools' to mix pupils of different faiths or none was an attempt to integrate Muslims into British society and to prevent them from segregating themselves too much. This attack on 'faith schools' was an attack on a very old and well-established part of British 'multiculturalism', since religiously denominated state schools have existed in Britain for many centuries, including for religious minorities (especially Catholics) since the nineteenth century.

The change in ideology culminated, on 8 December 2006, in Blair's speech on multiculturalism entitled 'The Duty to Integrate: Shared British Values'. The title said it all. It was a wholesale disavowal of the ideology of multiculturalism. Blair of course, like Kelly, said that multiculturalism and cultural diversity were good things. But his message was quite firm that this had limits. Blair did not beat around the bush – the problem lay with Muslims: 'The reason we are having this debate is not generalized extremism. It is a new and virulent form of ideology associated with a minority of our Muslim community.' Blair said that the Muslims must not expect the British to compromise on their values or on the rule of law, and that any introduction of sharia law or religious courts into British life (as has already happened in parts of the country) was to be firmly rejected.

Blair concluded using language which, hitherto, had existed only on the extreme right. Addressing immigrants, he said, 'Our tolerance is part of what makes Britain, Britain. *So conform to it; or don't come here.* We don't want the hate-mongers, whatever their race, religion or creed.' He concluded saying that immigrants has 'a duty to integrate' and denounced those opposed to such integration as 'racists and extremists'. His words were expressing exactly the same sentiment as had been expressed by Ray Honeyford and Norman Tebbit, only his language was far more brutal.

As both Blair and Kelly made clear, the collapse of the multicultural ideology was the fault of the Muslims. Their remarks came amidst a series of public controversies which underlined non-integration by Muslims, and after the Blair government had decided to introduce rather absurd 'citizenship tests' for new arrivals. This concentration on the specific problems associated with Muslim immigrants helps us to answer the question which inevitably arises in respect of this sudden and dramatic abandonment of the official liberal ideology of multiculturalism, that is, 'Why?'

There are two obvious candidates for explanation. The first is the fact in the rise in immigration. Since the mid-1990s, Britain has witnessed a huge influx of new migrants from all kinds of sources – asylum seekers, ordinary immigrants, migrant workers from the new EU states since 2004 (up to one million of these). The second most obvious explanation for the abandonment of multiculturalism as an ideology lies in the London bombings of 7 July 2005, when British Muslims conspired to blow up the London Underground, killing 52 people. When one puts together these two facts, is it any surprise that multiculturalism has been abandoned?

Yes. As far as the rise in immigration is concerned, this will not do as an explanation for the magnitude of the government's change on multiculturalism. Unlike France, there is no major anti-immigration party which poses an electoral threat to Labour. If there were one, then it would be entirely rational for an incumbent government to start to 'talk tough' on immigration to counter the danger of votes going to the other party. It is true that the British National Party has taken away some white working-class votes from Labour in certain areas but this is a micro-phenomenon, certainly not on the same scale as the loss of Muslim votes which has been massive following the war in Iraq.

Are the London bombings an adequate explanation for four or more decades of multiculturalism? Britain is a country which has dealt with terrorism for many decades in Northern Ireland. Yet when the IRA were letting off bombs in Northern Ireland and on the British mainland, there was no hand-wringing or self-examination about why 'British-born' people should turn against their own state. Instead, the policy of both Labour and Conservative governments throughout the entire period of the troubles in Northern Ireland was precisely to deny the IRA and its supporters any political legitimacy at all, and instead to deal with Irish terrorism as a purely criminal matter. The same goes for the very small-scale actions associated with certain 'animal liberation' extremists. The scale of Muslim terrorism in Britain is minute in comparison with the scale of Irish: thousands of people were killed over many decades in connection with the troubles in Ulster. Therefore, the appearance of a new form of Muslim terrorism is simply not an adequate explanation for the sudden abandonment of multiculturalism.

Instead, therefore, the following sombre conclusion imposes itself. The abandonment by Blair and his ministers of one of the liberal left's central tenets can only be understood as a cynical ploy to exploit people's fears about mass immigration to bolster support for the 'war on terror'. Exposed as a liar over Iraq, a war with which his name will remain associated in history for many decades to come, Tony Blair seems to have calculated that he can salvage something of his tattered reputation, and garner some support for the ongoing war in Iraq, by demonizing Britain's Muslim population and stating openly that Muslims represent a threat to western values. In other words, I submit, the abandonment of the ideology of multiculturalism is exactly what it appears to be – the abandonment of a tradition of tolerance and the instrumentalization of ethnic and religious differences within society for short-term, cynical foreign policy purposes.

John Laughland is the editor of *Geopolitical Affairs*.

NOTE

1   *Guardian*, 17 October 2006.

Italy: From a Country of Illegal Emigrants to a Country of Illegal Immigrants

VALERIA PALUMBO

On New Year's Eve 1962, Mario Trambusti, a 26-year-old baker from Bagno a Ripoli in Tuscany, fell to his death on the mountainous 'Death's Pass' (as it is called), that traverses the border between Italy and France above Ventimiglia. Mario was the eighty-seventh Italian victim of the lethal mountain crossing. He was trying to emigrate illegally into France and he may have gotten involved with a people-smuggler from Liguria or Calabria. Mario was a 'fenicottero', a flamingo, as they called Italians attempting to enter France illegally.

Less than forty years later something has changed. According to the Italian branch of Caritas, one million people entered Italy illegally between 2000 and 2004 and probably remained in the country. Again according to Caritas, the number of legal immigrants grew from 1,334,889 in 2001 to 2,800,000 in 2005, and now represent 4.8 per cent of the Italian population.

Let's go back to the Italian illegal emigrants. In the United States, they had many ways to insult people from Italy, for example, 'Wop' – 'without official papers', according to the historian Emilio Franzina.[1] Wop: they were not so wrong, we were often (and maybe we still are) illegal immigrants. And we had many ways to achieve our goal: to reach America, Antonio Margariti, one of thousands, went from Calabria to Milan, then to Chiasso, and later from Switzerland to Le Havre, and from there he set sail for America. From Le Havre, Marseille, Hamburg, Bremen, Glasgow: thousands of illegal Italian emigrants left for America.

Italians began very early on to leave the country, or rather, before 1860, the many little countries of which Italy was to be formed. Emigration without papers goes back to 1700. People went great distances: not only to France or Germany but also to Russia, North Africa, the Balkans and the Far East. These are exactly the same countries from which immigrants are now coming into Italy.

Every year 300,000 illegal African emigrants come to Europe. Eighty per cent of them pay criminal organizations to arrange their voyage. Human trafficking is worth €230 million a year, according to the UN. The Italian government estimates the trade to be worth €300 million a year.

Those entering Italy illegally generally do so via Libya. It costs between €1,525 and €1,990 just to cross the sea.² To cross the Sahara to reach the coast, from Niger for example, one has to pay between €1,700 and €3,400 more. The total cost is therefore between €3,515 and €5,390. At the end of their journey often lies slavery and even death.

At 1 a.m. on 26 December 1996, about three hundred illegal immigrants were offloaded from an old boat, the *Yiohan*, in the Straits of Sicily, to a Maltese fishing boat (F-174) which was supposed to land them. The two boats crashed. At 3 a.m. the fishing boat sank and 283 people drowned. They had all come from little villages in the Indian Punjab (160 casualties), Pakistan (31), Sri Lanka (92). Only twenty-nine survivors were rescued by the *Yiohan*. On 29 December, the *Yiohan* landed in Greece. The survivors reported what had happened, but they were not believed and were deported. Only two Italian newspapers, *Il Manifesto* and *La Repubblica*, began to write about the accident, the worst shipwreck in the Mediterranean Sea since the Second World War. Everyone else, the media and political authorities, said nothing. But in December 1998 the Prosecutor in Syracuse opened the trial against two of the thirteen traffickers accused of murder.

In June 2001, Giovanni Maria Bellu, a journalist at *La Repubblica*, 'rediscovered' the tragedy: a fisherman, Salvatore Lupo, had come across the remains of the shipwreck Other fishermen had found heads and rags but they had thrown them back into the sea. But Lupo discovered a body with an identity card. Anpalagan Ganeshu, born on 2 April 1979 in Chaukachceri, Sri Lanka, was on board with his elder brother, Arulalagan, who also died. *La Repubblica* paid for the soundings to be scanned by a robot, which found the ship and what remained of the 283 young people. Four Nobel Prize winners – Renato Dulbecco, Dario Fo, Rita Levi Montalcini and Carlo Rubbia – signed an appeal for the wreck to be brought to the surface. Their calls went unheeded.

On 29 April 2004, the Syracuse Court of Appeal acquitted all the defendants. The wreck had occurred only 19 miles from Portopalo but the Court rules that it had occurred outside Italian jurisdiction. The 283 dead had no justice and no grave.

In autumn 2004, Giovanni Maria Bellu published *I fantasmi di Portopalo* ('The Ghosts of Portopalo's').³ Later Mimmo Sammartino wrote *Un canto clandestino saliva dall'abisso* ('A clandestine song rises from the abyss').⁴ The Teatro della Cooperativa in Milan performed *La nave fantasma*, ('The ghost ship') by Giovanni Maria Bellu and Renato Sarti, together with Bebo Storti. I saw this impressive play and I very much hope that it will play elsewhere in Europe since 283 dead people are waiting for their voice to be heard.

At one point in the play, the fisherman, Salvatore Lupo, who has found the identity card and hidden it in a drawer, reads in the papers – he does not usually read the papers but that day his daughter had won a beauty contest – that the Greek captain of the boat had been freed. He thinks to himself:

> Behind this small piece of paper, there is a history. A history of poor people who died, human people, people like us. I think of the boy's family, I think of his father and his mother, I think first of all that this history is going too far. I take a decision. I have the proof: I have the identity card, I know where the shipwreck is. I call a friend in Rome. He puts me in touch with a journalist, Giovanni Maria Bellu of *La Repubblica*. We find the wreck. But the authorities do nothing. Yes, I know, that night there was a storm at sea, Gale Force 7, nobody could have been rescued. But the guilty men should have been punished.

Everyone knows what Ellis Island was. From 1892 to 1924, more than twenty million people passed through the (in)famous immigration centre. From 1880 to 1930, Italians were the most frequent arrivals, 4.6 million of them as compared to four million from the much larger Austro-Hungarian Empire and 3.3 million from the Russian Empire. From 1930 to 1965, there were 'only' 390,000 Italian immigrants. Coming from Southern Europe, Italians were less welcome than before. Ellis Island turned into a detention centre for people who were to be expelled: 62,000 people in 1931, 103,000 in 1932, 127,000 in 1933.

In Italy, Marco Rovelli has recently published *Lager italiani* ('Italian concentration camps').[5] The expression is perhaps too strong but it refers to the '*Centri di permanenza temporanea*' – 'temporary stay centres' – which have been set up all over Italy and Europe since the enforcement of the Schengen Agreement in 1995. The first immigration law in Italy dates from 1986; the second, the so-called 'Martelli Law' of 1990, sets down a procedure for deporting illegal immigrants and foreign criminals. People usually have 15 days to leave the country after a prefectural order to do so. The first such temporary camp was set up for Albanian refugees in Bari in 1991: after being put in Bari Stadium and other centres, they were all (or nearly all) deported back to Albania.

In 1998, Law No. 40, the so-called 'Turco-Napolitano Law', officially created these centres. It established entry quotas and expulsion procedures including the right of the police to detail illegal immigrants following an expulsion order. It is a good law, improving immigrants' rights and working conditions, but it makes it harder to enter Italy legally.

A further immigration law was passed on 30 July 2002, the so-called 'Bossi-Fini Law', No. 189. It extended the right of the police to detain people in temporary stay centres and created new ones. Illegal immigrants may now be detained in such centres for up to sixty days. If the detainee has not been deported within that time, they are released but still must leave the country within five days.

In recent years, around 15,000 people have been detained in these centres every year: 14,223 in 2003, 15,467 in 2004, 16,173 in 2005. Less than half were sent home in 2004, 68.6 per cent were sent home in 2005 and 18.5 per cent were set free (compared to 24.5 per cent in 2004). Most of those released remained in Italy illegally. Sixty per cent of the detainees have been in prison: as if their sentence were being prolonged.

As of the beginning of 2007, Italy has fourteen such centres. The most famous ones are in Bari, Brindisi, Crotone, Milan, Lampedusa and Rome. Two have been closed, in Agrigento and Lecce. The majority are managed by the Red Cross. Modena and Lampedusa are managed by the *Confraternità delle Misercordie d'Italia*; Lamezia Terme, Restinco and Gradisca d'Isonzo by co-operatives; Caltanissetta is managed by an *ad hoc* organization. In March 2007, the Minister of the Interior, Giuliano Amato, announced that, with the next Immigration Law, there will be fewer and smaller centres.

Law No. 40 of 1998 justified the creation of these centres in the name of the number of illegal landings along Italian coasts. Currently, immigrants without identity cards are in the minority. They come from all parts of the world. Between 10 and 26 July 2006, 2,111 people arrived in Lampedusa, the most southerly Italian island, from all over Africa: Morocco, Eritrea, Egypt, Tunisia. In the first ten months of 2006, 17,407 illegal immigrants were intercepted around Lampedusa, three times the local population. Among them, 988 were children and 639 women. According to *Medecins sans frontières*, there were 320 landings in ten months, compared to 127 in 2005. In that year 23,000 people were caught by the naval police.

But how many were *not* intercepted? How many of them landed? It is not easy to escape from Lampedusa: the island is small and well guarded. On 3 December 2006, fifty Iraqis were intercepted at Porto Cesareo in Puglia. Some days earlier, two men from Tunisia arrived in Pantelleria, a southerly island, on a jet-ski: it had run out of fuel but they managed to cross from Africa. Puglia has 800 km of coastline; the local Mafia and smugglers help the people traffickers, the famous *scafisti*. The controls are now much better than before but Italy has 8,600 km of coastline.

In 1995, between 30,000 and 37,000 illegal Italian immigrants lived in New York. According to the Immigration Office in Washington, there were a total of 67,000 in the country as a whole in 1992. This is the second largest group after Poles. In 1997, Berlin had at least 27,000 Italian illegal builders. Nowadays, out of the 120,000 people who work in the construction industry in Milan, some 60,000 are illegal immigrants according to Filea-Cgil, the building industry trades union. Foreigners account for 40 per cent of construction workers.

According to INAIL (the Italian National Institute of Insurance for Accidents at Work) there were more than 115,000 accidents reported by immigrants in 2004, of which 164 were fatal. The highest number of victims (three hundred a year) are in the construction industry. There were thirty-eight deaths in 2004, mostly in Lombardy and Piedmont. Half of the injured workers were aged between 26 and 35, and the majority came from eastern European countries (especially Romania and Albania).

In his wonderful book *L'orda* and Internet site <www.orda.it>, journalist Gian Antonio Stella of *Corriere della Sera* tells how hard it was to be an Italian emigrant.[6] Italians were perceived as too religious, too dark, too poor, too ignorant, too dirty and sometimes even too violent. Now, over 30 per cent of the prison population in Italy are immigrants, according to Caritas.

People think that illegal immigrants come from the sea. In fact only 10 per cent come by sea while 15 per cent come overland. Seventy-five per cent of illegal immigrants are people who have actually entered the country legally, usually on tourist or student visas, and who have stayed.

The reactions of Italians are often racist. Everyone knows the harsh declarations of Lega Nord members such as Mario Borghezio who, among many other attacks on immigrants, once sprayed disinfectant over foreign women. But I sometimes wonder: is this racism or powerlessness? Take the examples of the walls put up in Padua and Schio in 2006.

Padua has a big drug problem. The drugs trade is run by illegal immigrants. The authorities responded to the problem by building an 80-metre-long wall, 3 metres high, around a group of buildings where drugs were traded. The dealers were inside, the world was outside, the police were in between. According to the Mayor of Padua, Flavio Zanonato, the checkpoint at via Anelli was the only way to isolate the ugly 1970s buildings where a lot of illegal immigrants live. It is true that many of them are involved in drugs and prostitution but, if so, why not arrest them?

The same happened in Schio, a small town of 39,000 inhabitants with a long tradition of integrating foreigners and gypsies. But in October 2006, Mayor Luigi Dalla Via, a member of the centre-left coalition like Zanonato, decided to stop gypsies from coming to the town and built a wall in an industrial area in order to prevent them settling there. The inhabitants of the town supported this policy but this has more to do with desperation than with racism. Many Italians, even those who emigrated from South to North, or who have relatives who've emigrated, say they do not like immigrants. That is fear, that is prejudice. But it is also the result of the politicians' powerlessness to manage the new situation.

Europe and Italy have entry quotas, in an attempt to prevent illegal landings from countries like Libya. Usually those who are caught are sent back. In 2005, Italy deported 26,985 people; 35,437 in 2004; 37,756 in 2003 and 44,706 in 2002. In other words, the number of deportations is falling.

What is dangerous, because it creates both crime and racism, is that people do not actually leave the country after being ordered to do so. This is an increasing phenomenon: 40,586 did not leave in 2003, 45,697 in 2004, and 67,617 people in 2005. In 2005, for the first time since 1999, there were fewer people refused entry at the border, or deported, than those who managed to stay in Italy none the less. In the first six months of 2006, of the 62,545 illegal immigrants caught by police (2,869 more than in 2005) only 24,125 were in fact deported from the country.

Many people from Central Africa ask for political asylum. The least popular seemed to be the Romanians: of the 105,662 people who were obliged to leave Italy in 2004, 26,344 were from Romania. Now, as EU members, Romanians are allowed to enter the country without restrictions: problems have arisen with gypsy immigrants, first of all in Milan and in nearby Opera. People do not want them and living together seems to be very hard. Meanwhile, again in Milan, police had difficulty freeing the La Stecca district from illegal Senegalese drug pushers.

In 2005, 23,878 people were denied entry at the border, refusal generally occurring at airports. At Malpensa in Milan, 3.600 were refused entry; at Fiumicino in Rome, 2.351, while in Trieste, 3,923 were refused entry. There were also refusals at land borders n the Alps, Verbania-Domodossola (2,535) and Como-Ponte Chiasso. Fifty per cent of the people refused entry were from eastern Europe.

So for every ten illegal immigrants detected by the police, only four are actually sent home. According to the Italian Court of Auditors, it cost €115,467,000 to deport people in 2004, or €316,000 a day. Annually, €29 million are spent on projects to help immigrants.

In 2004, Italy counted 2,786,340 legal foreign residents. According to the immigration statistics of Caritas/Migrantes, there were three million legal foreign residents by the end of 2005: 44.5 per cent from Europe, 21 per cent from Asia, 18 per cent from America and 15.9 per cent from Africa. The same source estimates the number of illegal immigrants in 2005 at between 500,000 and 800,000 people. According to the research institute Eurispes, there were 800,000 illegal immigrants in 2006. As Caritas has said, one million people entered illegally Italy from 2000 to 2004, while 15,000 of them still work without regular wages and rights; 50,000 work as prostitutes. Meanwhile, 70,000 Italians a year migrate from South to North.

Since 1998, 1,811 illegal immigrants have died in the Straits of Sicily trying to reach the Italian coast. I think of them and of the thousands of illegal

Italian emigrants who died in the Atlantic Ocean in past centuries, as Gian Antonio Stella wrote in his book *Odissee*.[7] We know: they looked for a dream, sometimes just for a hope. Like the people who drowned off Portopalo, they had no justice. That is why Italy has the duty to remember and to help, to recognize the human rights of new immigrants.

Valeria Palumbo is a journalist with the Italian newspaper *L'Europeo*.

NOTES

1 Emilio Franzina, *Storia dell'emigrazione italiana* (Rome: Donzelli, 2001).
2 *Il Giornale*, 2 August 2006.
3 Giovanni Maria Bellu, *I fantasmi di Portopalo* ('The Ghosts of Portopalo') (Mondadori: Strade blu, 2004).
4 Mimmo Sammartino, *Un canto clandestino saliva dall'abisso* ('A clandestine song rises from the abyss') (Collana: Il divano, 2006).
5 Marco Rovelli, *Lager italiani* ('Italian concentration camps') (Biblioteca Universale Rizzoli: 2006 <http://bur.rcslibri.corriere.it>).
6 Gian Antonio Stella, *L'orda* (Milan: Rizzoli, 2002) <www.orda.it>.
7 Gian Antonio Stella, *Odissee* (Milan: Rizzoli, 2002).

Immigrants in Italy

LUIGINO SCRICCIOLO

2.8 million: that is the number of permanent foreign immigrants in Italy (4.8 per cent of the population). And that figure excludes half a million illegal immigrants. Italy comes third only to Germany (7.3 million immigrants) and France (3.5 million immigrants), and is on a par with Spain and Britain. Their geographical distribution is varied: if, once immigrants headed chiefly for the urban and industrial centres, today even provincial Italy and its small towns have a visible presence of immigrants who often live in their own worlds and have difficulty integrating. The province of Rome has the greatest number of foreigners (340,000), followed by Milan with 300,000; Turin and Brescia have around 100,000, while Padua, Treviso, Verona, Bergamo, Modena, Florence and Naples have around 50–70,000. Immigration is more concentrated in the North (59 per cent of immigrants), and stands at 27 and 14 per cent in central and southern Italy respectively. In 2005, the average influx stood at 190,000 stable entrants: 42,000 for work (besides 45,000 seasonal workers from outside the EU and 32,000 workers from the new member states), 67,000 for family reasons and 6,000 for religious motives.

The largest proportion of immigrant workers are Romanian (40 per cent); far behind them are Albanians, Moroccans and Poles, each with quotas between 15 and 10 per cent. Family reunions see Morocco and Albania at the top of the list (13,000 from each country), followed by Romania (8,000), China (7,000) and, at the lower end, India, Ukraine, Serbia-Montenegro, Bangladesh and Macedonia (3,000). As regards irregular migrations it should be stressed that the sea has caused many tragedies. It is now the Sicilian, rather than the Pugliese or Calabrese coasts, which is targeted. The main countries involved are African (Egypt, the Horn of Africa, Sudan, Sierra Leone, Burkina Faso and Nigeria) and Middle Eastern countries, although countries farther afield such as Pakistan and Bangladesh are also represented. The repatriation of illegal immigrants has remained almost the same for the past few years (approximately 105,000): a brake might be put on forced deportations by a new law, made necessary by a decision in which the Constitutional Court declared expulsions from Italy illegal until they had gone through the due legal process. According to Cesare Damiano, the Minister of Labour:

IMMIGRANTS IN ITALY

In our era migrations are increasing considerably because they are one of the most significant expressions of a globalized world. For Italy immigrants are above all a demographic and labour resource: thanks to them the population does not decrease and they provide an indispensible supplement to the labour force in various sectors. They represent an opportunity for, rather than a threat to, our well-being, culture, institutions and religious sense.[1]

Notwithstanding the attempts of the media, the government and those forces keeping it in power to avoid tension between immigrants and citizens, 'explosions' and conflicts arise constantly. The situation does not indicate a peaceful relationship between immigrants and citizens.

The Surrender in Padua: The Wall

Finally, the wall was erected: the answer to a situation that had been brewing for years and which had not been effectively dealt with by two centre-left coalitions, one centre-right coalition and a third centre-left coalition who were in power for two years under the same mayor, Flavio Zanonato, who had also governed for the previous two years as well.

A steel wall barely 4 mm thick, 84 m wide and 3 m high was erected in the Via Anelli zone of Padua, to isolate the surrounding area from the drug trade. Six small blocks thus became enclosed behind a kind of fence, the entrance being guarded by the police. On one side, gangs of drug dealers; on the other, Italian citizens and those immigrants wishing to live in secure surroundings. They said it would not become Beirut, that there would not be a state of siege. But an entire zone of Padua changed: beyond the wall is Via Anellli, which everyone used to call the ghetto and which now truly is one.

Here, five beehive-like blocks built in the 1970s to house members of the university had been gradually occupied by a population of immigrants, most of them illegal and involved in criminal activities ranging from drug dealing to prostitution. The urban geography changed in front of everyone's eyes: Via Anelli is a horizon of parables hung out from the terraces between lines of drying laundry, it is a camp among the cement, trashed cars parked everywhere, prostitution and drug dealing at all hours of the day and night. Controlled by highly profitable markets which already caused periodic battles, over the years ethnic divisions escalated to an uncontrollable degree. Inevitably, continuous frays occurred between immigrants in a furious battle to control the drug trade, with periodical intervention from special sections of the police.

Thus, eventually, the wall was erected. Of the five blocks, two remain inhabited. The remaining three were boarded up following an intense campaign led by Daniela Ruffini of the *Rifondazione Communista*, the commune's assessor of housing and migration. She proposed emptying the Via Anelli and rehousing some of its few permanent residents in other areas of the city and, where necessary, bringing the others up in front of the appropriate authorities. The citizenry's patience had worn thin and a document had already been produced inviting the population to set up groups to come to the defence of the Paduans. The wall and the subsequent emptying-out of the area decreased tensions. But the problem remains in the general consciousness of the city's inhabitants.

Cofferati in Bologna: Sanctions on Windscreen Cleaners and the Expulsion of Abusive People

Faced with growing discontent in the city and certain serious incidents (rapes in a park, nocturnal clashes between immigrants and Bolognese youths), Sergio Cofferati, centre-left mayor of Bologna and ex-general secretary of the CGIL (*Confederazione generale italiana del lavoro*, one of the oldest and largest trade unions) decided to impede 'urban illegality'. To be precise, this involved emptying camps of illegal nomads, removing conmen and windscreen cleaners from the historic city centre, and turning the screw on prostitution and drug rackets. The shantytowns on the banks of the River Reno were also emptied out as they were at risk of flooding, had no lighting and were in an extremely degraded state.

The parties of the extreme left (*Rifondazione Communista*, the Green Party and the Anti-Globalists) fought against Cofferati, leaving the intellectuals and the diocesan charity in a quandary. He was fully supported by the *Democratici di Sinistra* (reformed Communists), the Margherita centre-left coalition, and by the Bolognese church. Naturally, the entire centre-right applauded the fact that the 'red mayor' had suddenly become a supporter of law and order.

The question here is not to establish whether legality belongs to the left or to the right, nor to position oneself for or against Cofferati, but rather to see what the city thinks and to avoid a situation that causes friction between citizens and immigrants. A survey by the National Confederation of Artisans showed that 61.3 per cent of citizens believed that 'the rule of law should always be respected', 31 per cent 'agreed with the politics of solidarity and acceptance', while only 2.2 per cent were 'against Mayor Cofferati's choices'.

The Case of Treviso and the Mayor of the Northern League

The Northern League mayor of Treviso, Giancarlo Gentilini, lined up his own vigilantes against windscreen cleaners at traffic lights and got rid of park benches lest they encourage 'immigrant wastrels'. The local bishop, meanwhile, told his flock that 'helping the needy is the duty of every Christian'. It was zero tolerance: in Treviso there was no place for illegality. Accepting immigrant workers into the fabric of local life is another matter: regular, stable workers are needed by small businesses in the North-East.

Gentilini has reiterated his 'No' stance on immigrant windscreen cleaners point by point:

> I am talking, of course, about immigrants who work without any authorization and outside any organizational structure. The municipal police will intervene to deal with this kind of illegality, which is an affront to those citizens of Treviso who pay taxes, live peacefully and work hard. The 'Piave race'[2] is an honest breed, hard-working and law-abiding. I have said this at every level and I have come to be called a racist. I believe I represent the will of my citizens.[3]

In the case of Treviso, the reactions of the political world and civil society were unanimous in their condemnation: Gentilini had behaved like a 'xenophobic and racist' sheriff, especially when he referred to the inhabitants of the Veneto as the 'Piave race'. In the meantime, disturbing anti-immigrant graffiti appeared in the city and groups of citizens periodically patrol the city at night.

Turin: The San Salvario Area

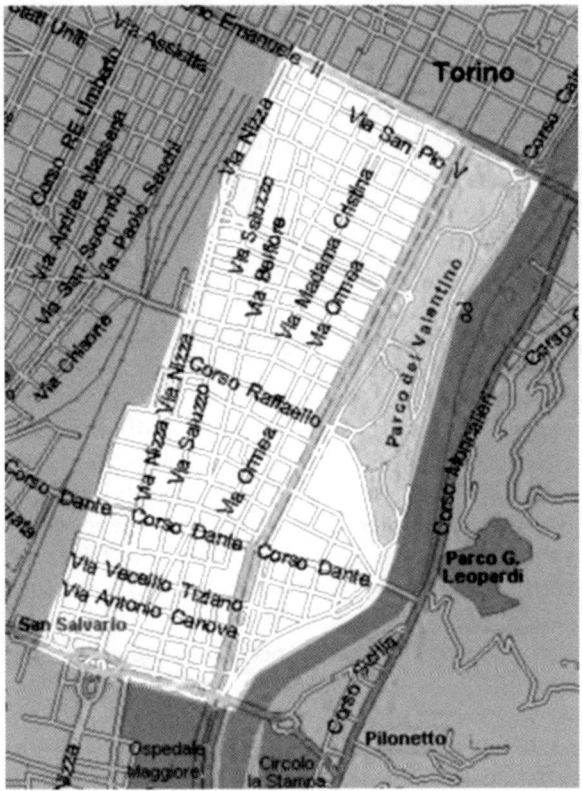

San Salvario district of Turin

Turin is repeatedly shaken by protests against immigrants. The names of certain parts of the city (San Salvario, Porta Palazzo, Pellerina, i Murazzi) are emblems of a new dissatisfaction or intolerance.

The San Salvario area is in the heart of Turin and is famous for the tensions and conflicts generated by the presence of immigrants. It is an area primarily inhabited by the petty bourgeoisie and the middle classes, who live in blocks further away from the railway station, while the immigrant areas (where the anti-immigrant demonstrations also take place) are to be found in the streets directly around the station. The streets adjacent to the station are infamous for their prostitution and drug and contraband dealing, both of which are controlled by bands of North Africans and Nigerians. Notwithstanding the shops run by immigrants, and the African and Asian restaurants which are also frequented by the local Turinese, the presence *en masse* of immigrants causes tensions and the price of property has plummeted. 'Block off San Salvario' has in the past few years appeared as graffiti on the city's walls (those wishing to block it off are obviously locals, and against the foreigners).

The percentage of residents born abroad, according to the official statistics, is equal to those born in the city (3.5 per cent), but is concentrated largely (5 per cent) in the northern part of the city. This zone houses the largest percentage of non-EU residents, most of them North Africans and Nigerians (see Figure 1).

San Salvario is also home to different religious communities (Catholics, Jews, Muslims and Waldensians), with a synagogue and two mosques housed in makeshift buildings. The area is host to various ethnic communities (apart from Italians, there are Sub-Saharan and North Africans, Asians, French and Anglo-Saxons).

In the heart of the area is the popular market of the piazza Madama Cristina, a place that could hardly be described as the stereotypical degraded or ghettoized periphery. The presence of foreigners here is very visible, as it is very concentrated. At night, drug-dealers take over certain streets. The gates of the via Nizza, opposite the station, are presided over day and night by unsavoury individuals. Fights and altercations occur, as do blood feuds.

The subjective perception of local inhabitants attributes this primarily to the density of population in the area, the poor living conditions in many buildings, the increase in micro-criminality and the close juxtaposition of different ethnic communities.

In these traditionally leftist areas, the centre-right now holds sway. The progressivists have lost because of people's growing sense of insecurity and competition between immigrants for the same places and streets. The vote seems go to whoever is most likely to take a firm hand with respect to those Sub-Saharan and North Africans with whom they are obliged to live.

Even in Porta Palazzo, another area which symbolizes Turin's racial conflict (though it is just minutes away from the Royal Palace and the Prefecture), the progressivist agenda has been ousted by the centre-right coalition in local elections. In short, it is the working classes who are most concerned about law and order, while the middle classes and the wealthy continue to side with the progressivist agenda. A constant stream of worrying signals continues to make people associate the cases of Porta Palazzo and San Salvario. Certain areas are controlled all day by groups involved in all kinds of illicit trafficking including drug trafficking and prostitution. Fights and blood feuds are common. Against these, the protests of the indigenous population is furious and fragmented. The town council and the police are inundated with demands that law and order be restored, that the black market be eliminated and housing tackled. There is a new wave of racism, as shown at the national level by the electoral success of the 'parties for order and against illegal immigration', and at a local level by an increase in tensions in multi-ethnic areas. Furthermore, the link established between crime and immigration gives rise to repressive measures: this means national laws limiting immigration and mechanisms at the local level for controlling territory and preventing certain kinds of behaviour designated as deviant. But the recently renewed spate of episodes of this kind in democratic countries, in which a climate of civil cohabitation did seem to have been established, makes one afraid that a new wave of intolerance and racism may be imminent.

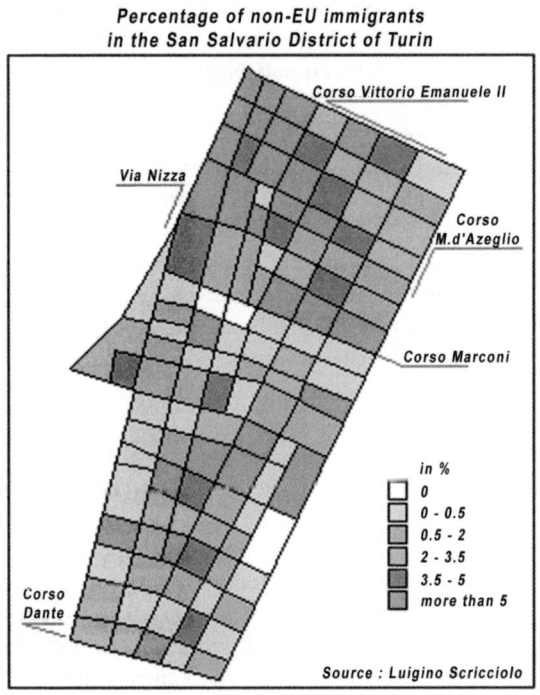

Figure 1

Rome, the Capital of Immigration

In the Trullo area of Rome, the latest punitive expedition against Romanians has taken place: one of their bars has been burned down. 'The same thing happens every night: they get drunk, they park in the middle of the street, they harass girls and the elderly. We are fed up with them,' say locals. Many

Trullo residents protest at the violence that is repeated almost every day. Twenty or so hooded youths armed with batons vented their fury against the bar on via Monte delle Capre in Trullo, and afterwards set fire to it. According to local residents, a true war has been going on for months between groups of local youths and Romanians. Fights between angry local youths and immigrants have become increasingly frequent.

But in Rome the number of favelas along the Tiber and the Aniene and in certain peripheral areas has increased: these are new ghettos where Albanians, Romanians, Macedonians, Kosovars and gypsies live. Even in the centre behind certain parts of the ancient city one finds camps where accidental death by fire is not unknown. The first ghettos are to be found near the Teatro Marcello, behind the Forum right between the Campidoglio and the Colosseum. In these places fifty or so non-EU illegal immigrants live, most from Asia and Eastern Europe. On the way up to St Gregory al Celio there are camps of Romanians, and on a street leading to Santa Francesca Romana, another camp of Asians has appeared. Tiburtina station has become home to a large number of homeless people: many live in train wagons. Others occupy disused factories in the areas near the EUR. But the truly hellish site is in the Tor Bella Monaca (a dormitory suburb), where one of the largest encampments is to be found, housing 1,500 homeless (from Muslim Korakane gypsies from Bosnia, to Kanjars and Orthodox Dazikaneche from Serbia, and, finally, Moroccans and Montenegrins). It is a kingdom of children with bloodied, scabrous faces. The other camp of despair is Casilino 900 where Kosovars, Serbs, Macedonians, Romanians, Albanians and a hundred or so Moroccans are forced to live together. This camp was created in the 1950s to house Calabrian immigrants; in the 1960s, the first gypsies arrived from eastern Europe and the Calabrians slowly but surely moved on. Now the camp houses about a thousand people of different ethnic backgrounds. Here filth reigns supreme and diseases that had long ago died out among us are a daily reality for them. But as long as the immigrants live apart or in favelas, run-ins and conflicts with the locals are avoided.

The Gypsy Question

Today in Italy there are between 60,000 and 90,000 'gypsies'. The largest proportion is made up of Sinti, 25,000 of whom live in nomadic camps; the others live in permanent housing (many of them are Italians born in Istria). The other large group and the latest to arrive is that of Roma from the former Yugoslavia, numbering between 10,000 and 12,000. To our farmers, these people are landless nomads; to city-dwellers, they are those marginalized on the city's outskirts; to factory workers, they are truants. To everyone, they are faithless and lawless. Every time people speak about gypsies, one is confronted with a sense of heavy frustration and of insecurity caused by the empty formality of the guarantees anchored in law. In this as in no other case, the mention of gypsies causes worry and embarrassment across the whole spectrum of public opinion, political parties and so-called civil society organizations. These camps of travellers, wherever they are located, give rise to protests and demonstrations by locals. The same thing happens in every city since, according to local residents, they are places of degradation, criminality and filth. The opening of Europe to Romania and Bulgaria has increased fear and unease among Italians, who are convinced that this will mean an influx of people from these countries into Italy, the EU's most permissive country.

Some Conclusions

One hears a lot about immigration. It divides public opinion and interests, above all, in the middle and lower strata of society, whose opinions on the matter tend to be similar regardless of their political positioning, be it left, right, or centre. The failure to give clear answers can pave the way for xenophobic movements and be a convenient scapegoat for those who wish to blame others for their own mistakes. In particular, cultural and religious differences provoke rejection. The working classes do not take well to competition with these 'foreigners in their own homes' in areas such as work and social support (housing, schools, health, and so on), particularly when an economic crisis means that their own living standards are falling. Islam is regarded with a great deal of suspicion, on both religious and cultural levels. When a high-ranking representative of the Union of Islamic Communities of Italy demands the right to polygamy, asks that crucifixes be removed from schools, or forces women to cover themselves from head to foot, tension increases. The demands made by certain sectors of the Muslim immigrant community and certain imams show a desire to impose on Italy their radical, male chauvinist, violent and anti-western version of Shiism. It is not acceptable to think of creating a charter of rights and values, as proposed by minister Giuliano Amato, to channel the Islamic question into the machinations of democracy. Italy has a Constitution which serves as a supreme charter for everyone and yet the Union of Islamic Communities of Italy and other Islamic communities have not even taken this into consideration as a basis for debate. One has the impression that the Italian political class has learned nothing from the experience of other European countries. The prevalent tendency elsewhere is to take a firmer line on accepting unqualified and uneducated immigrants, and conversely to try to attract well-qualified workers. They also invest in integrating immigrants, experimenting with new measures such as the French 'contract of welcome and integration'.

It sometimes happens that the existing laws in a recipient country are enforced on everyone with greater severity, in order to reassure the population that the immigrant does not present a threat and to avoid those feelings of insecurity that inevitably flare up into xenophobia and racist demonstrations. In Bologna, Mayor Cofferati has decided in favour of a more severe application of the law following some incidents of aggressiveness by (usually immigrant) windscreen cleaners. The event is emblematic of a presumed opposition of tolerance, seen as a value necessary for the acceptance of immigrants, against the rigorous application of the law, interpreted as the desire to exclude those who are different. It is not so. In my view, the position taken by the Bolognese mayor is much more effective as a policy of welcoming immigrants than the preaching of tolerance. It is those societies where one is certain of one's own rights which accept and favour the integration of different cultures, because they are less afraid of that which is different. Meanwhile, in those societies where that does not exist, and where an ambiguous 'tolerance' is in force, expressions of racism become more common, stemming from a fear of difference. If we want a large number of immigrants and their integration into our society, it is necessary that respect for law is enforced with a certain severity. An acceptance that is falsely tolerant generates fear of diversity which, in turn, generates ugly manifestations of racism which become difficult to eliminate. This is true both for Italy and for the rest of Europe. The appearance of the 'Identity, Tradition and Sovereignty' party in the European Parliament is a signal that needs to be confronted since fear, prejudice and hostility towards anti-western sentiments and against immigration (primarily Islamic immigration), cannot be left on the fringes indefinitely. It may be insignificant today but will grow if the questions of European identity and civilization are not properly confronted.

Luigino Scricciolo is an Italian journalist.

NOTES

1 ANSA, 31 March 2007.
2 The Piave river runs through Northern Italy.
3 La Padania, 22 March 2005

Migrations as a Factor of International Cooperation and Stability

GUIDO LENZI

Internationally, migration can be, should be and already is, to a certain extent, a issue requiring cooperation and stability. The world is indeed 'flat': communications and freedom of movement multiply among a host of international actors that have torn apart the intergovernmental framework established three centuries ago with the Treaties of Westphalia, at the end of the internecine intra-European Thirty Years' War.

It could well be said that international relations have been democratized. Such a development implies benefits and drawbacks for all nations concerned, big and small. There are more states now than ever in the history of humanity. It is not the first time that the world has undergone a process of globalization: the Silk Route, the discovery of the Americas, the Enlightenment, trade expansion in the nineteenth century, colonization and decolonization are among the many different previous expressions of it. As on those occasions, the new situation has radically altered the signposts. It has spread a feeling of precariousness typical of every transitional period, which should not induce panic but instead indicate the urgent need for all to adapt.

Among the many new phenomena, migratory flows must be seen as both the cause and consequence of this global upheaval. In order to deal with every such critical situation, international relations must move on from *ad hoc* emergency responses to a cooperative management of the underlying causes. Migrations in particular must be considered not as a pathology, which they obviously become if left unattended, but instead as a physiological development, potentially beneficial to the individuals concerned and equally to the countries of origin, both in terms of relieving their demographic pressures and of obtaining much-needed financial remittances (which in some countries account for 70 per cent of the national revenue). Historically, Italy has acquired both a very wide experience and a special sensitivity in the matter.

In political terms as well, migrations have become one of the most relevant 'human security' issues that have come to the fore in recent times, requiring an array of cooperative methods. These are needed for various reasons: in order to move from bilateral relationships to a network of involvements, and from an economic and social approach to a more political one, which should combine the 'pull' and 'push' factors, stamping out the criminal connections that go with migration, in terms both of people smuggling and of straightforward trafficking of human beings. This latter situation is akin to the slave trade of yore. In my country, Lampedusa has long been witness to an unbearable constant stream of modern-day 'Rafts of the Medusa', so dramatically rendered by Géricault over a century ago. Over and above security considerations, it is therefore finally a matter of preserving human dignity.

Three million legal immigrants come annually into OECD countries (with fewer asylum seekers and more family reunions). Successive Interior Ministers have stated that, contrary to other European countries, Italy needs an inflow of unskilled labour: 2.6 million immigrants (1 million of whom are Muslim) already reside legally with us, accounting for 5 per cent of our population, 6.1 per cent of our GNP and 2 billion Euros of remittances to their countries of origin.

Nationally, improved integration and intercultural dialogue must counterbalance the instinct of self-segregation and separation, which our open and pluralistic societies do not justify, and the eventual naturalization process excludes. True, our welfare states are less generous than in the past, but the resulting additional effort is required from every individual citizen, both current and would-be. Every immigrant must be encouraged not to assimilate and thereby lose their national identity, but to integrate in the new social environment, which does not imply cutting ties with their country of origin, on the contrary. For over a century, Italian immigration to the four corners of the world has shown the way. As Ernest Renan said in those days, all that is required is a will to live together.

Internationally, multilateralism must provide the broader framework in which to insert bilateral cooperative agreements. Partnership is the buzzword used nowadays to indicate the usefulness of sharing responsibilities, tasks and benefits, while protecting in the process individual national interests. International organizations such as the International Organisation for Migration (IOM), the United Nations High Commissioner for Refugees (UNHCR), the Organisation for Economic Cooperation and Development (OECD)and the United Nations as a whole with the 'high-level dialogue on migrations' launched in September 2006, provide the general framework. Among Mediterranean countries, the 5+5[1] and the Mediterranean Forum[2] contexts are finding their way, supplementing the Euro-Mediterranean

'Barcelona process',[3] which can and must proceed on specific issues, such as the implications of migratory flows, even when (especially whenever) its comprehensive political outer shell tends to falter.

The European Union has clearly restated its political intentions, by including its Mediterranean partners in its 'neighbourhood policy', and most recently by giving it further substance with the Rabat ministerial conference held on 10 and 11 July 2006 and the Tripoli Summit of 22 and 23 November 2006, dedicated to migration issues. The joint involvement of the countries of origin, transit and destination that resulted from these gatherings have put their management on a new footing, addressing at the same time the causes and consequences of migratory flows, addressing comprehensively and coherently (in a 'global approach', as has been stated) their economic development, social integration and law enforcement aspects.

The Tripoli meeting, resulting from a persistent Italian initiative, proved ground-breaking in that respect. Its conclusions should now constitute the much-needed essential common terms of reference for all countries concerned.

Guido Lenzi is a diplomatic adviser to the Italian Ministry of the Interior.

NOTES

1 The 5+5 is an informal western Mediterranean cooperative group of the Ministries of the Interior of France, Italy, Malta, Spain, Portugal, as well as Algeria, Libya, Morocco, Mauritania and Tunisia.
2 The Mediterranean Forum includes additionally Egypt, Greece and Turkey for informal political consultations supportive of the Euro-Mediterranean (that is, 'Barcelona') process (see n. 3 below).
3 The Euro-Mediterranean Conference of Ministers of Foreign Affairs, held in Barcelona on 27–28 November 1995, marked the starting-point of the Euro-Mediterranean Partnership ('Barcelona Process'), a wide framework of political, economic and social relations between the member states of the European Union and partners of the southern Mediterranean.

Immigration, Radicalization and Islamic Discourse in Europe: An Examination and a Sufi Response

NIR BOMS

I have a more technical point to make today. I stand here as a French citizen. I want to make clear that I am not French and have no relation. I'm a sworn enemy of France. So I want to make this in the record that I'm not French, okay? I tell you I am a Muslim, and I have nothing to do with a nation of homosexual Crusaders. And I am not a frog. That's the first thing ....

Zacarias Moussaoui, the twentieth hijacker, to the US District Court for the Eastern District of Colombia

I am not only British. I am English – because England is the only home I have ever known – even though I am a Muslim and I have an Indian name.

Munaf Zeena, Director of the North London Muslim Community Center

Many great and heated debates are hiding between the lines above, that represent only a fraction of the many voices expressed by Muslims in Europe.

In recent years, I believe that many of us have become increasingly aware of the voices of extremism such as that of Zacarias Moussaoui quoted above. These voices often fall into a stereotype that links Islam, immigration and extremism.

Moussaoui could easily fall into that very rubric. He was born in France to a mother that was married at the age of 14 in Morocco. Like many other poor immigrants, Moussaoui endured domestic violence at home and racism and discrimination outside. His family moved frequently between different parts of France and eventually to Britain. There he appeared to have made it. He earned a master's degree in iInternational business from South Bank University in London. He made money. He could have been a success story.

But he was still angry. Consequently, he took his initial steps into the world of radicalism at the Brixton mosque and later with members of the Finsbury Park mosque, where the extremist Abu Hamza al-Masri held lessons. In 1998, he attended the Khalden training camp in Afghanistan. Consequently, he joined Al Qaeda and participated in the planning of the September 11 attacks on the US.

Munaf Zeena, the 45-year-old director of a North London Muslim centre, has a very similar background. Like Moussaoui, he too suffered racism from British nationalists while growing up. He also has been to Pakistan and felt the enormous gulf between his own, British-born generation and that of his parents. But Zeena took a different path and in the last twenty years he became an educator and community activist who preach tolerance and moderation. Zeena found his own way to fight radicalism. He believes that in order to defeat extremism a sense of community must be created.[1]

'What turned *them* into terrorists and what turned me into a man of peace?' asked Zeena in a Muslim gathering in London a week after the London Underground terrorist attacks of 7 July 2005. This article attempts to highlight a partial answer to this question by addressing the issue of political Islam and its radical and 'moderate' discourses.

'Political Islam' is usually linked with Islamist tendencies that are often associated with the most radical voices of Islam. Statements, such as those given by radical markers such as Osama bin Laden of Al Qaeda or President Ahmadinejad of Iran stress the idea of the 'clash of civilizations' in which Islam should find itself fighting with the West. Some scholars tend to use similar language as they attempt to remind Europe and the West of a rapidly changing reality that requires a policy shift *vis-à-vis* the increasing Muslim population in its midst. Since Islam appears to be fighting the West, they would say, the West should be prepared to fight Islam.

I will argue that these extreme responses are not constructive, as they contribute to a radicalization process of newcomer Muslim immigrants to Europe and elsewhere. Further, I will argue that the 'clash of civilizations' terminology – that seeks to paint Muslim migrants as isolationists, radicals and extremists – fails to capture the real dimensions of the problems associated with Muslim immigration to the West. As this article will show, many Muslims are equally worried about these growing trends of radicalism among their fellow believers. More so, as the Sufi case further demonstrates, there is a solid Muslim leadership that is willing to make significant efforts to battle the phenomenon of Muslim radicalism.

* * *

Not surprisingly, the western discourse surrounding Islam in Europe often chooses to focus on the radical voices of Islam, on its dangers and on a 'European Awakening' that is allegedly necessary as a response to a threatening Islamic wave. Indeed, there are a few good reasons for that view.

As has been the case in the past, the world of Islam may do more to define and shape Europe in the twenty-first century than the United States, Russia, or even the European Union. Islam is at least the second largest religion in sixteen of the thirty-seven European states. In 1996, there were 200,000 Muslims in Belgium; 2.5 million in France and about 800,000 Muslims in Britain.[2] Today, in 2007, most estimates point to 400,000 Muslims in Belgium; 5–6 million Muslims in France and 1.6 million in Britain. These numbers, which represent a 100 per cent population growth within ten to fifteen years, reveal the vast challenges of immigration and integration that are often juxtaposed with the questions of communal influences. Will the Muslim community in Europe push its new members towards integration and moderation? Or, will the community's leaders push its members towards the opposite direction, that is, towards isolation and extremism?

The opening of the twenty-first century brought religious tensions to a new peak in Europe and brought new and unprecedented events to a European soil that had thought itself immune from the influence of radical ideologies. Events such as the murder of Theo van Gogh by an Islamic radical; the involvement of European residents and citizens in terrorist attacks that took place on European soil; the violent riots and the waves of demonstrations that accompanied the printing of cartoons in a Danish newspaper and, finally, in 2005, the wave of riots in immigrant-dominated areas near Paris, during which nurseries, schools, shops and over nine hundred cars were burned– these are just a few examples that illuminated the scope of an ever-widening gulf between a growing immigrant population and the European countries in which they live. A series of social, economic and political issues can be used to explain some of these events. But the phenomenon of radicalization, as the following examples will further illuminate, still stands out alarmingly.

A Nixon Center study which analysed interviews with 375 suspected or convicted terrorists in western Europe and North America between 1993 and 2004 found more than twice as many Frenchmen as Saudis and more Britons than Sudanese, Yemenis, Emiratis, Lebanese, or Libyans combined. In fact, fully a quarter of the jihadists it listed were western European nationals.[3]

The fact that European Muslims were willing to join the ranks of terrorist groups who vouch for Osama bin Laden appeared shocking to many Europeans. But, as repeated polls have shown, these individuals were connected to a broader trend of radicalization that can be found outside Europe as well. Recent finding by the Pew Global Attitude project affirmed that the rise of Islamic extremism is indeed an issue of global concern. Most respondents in countries as different as Germany, India and Britain, Pakistan, Indonesia and Egypt, expressed concerns about the rise of Islamic extremism. Similar concerns were expressed by the majority of British, French, Spanish and German Muslims. But at the same time, the Muslim respondents in Britain, French, Spain and Germany reported 25 per cent support to Islamic extremists such as Osama bin Laden. In another recent survey, 16 per cent of Muslims in France and Spain and 15 per cent of Muslims in Britain said that they were willing to support suicide bombing in order to protect Islam. Thirteen per cent of Muslims in Britain felt the terrorist attacks in London on 7 July 2005 were justified.[4]

These alarming findings may have not yet sunk in in Europe, where people are slow to admit that 'ideology matters'. Scholars and pundits will often brandish these alarming figures to support their call for a radical change to a whole set of policies concerning immigration, national security and education. But while Europe further engages in this important debate on the limits of tolerance and on the meaning of religious freedom, I believe it is failing to focus on another part of the Islamic discourse in Europe. This part of the debate is led by Muslims whom I will call 'integrators' and who are likewise alarmed by the growing influence of radical clerics and by the radicalization process that they see as fundamentally dangerous to their own interpretation of Islam.

The 'integrators' believe that these radical clerics, many of whom immigrated to Europe in the last fifteen years, do not speak the language of Islam which talks of tolerance, peaceful worship and to 'live and let live'. In fact, these clerics do not speak the languages of Europe rather, they speak in Arabic or Urdu to a crowd of Muslim immigrants that is more likely to be swayed by the clerics' arguments. However, this 'imported' radical discourse finds its own adversaries within the Muslim community in Europe. In some ways, this struggle can also be framed as one between 'European Muslims', who have been in Europe for three or four generations, and the 'newcomers', radical elements that succeed in engaging the first- and second-generation immigrants who suffer from the difficulties typical of many immigrants and who are more prone to find an answer in a radical Islamist message.

The Debate on Moderate Islam

All around the world, Muslims and non-Muslims are engaged in a debate about the nature of Islam. This debate knows countless slants and perspectives but its core revolves around the link between the noticeable trend of Islamic radicalism and the question of Islam itself as being the source of it. Abdel Rahman al-Rashed, the director of Al-Arabiya television news channel, eloquently put it when he wrote 'It is a certain fact that not all Muslims are terrorists, but it is equally certain, and exceptionally painful, that almost all terrorists are Muslims.'[5]

This debate touches the very soul of Islam and exposes the problems of discussing the issue of reform within Islam which by itself presents a set of theological problems. The doctrine of *Tawhīd* ('unification', the Oneness of God) and its indisputable message makes the discussion of 'reform' in an Islamic context difficult since it needs to reconcile sets of what appear to be indisputable truths.

The strong language used in various Islamic texts appears to leave little room for manoeuvre for further interpretations: Islam is divine, unified and indisputable. As it is taught in the Koran: 'There is no right on Him that is binding, and no one exercises rule over Him. Every endowment from Him is due to His Generosity and every punishment from Him is just. He is not questioned about what He does, but they are questioned.'[6]

Further, the implicit equation of 'reform' (*Islach*) with 'improvement' has led many Muslims to reject the applicability of the concept of reform in Islam, for Islam to them is perfect by definition and is not susceptible for improvement or reform.[7] Some Muslims also seek to characterize that debate as an attempt by non-Muslims to meddle in internal Muslim affairs and to inject 'western' ideas into Islam and subsequently divide the Islamic camp. In their view, 'reformist Islam' is a political invention that is essentially alien to true Islam.[8] The Central Council of Muslims in Germany, for example, recently stated that 'Islam is faced with the threat, due to political and governmental pressure, of being split into two denominations: Islam and Reformist Islam.'[9]

Of course, Islam was not immune from disputes, starting from the murder of Imam Ali in 661. Attempts to shape, reform and reinterpret Islam have soon followed and helped create the different schools of Islamic jurisprudence along with the lines of authority that influence Muslim tribes, orders, states and individuals until this very day.

Attempts to adopt and incorporate 'foreign' or 'western' influences into Islam are likewise not new. Islam, which in turn influenced Judaism, Christianity and later religions that emerged from it, was influenced by non-Muslim ideas in a similar fashion. Early thinkers like al-Farabi (870–950 CE), Ibn Sina (980–1037 CE); Ibn Rushd (1126–1198 CE) and Ibn Khaldun (1332–1406 CE) were influenced by Aristotelian and rational thought.[10] Thinkers such as Jamal ad-Din al-Afghani (1839–97), Muhammad Abduh (1849–1905) and his disciple Rashid Ridda (1865–1935), drew inspiration from western writers and philosophers and began to promote the idea that not Islam itself, but rather its archaic interpretation and its obsolete system of norms were to blame for the backwardness of Islamic societies. Here, for example, Muhammad Rashid Ridda, the founder of the Salafiyyah movement in Egypt, offers his criticism against Muslim leaders who abused Islam for the wrong political purposes:

> Muslims have lost the truth of their religion, and this has been encouraged by bad political rulers, for the true Islam involves two things, acceptance of the unity of God and consultation in matters of State, and despotic rulers have tried to make Muslims forget the second by encouraging them to abandon the first.[11]

Responding to foreign influences in Islamic lands, Khairuddin At-Tunisi (1810–99), the leader of the nineteenth-century reform movement in Tunisia wrote:

> Kindling the Umma's potential liberty through the adoption of sound administrative procedures and enabling it to have a say in political affairs, would put it on a faster track toward civilization, would limit the rule of despotism, and would stop the influx of European civilization that is sweeping everything along its path.[12]

Reviewing the background of some of these attempts to reform Islam is relevant to the issue at hand. However, it is beyond the scope of this article. It will suffice to point out that the current debate is not unprecedented, that it has deep historical roots and that it had led to fractionalisation within Islam in the past. The forces of reform and moderation existed and engaged the more radical elements of political Islam throughout Islamic history. Like today, they were often sidetracked by them.

Last but not least, it is important to point out that this discourse on 'moderation' and reform is indeed fundamental as it touches the very soul and nature of Islam itself. While the 'moderates' will likely accept the idea of 'different interpretations', the radicals will reject that notion wholeheartedly.

The Sufi Case

The alarming findings that were quoted above did shake a number of moderate Muslim leaders who understood that they may have already missed one of the last calls for action. In particular, these developments triggered an interesting shift in the traditional position of some Muslim Sufis who subsequently decided to join the political game.

I have found this shift of particular interest and relevance due to the following five reasons:

- It can, in a modest way, be considered an historical precedence.
- It is one of the strongest Muslim responses to radicalism that can be seen today.
- It is an organized and growing effort that is making headway in Asia, the Americas, Europe and the Middle East.
- An important portion of it is centred in Europe and aims to influence the European Muslim discourse.
- It is not a case of 'moderate Islam' (a definition that is very problematic by itself) but of political Islam of a different sort. It is also not the only example of the phenomenon I am about to describe.

Sufism, or *tasawwuf* in Arabic, is the name by which Islamic mysticism came to be known in the eighth or ninth century. The term 'Sufi' derives from the Arabic words 'suf' ('wool') and 'safa' (purity) and was applied to Muslim ascetics and mystics because they wore garments made out of wool. According to another view it is derived from the Arabic verb 'safwe', meaning 'those who are selected' – an idea that is frequently quoted in Sufi literature.

Sufi Islam is less rigid in its approach to Islamic law (*Sharia*), stressing the internal devotion to God and the pursuit of peace, equality and tolerance.[13] These moderate teachings were often at odds with both Sunni and Shiite Islam, which saw Sufism as a deviation from the true teachings of the Koran. Early Sufi mystics like Al Hallaj of Basra were charged with witchcraft and were persecuted for preaching the Sufi way. Consequently, their leaders learned to stay away from politics and did not always seek to challenge the ruling Muslim leadership.

But recently, some Sufi leaders have understood that a higher calling may be at stake. 'It is time for the middle ground to stand up and be counted before it is too late', wrote Haras Rafiq, the founder of the Sufi Council in Britain. 'Why is it that the more radical minority seems to have "taken over the microphone", and is ever increasingly becoming viewed by many as the mainstream version of Islam?' he asked in an editorial that appeared in a new publication, *SPIRITthemag* that is aimed first and foremost towards British Muslims.

The Sufi Muslim Council soon established its position against any form of radicalism:

> The Nazis – oppressors of many nations and of the Jews – stand condemned. Christian-Irish extremists engaged in fratricide stand condemned. Christian-Serbian extremists, oppressors of innocent Muslims in Bosnia and Kosova, stand condemned. Extremist Jews

> attacking innocents, stand condemned. Similarly, Muslim extremists, like bin Laden and his affiliates – mst be condemned. Therefore, we stand up as Muslims in the UK, declaring that we are not supporting any of these extremists, nor do we have anything to do with them.[14]

Rafiq and his colleagues followed up with a series of educational programmes, conferences and public speeches that aimed at attacking the radical elements of Islam within Britain (including the main Muslim establishment) which are considered, in their eyes, to be extreme and pro-Wahabbi.

Rafiq is not the only Sufi who has adopted such a position. When Sheikh Abdullah Algharib Alhamad Altamimee, a Syrian Sufi scholar who has been teaching Islam for thirteen years, decided to speak out against the Assad regime, he also broke with a thousand years of Sufi tradition. When he decided to travel to Washington and officially join the ranks of the Syrian opposition, he told his wife that his life was no longer in his hands. 'There are two million eligible young women who are not married mainly due to the fact that their potential husbands, two million eligible men, are too poor to support a future family,' he told me when I asked him about the reasons for his uncommon move. He explained that this is his own Islamic call and added that if he is to sacrifice his life for his people, he will receive his reward from God.

In Indonesia, a place influenced by Sufism from the time of Sultan Malik Al-Saleh in the thirteenth century, similar voices are heard. Like other parts of the Muslim world, Indonesia has experienced an Islamic revival since the 1970s.[15] Growing numbers of mosques and prayer houses, the increasing popularity of head coverings among Muslim women and schoolgirls, the more common sight of Muslims excusing themselves for daily prayers and attending services at their workplaces, the appearance of new forms of Islamic student activity on university campuses, strong popular agitation against government actions seen as prejudicial to the Muslim community, and the establishment in 1991 of an Islamic bank are all signs of that Islamic revival. But aside from these 'outer' (*lahir*) expression of Islam, Indonesia also experienced the increasing popularity of Islam's 'inner' (*batin*) spiritual expression. Sufism, according to Howel:

> ... has inspired new enthusiasm, even in the sectors of Indonesian society most intensely engaged in modernisation and globalisation: the urban middle and upper classes. This interest is expressed through the participation of urbanites in the long-established, rural-based Sufi orders, the *tarekat*, but also through novel institutional forms in the towns and cities.[16]

As with the Sufi Muslim Council and Sheik Tammimi, Sufi leaders in Indonesia have also began to use Sufi principles in a political way and in a struggle against a more radical interpretation of Islam.

Ahmad Dhani, a Sufi and a rock star, is one example for these new voices. In 2004, he issued his *Laskar Cinta* album, which means 'Warriors of Love'. The title was a very deliberate choice: Laskar Jihad ('Warriors of Jihad') is a violent militia that was led by Jafar Umar Thalib, a veteran of the Afghan jihad who claims to have met Osama bin Laden. In 1999, following an incident with a Christian bus driver and a Muslim passenger, Thalib's militia shipped thousands of fighters to the Maluku Islands into the region by boats to 'wage jihad'. The conflict lasted three years – an estimated ten thousand people perished on the island of Ambon alone, and around half a million Indonesians were driven from their homes.

The *Laskar Cinta* album was designed to provide Indonesian youth with a choice between joining the army of jihad and joining Dhani's army of love. It sold hundreds of thousands of copies. And it brought some harsh responses from elements like the Islamic Defenders Front, a radical group in Indonesia that is affiliated with Hizb-ut-Tahrir. As a result, Dhani and his wife, Indonesian pop star Maia, and their children went into hiding last year.

Despite the furore created by their music, Dhani and his group emerged last December with a new song, also with the title 'Laskar Cinta,' that soared to No. 1 on Indonesian radio and MTV Asia. 'Laskar Cinta' is the first track in Dewa's latest album, *Republic of Love*:

> Watch out, watch out and be on guard –
> for lost souls, anger twisting their hearts, for
> lost souls, poisoned by ignorance and hate ...
> Warriors of Love, teach the mystical science of love,
> for only love is the eternal truth and the shining path for all
> God's children everywhere in the world.

Abdulrrahman Wahid , a former Prime Minister of Indonesia, a Sufi leader and a patron of Dhani , is now working with a group of Arab musicians and producers in order to bring these Sufi messages to mainstream Arabic poetry world-wide.

I realize that pop songs do not usually make their way to academic journals but I have chosen these examples to demonstrate that this important discourse – that can also be described in the context of a 'war of ideas' – is taking place throughout the Muslim world and in some untraditional ways.

Conclusions

This article has focused on the connection between Muslim immigration, radicalism and Islamic discourse by showing two dimensions of the Muslim expatriate community: a radical dimension, that was demonstrated by a brief examination of the European case, and a counter-reaction to these radical influences, illustrated by the Sufi case in Europe and elsewhere.

The influence of non-state players in the international area, chief among which are terrorist groups, is already proven. Of course, radical groups, who are willing to operate without considering the consequences, have an advantage in this area. But moderate groups, who have also become much more active since the turn of the century, can have an influence as well. Current research appears to provide only a partial perspective as to nature of the 'integrators', their work, their potential and their possible contribution to a new Muslim discourse.

It is assumed that radicalism came (or was imported to) Europe from outside its peninsula. It is also assumed that ongoing influences of Muslim and Arab lands on Europe will not likely bring radically different messages to Europe in the near future. In that sense, we can further assume that 'moderation' will not be likely to emerge in Europe in a similar way to the emergence of racialism. If it were to emerge, it will have to emerge from within.

Radicalism can be tackled in a number of ways but governmental policies can make a difference. France and England adopted different immigration policies and different approaches when dealing with the absorption of their minorities. In a Pew Research poll of Muslims conducted in spring 2006, 81 per cent of British Muslims said they were Muslim first and British second. But 'only' 46 per cent of French Muslims were saying the same thing.[17] Government can help to shape a more positive or negative climate that can in turn encourage or discourage isolation, integration , radicalism and moderation. And governments may have some dedicated allies that are committed to these very objectives. These allies, as the Sufi case illuminates, have a direction, a guiding force and even a growing 'camp' that slowly bring together Muslim individuals and groups who seek to advocate a different Muslim agenda. These groups must be taken into consideration.

Muslims in Western lands are becoming increasingly organized. For some, this is a threatening development that reflects much of the problem associated with radicalism rather than its 'solution'. But, as was the case for radical groups, for expatriate groups and for immigrant diasporas world-wide, these groups quickly develop lives of their own and agendas that can be markedly different from those originally represented by their 'founding fathers'. Judaism and Christianity provided many examples for religious doctrines, practices and perspectives that have changed as a result of communal dynamics in foreign lands. Islam can provide a few examples of its own, including the recently developed framework of Minority Jurisprudence (*Fiqh al-Aqalliyyat*) that was created to address the unique religious needs of Muslims living in the West.[18] It is too early to say whether theological and/or other major developments will emerge as a result of this process. But what can be said is that this is a living dynamic that has a strong potential of influencing the further segregation or integration of Europe's Muslims.

Finally, it is important to recognize the 'integrationist' not only because of their merit and because their work makes them a better partner for dialogue, but also because that recognition helps to frame the real challenge that lies ahead: not a *war of civilization* that puts Islam against the world (or the world against Islam) but rather a *war of ideologies*, in which Muslims, Christians and Jews can fight shoulder to shoulder against radicalism and extremism worldwide.

Nir Boms is Vice-President of the Center for Freedom in the Middle East, Washington, DC.

NOTES

1. John Daniszewskim, 'Moderates Raise Voices to Influence the Young', *Los Angeles Times*, 18 September 2005.
2. J. Rath, R. Penninx, K. Groenendijk and A. Meyer, *Western Europe and its Islam: The Social Reaction to the Institutionalisation of a 'New' Religion in the Netherlands, Belgium and the United Kingdom*, International Comparative Studies Series, 2 (Leiden/Boston/Tokyo: Brill, 2001).
3. Robert S. Leike, 'Bearers of Global Jihad: Immigration and National Security after 9-11', (Washington, DC:Nixon Center, 2005).
4. 'The Great Divide: How Westerners and Muslims View Each Other', Pew Global Attitudes Project, June 2006.
5. Abdel Rahman Al-Rashed, 'Innocent Religion is now a Message of Hate', *Al-Sharq Al-Awsat*, 3 September 2004.
6. Qur'an, Surat an-Nur, 21: Al-Anbiya', 23; Surat al-'Anbiya', 23.
7. Andreas Jacobs, 'Reformist Islam: Protagonists, Methods, and Themes of Progressive Thinking in Contemporary Islam', Brochure series, September (Berlin/Sankt Augustin: Konrad-Adenauer-Stiftung, 2004), pp. 3–4.
8. Ibid., p. 5.
9. Ibid.
10. Stephen Frederic Dale, 'Ibn Khaldum: The Last Greek and the First Annaliste Historian', *International Journal of Middle East Studies*, 38 (2006), pp. 431–51, p. 432.
11. Cited in Albert Hourani, *Arabic Thought in the Liberal Age 1798–1939* (Oxford: Oxford University Press, 1962), p. 228.
12. Khairuddin At-Tunisi, *Aqwam Al-Masalik Fi Taqwim Al-Mamalik* (The Straight Path to Reformation of Governments) (Al-Dar Al-Tunisiyah, 1972), p. 185.
13. It would not be accurate to characterize all Sufis movements as 'moderate' or to draw broad-brush characterizations on Sufism overall. Sufis in general are complex, and cover many different 'stripes'

of Islam, for example, Sufism started out as a Shia movement, but now is mainly a Sunni movement. Hanbalis, Shafis, Malikis and Hanafis can all belong to different Sufi *'tariqas'* (brotherhoods). In fact, both the Muslim Brotherhood in Egypt, as well as Al Qaeda, have Sufi roots.

14 Declaration of the Sufi Muslim Council, *SPIRITthemag – The Voice of the Silent Majority*, 1 (June 2006), p. 14.

15 See, for example, Robert W. Hefner, 'Islam in an Era of Nation-States: Politics and Religious Renewal in Muslim Southeast Asia', in Robert W. Hefner and Patricia Horvatich (eds), *Islam in an Era of Nation-States: Politics and Religious Renewal in Muslim Southeast Asia*, (Honolulu: University of Hawaii Press, 1997), pp. 3–40.

16 Julia Day Howell, 'Sufism and the Indonesian Islamic Revival', *Journal of Asian Studies*, 60, 3 (August 2001), pp. 701–29.

17 Jonathan Paris, 'Europe and Its Muslims', *Foriegn Affairs*, 86, 1 (February 2007).

18 *Fiqh al-aqalliyyat* was developed as a means of assisting Muslim minorities in the West. The *fiqh*, or jurisprudence, for Muslim minorities, is a legal doctrine introduced in the 1990s by two prominent Muslim religious figures, Shaykh Dr Taha Jabir al-Alwani of Virginia, and Shaykh Dr Yusuf al-Qaradawi of Qatar. It may, however, also be applied in other parts of the world with large Muslim minorities, such as India. Shaykh Dr Taha Jabir al-Alwani, founder of *fiqh al-aqalliyyat*, gave a video lecture at a 'Jurisprudence Workshop' held in New Delhi (<www.asharqalawsat.com>, 14 February 2004). Dr Yoginder Sikand, an Indian Muslim intellectual, regularly includes articles about *fiqh al-aqalliyyat* in his online journal *Qalandar* (<www.islaminterfaith.org/dec2004/article4.htm>). For a complete review see Shammai Fishman, *Fiqh al-Aqalliyyat: A Legal Theory*, Research Monographs on the Muslim Series No. 1 (Washington, DC: Center on Islam, Democracy, and the Future of the Muslim World, Hudson Institute, October 2006).

The Migration Policy of the City of Athens

KALLIOPI BOURDARA

A cohesive community is the one that has a common vision and a sense of belonging for all, where the diversities of their citizens are appreciated and positively valued, where those from different backgrounds have similar life opportunities, and where strong and positive relations are developed between people from different backgrounds in workplaces, in schools and within neighbourhoods.

Community cohesion cannot be achieved and sustained with single measures. It is necessary to display a number of activities: listen to the community's problems, develop a strategy for delivering outcomes, continue to improve equal services, develop common bonds and friendship, share culture, respect differences, encourage community participation, and use sporting, cultural and interest activities to promote social cohesion.

We believe that society learns to recognize and accept differences on a daily basis. This is the result of being together at schools, in neighbourhoods, in the market, at work, in places of entertainment. In this light, we established the first Municipal Intercultural Centre, organized seminars on interculturalism, and operated a reception office, offering information and consultation to immigrants. We also established the first multilingual municipal radio station, hosted events and invited migrant communities to attend city events.

Unlike the prevailing trend elsewhere, the city of Athens treats immigration not as a 'problem' but as an asset.

This is a reality very much connected to globalization, to Greece's economic development and participation in the EU's economic zone, to the higher levels of literacy among the population (creating a surplus in the professional workforce and a deficit of semi-skilled and manual workers) and to the fact that many of Greece's immediate neighbours are former communist countries. According to the latest census in Greece in 2001, migrants amount up to 22 per cent of the total population of the city of Athens.

Over the last fourteen years, Greece has experienced a massive influx of migrants from neighbouring countries, as well as from Africa and Asia. In this period, the density percentage per square meter of the migrant population in the city of Athens has reached the levels which it took the United States of America 150 years to reach.

The city of Athens, like many other cities in Greece and elsewhere in Europe, must face the following paradox: migration policies are being adopted at the European Union or the national level, yet it is the cities that are called upon to implement them on a micro-scale. Very often it is the case that funds for necessary actions are only available centrally and are allocated on the basis of a top-down philosophy.

Athens has accepted this challenge and has adopted a strategic framework for implementing integration policies for the migrants who chose Athens as a safe environment for themselves and their families.

In the framework of this policy, the city of Athens:

- is in direct line of communication with agencies implementing relevant programmes, migrant associations and non-governmental organizations assisting immigrant and refugee populations;
- encourages the participation of immigrants in events and activities in the city, such as the Athens' Volunteer Programme during the 2004 Olympics and Paralympics;
- offers space and assists in the communication to the public of activities, promoting xenophilia and organized by immigrant communities;
- has upgraded the infrastructure and the services provided by the Immigration Office with additional, qualified personnel, able to communicate in basic migrant languages and, relocating the Office in newly renovated spacious and more functional premises;
- has established the multilingual municipal radio programme 'Athens International Radio 104.4', airing programmes in twelve languages (English, French, German, Spanish, Italian, Arabic, Albanian, Bulgarian, Polish, Romanian, Russian and Filipino).

In the framework of EQUAL Programme (Development Synergy, 'Forum for the Social Cohesion'), the Centre for Employment and Entrepreneurship of the City of Athens established an office for the reception of immigrants, offering information and consultation on issues regarding employment and procedures for the issuing for residence and working permits. The number of beneficiaries reached almost 6,000 in eighteen months' time. Also within the framework of EQUAL Programme, the Centre for Employment organized, in November 2003 and in December 2004, seminars on interculturalism, targeting employees working in Athens and other cities in Attica.

We are convinced that the mere management of bureaucratic issues is not enough. The city must show itself to be active, by assisting, contributing to and activating all dynamic human resources in order to provide solutions to the major and substantial problems of our fellow citizens who come from abroad. Local governments are the main actor for cultivating an atmosphere of trust among citizens, for achieving creative interaction between cultures and civilizations, as well as social cohesion – elements crucial for the formation of viable future multicultural societies.

For the city of Athens, immigrants are a part of the city, a part of its human and cultural wealth. Our wish is to give to all Athenian residents the opportunities that are necessary for pursuing and fulfilling their dreams and aspirations.

Dr Kalliopi Bourdara is the Deputy Mayor of Athens.

The Social Integration of Immigrants in Greece

MARKOS PAPAKONSTANTIS

Until the end of the 1980s, the word 'immigrant' in Greece referred to distant relatives of Greeks who had left their country in search of a better life[1] in two waves: one at the beginning of the twentieth century,[2] and another in the 1950s and 1960s.[3] There were only a few immigrants in Greece itself.[4] These were mainly manual labourers from Egypt, Ethiopia, Pakistan and the Philippines.[5] Small wonder, then, that Greece saw little need for a large-scale legislative programme to regulate immigration before 1991. Foreign immigrants were governed by Law No. 4310, passed originally in 1929,[6] and revised in 1948, a text which dealt above all with migratory flows.

At the beginning of the 1990s, the first waves of immigrants arrived, mainly from the Balkans following the collapse of the Eastern bloc and the socio-economic changes which flowed from that. For the immigrants from these countries, Greece offered a number of attractive characteristics, including a prosperous economy with job opportunities and high salaries (relative to those paid elsewhere in the Balkans), the rule of law with stable institutions and economy; membership of the EU; and the fact that moving there or being deported from there was relatively cheap and stress-free in comparison to emigration to more distant countries.

At first, the large-scale arrival of these immigrants surprised the Greek state. Greece was confronted with a completely new situation for which it was unprepared. The first law, No. 1975, passed in 1991, was entitled 'The arrival, departure, residence and labour rights of foreigners, deportation, procedures for the recognition of foreign refugees and other measures'[7] and was aimed principally at restricting immigration, at better border control, and at the expulsion of illegal immigrants. From 1991 to 2001, the situation changed rapidly; whereas there were 167,000 immigrants registered in 1991, their numbers exceeded 800,000 ten years later. This new reality forced the government to adopt a new immigration policy, which it did in Law No.2910/2001 entitled 'The entry and residence of foreigners on Greek territory; acquisition of Hellenic nationality by naturalization and other measures.'[8] In the meantime, two presidential decrees (358/1997 and 359/1997)[9] were issued in order to put in place the first programme for regularizing those immigrants already in the country.[10] When it submitted the new law, No. 2910/2001, the Greek government had two objectives: on the one hand, it wanted to put in place a second programme for regularization, in order to deal with the massive arrival of immigrants without permits; on the other hand, it wanted to put in place measures which could deal with immigration in the medium term, including better border controls, the conditions under which foreigners could come to Greece to work or to study, and finally, naturalization of resident immigrants.

Law 3386/2005, 'The entry, residence and social integration of nationals of third countries on Greek territory'[11] represents the third key effort by the Greek legislature to regulate unresolved and badly managed migratory issues and to give illegal immigrants a third chance to be regularized. After fifteen years in which there had been immigrants in Greece, the Greek government had to take into consideration a number of facts:

- the permanence of the phenomenon of immigration;

- the real magnitude of the question both in qualitative and quantitative terms, that is, the number of immigrants and the consequences of their presence for Greek society and the economy;

- the need to regulate large-scale entries and residence by means of a programme of social integration above and beyond simple police measures;

- the need for society to adapt to the continuously evolving needs and obligations created by the permanent residence of foreigners;

- the weaknesses and gaps in the previous legislation, and

- the experience of the country and of the rest of the EU, and the EU's own powers in the matter.

Some of these matters were taken seriously by the Greek government and influenced the drafting of the law. Others did not. This law was completed by two other decrees, Nos 150/2006[12] and 131/2006[13], the adoption of which was necessary in order to harmonize Greek law with EU Directive 2003/109/CE on the status of long-term foreign residents[14] and with EU Directive 2003/86/CE on family regroupment.[15]

In order for integration to work in Greek society, it is necessary to create appropriate conditions, both in order to promote peaceful coexistence between Greeks and foreigners, and in order that Greece fulfil its EU obligations. In the first case, the creation of a framework in which immigrants can integrate and function as an integral part of Greek society is necessary in order that ghettoes do not develop and to prevent outbreaks of xenophobia and racism on the part of Greek nationals, reactions which are the result of the creation of barriers in society. In the second case, the changes in Greek immigration policy which have occurred in recent years, above and beyond the changes which have occurred on the ground, and Greece's transformation from a country of emigration to a country of immigration, reflect developments at EU level. Faced with similar problems created by the continuous influx of immigrants, Greece's EU partners decided to try to put in place common rules to facilitate the social integration of this sensitive segment of the population.

In this article, I first research the juridical framework which Greece has adopted to favour the integration of immigrants by means of Law 3386/2005. I use the word 'research' deliberately because the provisions for integration are few and far between, while some of them have had the opposite result from that desired. Then I will look at how these provisions have been applied in the administrative domain and how they thus become real policy. Finally, I will present issues which remain to be resolved if we want to speak of a real desire to integrate immigrants.

The Legal Framework for the Social Integration of Immigrants

Before discussing the provisions of Law 3386/2005 which deal with the integration of immigrants, four observations need to be made. First, the integration of a person into society is permanent. It starts with birth and finishes with death. During the process of socialization, some factors play a preponderant role in deciding whether or not integration succeeds. This continuous procedure is not unilateral. It can take different forms depending on whether the society in question promotes or legislates in favour of integration.

Second, if we distinguish between the notions of integration and incorporation, we will come to the following conclusion: all areas of daily life which concern public life constitute factors of social integration. By contrast, those which concern private life concern social incorporation.[16] Employment, housing, language, education and civil rights are in the first category, while family regroupment, religion and the respect for cultural differences are in the second. Respect for the private life of foreign immigrants by the host country is a major contributory factor to integration. A state which preaches that it has a migration policy with a heavy emphasis on social integration must not neglect policies which favour incorporation.

Third, the law divides immigrants into two categories: legal and illegal. For those in the first category, the law provides for rights and duties. For the second, there are only prohibitions. Consequently, integration concerns only legal immigrants.

Fourth, immigrants constitute a sensitive segment of the population which, in order to integrate into society, must pass through a procedure of resocialization. This procedure presents difficulties considering that the foreigner is often called upon to adapt to a society whose characteristics are quite different from that in which he or she grew up. Because of these concrete difficulties, the creation of conditions which will facilitate this effort is essential for any state which understands that harmonious coexistence produces only beneficial results.

Before discussing the provisions of the law on the matter in hand, we must briefly describe the procedure of regularization which is the starting-point for social integration.

The Procedure for Regularizing Immigrants in Greece

Delivery of a residence permit is the essential prerequisite for any state policy on integration. From the moment when the foreigner has a residence permit, they can legally remain and work in the country.

The law provides for two categories of applicants for residence permits: those who have the right to enter Greece and those who enter without having the required papers. For the latter case, there are transitional arrangements which provide the starting-point for regularization. This concerns only those who can prove that they entered Greece before 2005.

In the first category, nationals of third countries who want a residence permit must fulfil the following conditions: they must have entered the country legally and must be able to prove that they have resided in Greece either by means of a passport or using any other document recognized by international conventions. There must be no record of any threat to public order and no danger of any threat to health. The applicant must also be able to show that they have the necessary resources to be able to return to their country of origin.

In the second category, nationals of third countries have the right to apply for a residence permit as soon as one of the following conditions is satisfied: that they have entered and resided in Greece before 31 December 2004 ; that they represent no threat to public order or health; that they have registered with the tax authorities; that they have paid their social security contributions; that they swear a formal attestation about their employment in Greece, the members of their family who live with them, and that they have no criminal record.

People may apply for residence permits if a) they have a legal visa for entering the country; b) if they are applying for humanitarian reasons; c) if their applications for asylum have been rejected; d) if they are relatives of a resident who is abroad; e) relatives of an illegal resident and adult children of legal residents if they have a visa on the family passport, or if they are listed on it as being protected members of the family.

The procedure which legal immigrants must follow in order to obtain a residence permit is as follows: the foreigner submits their application with all the relevant documentation to the local authority where they presently live. The decision on whether to grant the residence permit is taken by the secretary-general of the local authority. Permits are issued for one year and can be renewed for a two-year period after that. The procedure is essentially the same for illegal residents, who must prove that they have resided in Greece before 31 December 2004.

Provisions in Greek Law for the Integration of Immigrants

Law 3386/2005 lays out the objectives of social integration and explains how it is to be achieved. It provides that integration entails both rights and obligations for foreigners. They are allowed to participate in the economic, social and cultural life of the country, and they are obliged to respect the fundamental values of Greek society. The principles on which integration is based are: a) avoidance of all forms of discrimination; b) equality; c) respect for fundamental rights; d) support for their own contribution to the social, economic and cultural life of the country; e) support for the immigrant's family; f) participation in the social integration process. These various principles are embodied in the Integrated Action Programme. This requires: a) a certified knowledge of Greek; b) a course on Greek history and civilization; c) work in the Greek labour market; d) active social participation. The requirement that the immigrant know Greek and Greek culture applies only to those who wish to apply for long-term residence permits. The decision on this was taken by three ministries acting together: the ministries of the Interior, of Public Administration and of Religious Affairs.[17] These ministries run and oversee the certification procedure and organize exams to text the applicants' knowledge.

Law 3386 has no provisions governing insertion into the labour market. Indeed, it creates a limited system for immigrants, reducing their mobility, flexibility and right to work. Work permits are issued for specific kinds of jobs and are not necessarily transferable to other jobs: the restrictions on immigrants are of course far more onerous than on Greek nationals. Immigrants wishing to start up their own businesses are also subject to special rules which require them to put up tens of thousands of euros first.

Education and training play an important role in the social integration of foreigners. The law provides that education is a right enjoyed without restrictions by minors who are nationals of third countries residing in Greece. They need the same documents for enrolment in schools as Greeks do. Those without such documentation may also be accepted if they are refugees, or if their regularization is in the pipeline. Adult foreigners may also have access to the tertiary education system on the same basis as nationals. However, there are no provisions for the recognition of degrees from foreign universities.

The Political Framework of the Social Integration of Immigrants

The letter of the law will remain hollow if the administration is not prepared to implement its provisions. This is the problem in Greece where the problem is not the lack of law but the failure to apply it.

Moreover, integration is a process which goes beyond mere formal regularization. It requires the participation of the state, which creates the necessary infrastructure, and of society which consolidates the state's efforts. Society is a living organism which can repel foreign bodies if they threaten to upset its internal balance. At this point, the role of the state becomes decisive. Beyond infrastructure, it needs to create the conditions which are necessary so that the blending of civilizations can be of benefit to all Greek society, that is, including its foreign immigrants. It must ensure that integration and coexistence are not threatened by individual cases but that any unpleasant events, having been condemned, should instead constitute the starting-point for thinking about immigration matters and setting off in the right direction.

Laws need to be the foundation stone for policy and they need to be flexible and durable. Unfortunately, the provisions for integration in Law 3386 are limited. The Integral Action Programme remains simply a written project. Of the four initiatives it lays down, only the certification of knowledge of Greek language and history can be implemented. But even the inter-ministerial decision which lays out the certification process makes it difficult to get the certificate. A foreigner who wishes to obtain a long-term residence permit is obliged to have a hundred hours of Greek-language lessons, and twenty-five hours of lessons in Greek history and civilization. At the end, they must take written and oral exams. It is difficult for immigrants to fulfil these conditions because the lessons must be taken in an adult education centre and not all regions of Greece have them. The immigrant also must write an essay on a matter of topical interest and also take comprehension and grammar tests, and a recorded oral exam. This alone is enough to deter a lot of immigrants and thus deprive them of the possibility of obtaining the long-term residence permit. Immigrants therefore remain in a situation of uncertainty with two-year residence permits, and this uncertainty does nothing to foster a sense of integration.

The law therefore lacks dynamism and a sense of perspective. These are serious flaws for a law which aims at solving such a serious social question. It also fails to make any provision for second-generation immigrants. There is a lack of foresight as to what to do with the children of immigrants who have either been born or who have grown up in Greece. These children cannot obtain Greek citizenship an they are even denied birth certificates. This puts the children at constant risk of expulsion, even though they are called on to pay taxes. It is obviously impossible to integrate them under such conditions. Like their parents, they are second-class citizens with limited rights, deprived of the equality of opportunity. The state bears a heavy responsibility for this.

Conclusion

Greece evidently does not have a long tradition of immigration policy but this is no justification for the failure to pursue an overall policy of integration. The failure is the result of the weakness of the state to implement its plans, a reluctance to spend money and a desire to avoid the political cost of pursuing a policy which might provoke a counter-reaction in Greek society. The state wants to make integration difficult so that immigrants will be encouraged to leave. This is not a policy but the lack of a policy which hinders integration.

The reference in the title of the law to 'the social integration of nationals of third countries on Greek territory' therefore cannot be justified by the actual content of the law. On the basis that social changes need time and that integration is such a social change, the limited reference to certain provisions in the law could be the seed which will put down roots and grow, providing that the state cultivates it and protects it from natural catastrophes. However, in its practices, the state has shown that it has no intention of playing this role. It is a source of consolation that Greek society has shown itself, in its overwhelming majority, to be tolerant and hospitable towards immigrants and that it helps solve the problems which arise from coexistence with foreigners.

Markos Papakonstantis is an advocate and research fellow at the Institute for Strategic and Development Studies (ISTAME) in Athens.

NOTES

1   The three main reasons which led to Greek emigration were: a) the rumour that there existed huge amounts of uncultivated land in America; b) the development of transport; c) personal, political and economic reasons. See L. Dollot, *Les migrations humaines* (Paris: PUF, Coll. Que sais-je, 1958), pp. 74–6.

2   In the period 1900–1920, Greece lost 8 per cent of its total population. Some 25,000 Greeks left the country every year for the 'Promised Land', leaving behind a country which was financially destroyed and wracked by political uncertainty.
3   In the 1960s, the flight of the population, above all to West Germany, reached 459,000, that is, 5 per cent of the population at the time. This flow differs from that of the period 1900–1920, in that it was intra-European and not overseas.
4   According to Eurostat, the number of foreigners in Greece in 1981 was 171,424 and 797,093 in 1991. Foreigners represented 1.4 per cent of the population in 1990, and 8.1 per cent in 2004.
5   See L. Stergiou Nicolaou, 'Problématismes de la politique migratoire en Grèce et en Europe', *Ekpedeussi & Epistimi*, 1, 3 (April 2006), p. 276.
6   The first law on 'The installation and movement of foreigners in Greece' was voted in 1925 and was in force for two years from 1927 until 1929. Law 4310 of 1929 had the following title: 'Law on the installation and movement of foreigners in Greece, police control of passports, expulsions and deportations'. In addition to security matters, this law also dealt with employment and repatriation. It forbade foreigners from entering Greece without a work permit. There was a debate at the time on whether the restriction on the liberty of movement of persons were compatible with the terms of the 1927 constitution.
7   Official Journal of the Greek Republic, A 184 /4.12.1991.
8   Official Journal of the Greek Republic, A 91/2.5.2001.
9   Official Journal of the Greek Republic, A 240/28.11.1997.
10 371,641 immigrants applied for a 'white card' (residence permit for indefinite period), of which 212,860 were foreigners seeking regularization.
11 Official Journal of the Greek Republic, A 212/23.8.2005.
12 Official Journal of the Greek Republic, A 160/31.7.2006.
13 Official Journal of the Greek Republic, A 143/13.7.2006.
14 Official Journal of the European Communities, no L 16 du 23 janvier 2004.
15 Official Journal of the European Communities, no L 251 du 3 octobre 2003.
16 Institut de Travail de GSEE/ADEDY, Circoscription d'Attiki, *Recherche sur les formes d'intégration sociale des immigres économiques dans la circonscription d'Attiki 2003–2004*, pp. 37–8.
17 See Ministries of the Interior, of Public Administration and of Religious Affairs, *Certificate of knowledge of the Greek language and of elements of Greek history and civilization for nationals of third countries desiring to obtain a long-term residence permit*, 23 November 2006.

Are the Spanish For or Against Immigration?

MARISA ORTÚN RUBIO

In order to understand how the Spanish react to immigration we need to take several factors into account:

- *The sudden and unexpected growth of population* – Spain's population has grown by more than three million people from 2000 to 2005, that is, more than during the previous twenty years. Considering Spain's low reproduction rate, we know that the main reason for this increase is immigration. In 1981, there were 200,000 foreigners living in Spain; in 2001, there were more than one million. On 1 July 2006, according to the Instituto Nacional de Estadística, the number of people living in Spain was estimated at 44.07 million, of whom 4.27 million were foreigners. For the period 2001–06, in some regions the population of immigrants could be multiplied by a factor of four or five.

- *Spain is considered an immigrant host country* – It has become a destination country for African and Latin American immigrants who are drawn to the country because of historical, cultural and linguistic links.

- *Legal instability in migration matters* – José María Aznar of the Partido Popular led a government which was very restrictive in immigration matters – foreigners had very few rights, and the rhetoric focused primarily on border security and cracking down on people smugglers. The year 2004 saw a change in government to that of José Luis Rodríguez Zapatero of the Partido Socialista Obrero Español (PSOE) and a shift in focus to the more humanitarian aspects of immigration and its problems. This shift encouraged the regularization of immigrants as well as seeking more permanent solutions. Since 1985, there have been five laws enacted regarding foreigners, the last on 20 November 2003, which came into effect on 30 December 2004. Between 1991 and 2006, there have been eight regularization procedures (five from the Partido Popular, and three from the PSOE) in an attempt to deal with the increasing number of immigrants.

- *The social visibility of immigrants has increased* – From 2001, events have been organized by the illegal immigrants and asylum seekers, for example, hunger strikes, and the occupation of churches. Previously, there had been only minor activity and people had not yet realized the extent of the problem.

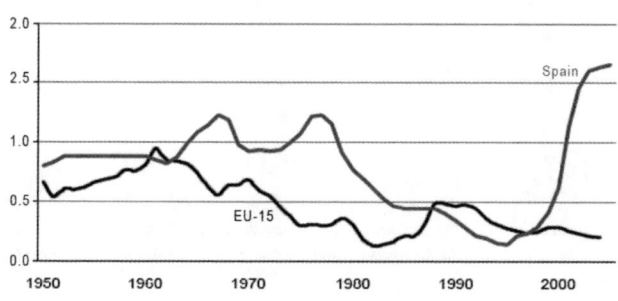

Annual Growth Rate of the Population in Spain and in the Member States of the EU.

Sources : Maddison historical series & INE

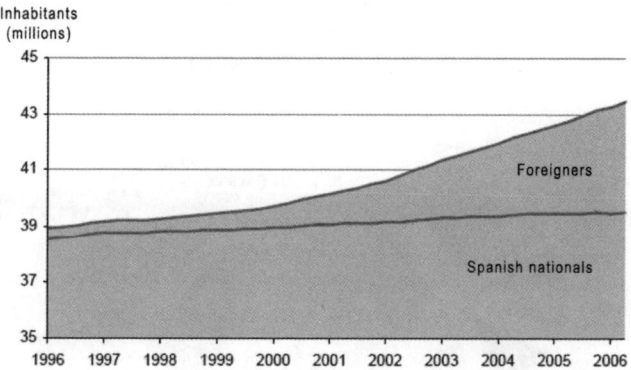

Development of the Spanish population

Source: poll of active population, INE.2006

This graph compares the numbers of immigrants in Spain and France and classifies them by their region of origin:

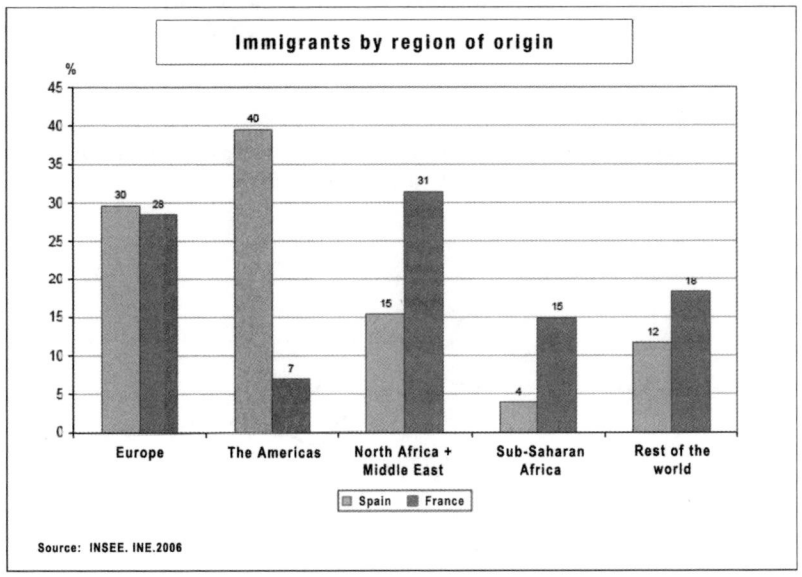

This graph shows immigrants in Spain by their country of origin:

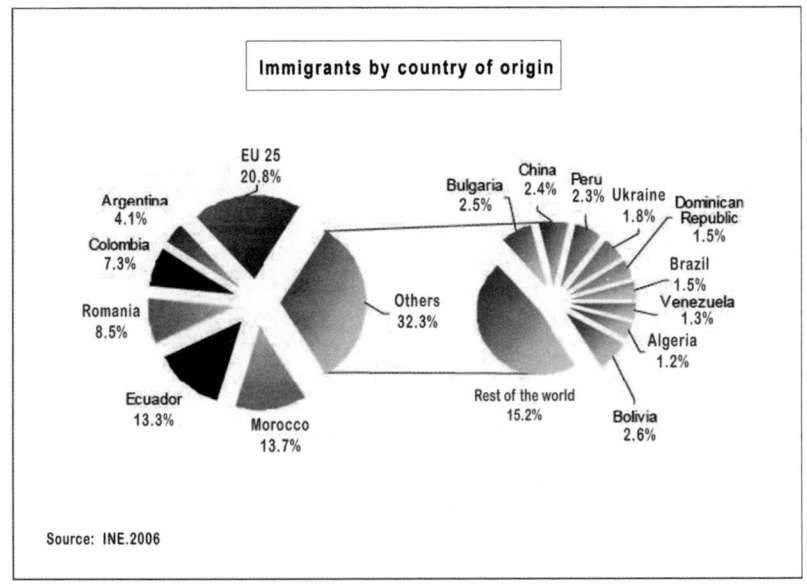

ARE THE SPANISH FOR OR AGAINST IMMIGRATION?

Immigrants from Morocco account for 13.7 per cent of the migrant population; 13 per cent come from Equador. In 2001, immigration from Latin America increased by 580 per cent due to a catastrophic drop in GNP in 1999 with resulting sudden and widespread rise in poverty levels, the greatest ever experienced in the region. Romanians account for 8.5 per cent of immigrants, and Columbians 7.3 per cent from Columbia.

A note on methodology: the numbers regarding immigration are problematic because of overlap. Registration for the census takes place at the civil council level in the community of residence. However, if someone relocates to another area, they are not necessarily deleted from their previous community's census rolls, as government grants are allocated based on the number of inhabitants in each council. Thus it is in the interest of the civil councils to revise upwards their census rolls, but this skews attempts for determining the true number of immigrants in any area, especially when attempting to discuss illegal immigration. Estimates have made of 700,000–1.6 million immigrants, based on those registered in civil councils, or those who have not yet obtained a residence permit.

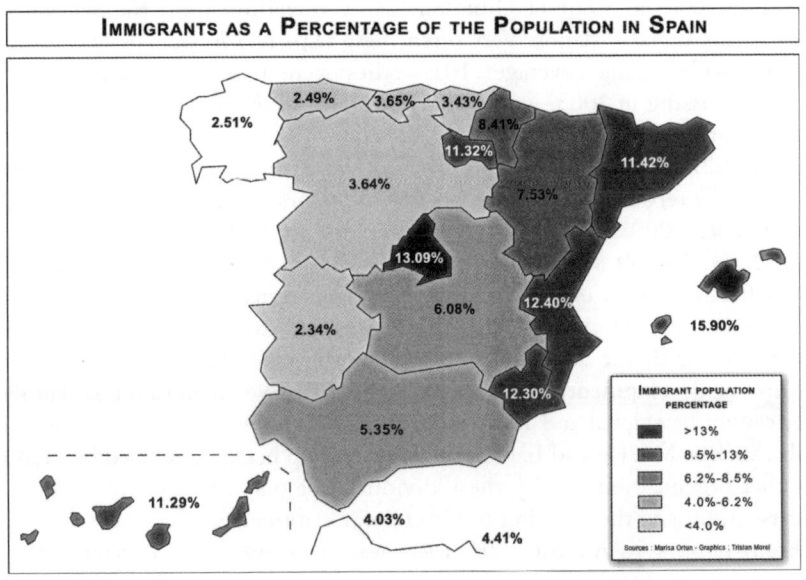

Some regions present high percentages of immigrant in their populations: the Balearic Islands have 16 per cent, Madrid 13 per cent, Murcia, Catalonia and Valencia approximately 12 per cent; other regions have lower rates: Asturias, Extremadura and Galicia have 2.5 per cent. The highest rates in immigrant population density are found on the Mediterranean coast and in Madrid, that is, in the economically dynamic zones, part of whose vitality is due to input from the 'underground' economy, which attracts cheap labour.

The case concerning Catalonia is a specific one. According to census findings released 1 July 2006, the number of immigrants had been multiplied five-fold in the past six years. Of these, 34 per cent are estimated to be illegal immigrants. This is a new phenomenon concerning Catalonia's population, 13.13 per cent of whom are foreigners. This has a structural impact on the population as its average age is lowered, by the influx of younger immigrants, as well as an impact on the 'character' of the region – one-third of Catalonia's immigrants originate from Latin America.

A study published by the Ministry of Work in February 2002 classified foreigners according to the way they entered the country: 62 per cent arrived by plane; 28 per cent by ship; 5 per cent by car; 2 per cent by train; 2 per cent by small water-craft, and 1 per cent by 'other means'.

What is striking here is the distance covered by a small minority of migrants who travelled by *pateras* or *cayucos* (small boats, or canoes which carry around ten people). The migrants leave the Maghreb or Sub-Saharan Africa, passing across the Strait of Gibraltar, or by travelling from the Senegalese coast to the Canary Islands. Very often these trips end in tragedy, with some gaining wide media coverage: 101 bodies were reported found and 109 reported missing in 2003; in 2004 there were 81 found and 60 missing. many more go unreported.

In 2002, 12,600 Moroccans arrived in Spain via *pateras* but were immediately repatriated, in accordance with Spain's agreement on readmission with Rabat; 9,000 Sub-Saharan Africans arrived in the Canary Islands.

In 2003, 19,000 Moroccans crossed the Strait of Gibraltar. Thousands of Sub-Saharans arrived on the Canary Islands and Aznar's government ordered the transfer of those people to the Peninsula, the NGOs signing some agreements with the authorities to finance the task. These arrivals mark a disturbing development: many are minors who arrive having no host family; they cannot be evicted and are attracted to cities well known for their football clubs, such as Madrid and Barcelona. Some end up being supported by NGOs or other organizations, or by the individual Regions, or they end up in the streets, adding to the growing levels of petty criminality.

From January to August 2006, there was extensive media coverage of the dramatic arrival of 28,000 people via *cayucos* along the Canary Island coastline, which inundated the country's infrastructure. Nearly 12,000 people were transferred to the Peninsula, 10,000 were fed and sheltered by the NGOs, and only 20 per cent were attended by the individual Regions. In general, if there are no agreements on repatriation with their country of origin, migrants are transferred to Regions of the Peninsula which have given their authorization.

In August 2006, Catalonia protested when planes started to arrive there with large numbers of immigrants that the community was unable to accommodate. There had also been a refusal from the Partido Popular-run Region of Rioja.

It is likely that the rhetoric of the political parties and official institutions had for a long time been marked by a certain mix of tolerance and good-natured recognition – after all, wasn't Spain a country of emigrants thirty years ago? But the numbers given by the NGOs for hosting and integrating immigrants are astonishing. In 2006, the NGOs received €6.5 million to help with the tide of immigrants; in 2007, the budget increased by 100 per cent. The political parties voted for an increase of 50 per cent in grants to the Regions: €15 million to humanitarian relief; €10 million to minors; €15 million to integration and educational actions.

The Representation of the Citizens

With all this, there is a crucial question: how do Spanish people feel about this policy of facilitating integration? During 2001–02, some studies showed that people were harbouring doubts. The impression given was that people were reluctant to reveal their feelings of hostility, which are masked by declarations of altruism and generosity.

The opinion polls of the Centro de Investigaciones Sociológicas (CIS) shown below indicate some very distinct realities: unemployment, immigration and terrorism seem to be the three major issues that worry Spanish citizens.

OPINION POLL 2005: IMMIGRATION – SEEN AS ONE OF THE MOST IMPORTANT PROBLEMS FACING SPANISH SOCIETY

2005	Percentage	Position
January	21.4	3
February	23.6	3
March	18.7	4
April	29.5	3
May	27.7	3
June	23.1	3
July	22.7	3
September	32.8	3
October	37.4	2
November	40.0	2
December	29.4	2

In 2000, 43 per cent considered immigration as a positive feature, 24 per cent thought differently; 51 per cent believed that the increase of delinquency was linked to immigration, as against 35 per cent who did not.

According to the opinion polls of the Centro de Investigaciones Sociológicas, in May 2006, 27.7 per cent of people interviewed considered immigration to be the country's biggest problem; by July, that number had increased to 38 per cent, and by September, 59.2 per cent. Everyone agreed that the images people were being fed on a daily basis by the media, of waves of Sub-Saharan Africans arriving on the shores of the Canary Islands, had a negative impact on opinions.

In May 2006, a Gallup opinion poll for *El Mundo* reported that people thought:

- There are too many immigrants: 69.1 per cent.

- The massive regularization of immigrants in 2005 made people react 70.2 per cent against, with 18.1 per cent considering that the number was a satisfactory one, while 2.5 per cent said there should be more immigration.

- Those in favour of a quota system, dependent on the needs of the labour market: 83 per cent.

In July 2006, a survey was conducted in Catalonia, which was published in *El Diario de León* in July 2006, and extrapolated for all of Spain. It revealed that:

- Nine out of ten people interviewed believed there are too many immigrants.

- Regularization should continue: 16 per cent.

- There should be stricter control at the borders: 59 per cent.

- Smugglers should be repatriated: 20 per cent.

In an Internet survey conducted in September 2006 for the daily newspaper *La Vanguardia*, 58.6 per cent were against imposing stricter legislation on immigration, while 40.2 per cent were in favour (with 37,655 people participating). By early December 2006, this number had changed, with 58.4 per cent voting for more strict controls (with 38,232 people participating). However, according to the postings on the Internet forum which the newspaper had started, people had lost interest in the debate surrounding this subject.

According to the responses received by the market research group APPEND, and published by *La Vanguardia* on 22 October 2006:

ARE THE SPANISH FOR OR AGAINST IMMIGRATION?

- Immigration does not contribute to economic growth: 43 per cent agreed.
- There is a correlation between immigration and a breakdown in law and order: 73 per cent.
- Immigration is perceived as a danger for the country: 32.5 per cent
- People should be able to live in any country no matter what one's origins: 50 per cent.
- Spanish people are worried about mixed (Spanish–immigrant) schooling: 27.2 per cent.

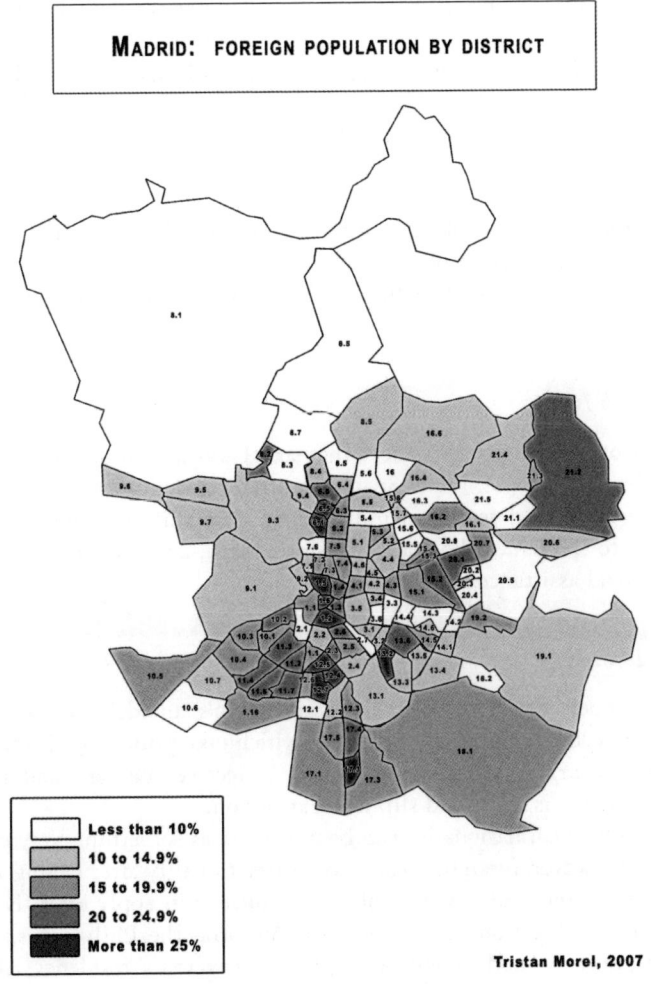

The figures published by the Observatorio Permanente de la Immigración, on the basis of interviews and meetings with the inhabitants of neighbourhoods which have a significant number of immigrants, are even more interesting. Immigrants tend to settle down in town centres, that is, in old and decaying flats and other properties; in this way, they contribute to maintaining a certain dynamism in the property market: the owners tend to sell to immigrants in order to buy better properties elsewhere. This explains the concentration of immigrants in the centre of Madrid and in Barcelona.

The interviews and meetings that took place in 2000–04 show that immigration was perceived to be a problem by all the survey participants. This geopolitical concentration in the town centres is likely to have consequences, such as the emergence of 'populist' anti-immigrant groups and organizations.

Interviewees' concerns were expressed as falling into the following categories:

Feeling Invaded

Even if statistics do not make the difference between 'permanent stock' and a flow of immigrants (or people simply passing through or returning), the population has the general feeling that waves of people are invading the country.

Public Health Issues

A majority of Spanish people believe that social and welfare services should be reserved for local people. Many Spaniards often condemn the fact that social health care is granted to illegal immigrants, that children from immigrant families go to the same school as Spanish children, and that they receive any available social assistance from the state.

National Identity

Some newspapers report on what they call 'Spain's suicide'. Despite the fact that Spain is a conglomeration of Regions which take pride in their individual identities, the argument here concerns a collective memory and national cohesion which it is felt could slip into extinction.

Access to Spanish nationality can be perceived as something very difficult to achieve. However, foreigners can apply for nationality after having lived in the country for ten years, while political refugees can apply after five years. Immigrants coming from Latin America, Andorra, the Philippines, Guinea Equatorial, Portugal and Israelis can apply after two years' residency. After one year, children born in Spain of foreign parents, and foreigners married to a Spanish citizen can apply for Spanish citizenship (Legislation has become more strict in order to prevent someone gaining Spanish citizenship through an unconsummated marriages, that is, marriages of convenience.)

In some sectors, public opinion claims that nationality is granted by a simple administrative act, without taking into account the origins of the individual seeking citizenship, their culture, or their attitude towards Spanish history and institutions, with resulting doubtful consequences for national cohesion.

Immigration: A Time Bomb

For the time being, circumstances are favourable for immigration and the report of the government's Economic Bureau (15 November 2006), as well as the La Caixa Report (August 2006), underline the positive impact of immigration on the Spanish economy. However, if Spain were to suffer a recession, the crisis would be considerable. Indeed, though presently there is hardly any competition concerning low-skilled jobs, the media and the press declare that the job market can't possibly absorb any more immigrants. Others believe the contrary.

The Fear of Islam

Spain is a non-religious state, as stipulated in its Constitution of 1978. However, Catholicism is very important in Spain and consequently a question arises here concerning the relation between Catholicism and Islam. In 1979, Spain signed the Saint-Siege Agreements concerning Spanish law, education, culture, armed forces, and the financing of the Church. Latin American immigrants, traditionally practicing Catholics and wanting Catholic education for their children, reinforce this state support of the Church.

Spanish Muslims are still a minority – only 2 per cent of the population; however, the image the Spaniards have of Muslim immigrants from the Maghreb is a very negative one. Historically speaking, Islam is considered to be a conquering religion. When Granada fell in 1492 and consequently the Muslims lost al-Andalusia, this was felt as a collective trauma; Muslims, wanting to reintegrate Spain into the *umma*, attempted to regain their Spanish territories and restore their hegemony.

Public awareness of this history has increased, especially since the Madrid bombings of 11 March 2004. Some writers pronounce that Muslim immigrants will become instruments in the hands of the jihadists; these ideas prevail against all the associations which fight for tolerance and integration which wish to distinguish themselves from the fundamentalist imams. Some forums for citizen participation indicate the rise of a new phenomenon: Islamophobia, and the accompanying conviction that Muslims want to start a war against Spain. This fear of Islam leads to xenophobia and to racism, though there are as yet no official data on the spread of such a phenomenon, though there are often articles in the press about this. There has been some local resistance to the spread of Islam, for example, in Andalusia and in Catalonia, where there has been some outcry against the construction of mosques.

In Madrid, the association 'Iniciativa Habitable' is attempting, via its Internet website, to mobilize the inhabitants of 'various communities'; it plans to put up candidates in the municipal elections of 2007, though the majority of citizens would not identify themselves with the group's rhetoric and institutional policy. The programme of the association includes the following measures:

- In general, stricter control of immigration;
- Eviction of immigrants who have no residence permit;
- Abolition of any legal assistance for smugglers;
- The protection of citizens and their interests, as well as support for small and medium sized enterprises;
- National preference in terms of social rights;
- Benefits granted to Spanish families who are in need, provision of social assistance which is supposedly 'only granted to immigrants';
- Expulsion of violent children from schools and an end to the policy of grouping violent children in special centres;
- A limit on free health care for immigrants;
- Special taxes on money transfers abroad, in order to avoid 'the flight of capital towards the immigrants' countries of origin;
- No right to vote for immigrants;
- No automatic granting of nationality, whether after years of residence or because of birth in Spain;
- Control of marriages of convenience.

The future will reveal if the demands of this programme, which highlights frustrations among the Spanish population, will meet with any success; on the other hand, what is certain about these measures is that at least some of them are likely to appear in other parties' political programmes in order to gain votes. Basically, 'indigenous' Spaniards consider themselves entitled to certain social rights, such as religion, Spanish land, culture, and so on; certainly they are indeed part of, and will contribute to, this future. The rejection of immigration may act as a unifying principle.

Marisa Ortún Rubio is a historian with the Spanish Consular Service in Paris.

The Revival of Islam in Spain

ROSA MARÍA RODRÍGUEZ MAGDA

Migrations into today's Spain are a fairly recent phenomenon, but they became increasingly important over the past few years. Close cultural and linguistic relations led to a high level of immigration from Latin America, particularly from Ecuador, Bolivia, Columbia and Argentina. Similarly, the fact that Spain is geographically close to North Africa has made it easier for more and more people from the Maghreb and Sub-Saharan Africa to immigrate.

There are 44,390,000 residents in Spain, among whom 3,880,000 are foreigners – 8.7 per cent of the population. Moreover, there were 1,650,000 illegal immigrants in Spain in December 2006. Illegal immigration, organized by mafia groups, has recently increased dramatically in the Canary Islands: around 30,000 people in 2006 came from North and Sub-Saharan Africa, five times the numbers recorded in the previous year.

In previous years, successive governments decided to regularize immigrants: according to the *Partido Popular*, in opposition at the moment, this created a 'call effect' on illegal immigration that neither Spain nor Europe can sustain.

Each nationality should be considered as a specific case, but the biggest cultural gap concerns Muslims, despite the historical coexistence of customs and religions between the Spanish cities of Ceuta in Spain and Melilla in North Africa. In Ceuta, 27,000 out of 71,500 inhabitants are Muslims; in Melilla, Muslims number 26,400 out of a population of 64,400. But the Muslim population is constantly increasing; according to some estimates, Muslims will represent the majority of the population of both cities within the next decade. Thirty per cent of the armed forces in Ceuta and Melilla are Muslims.

There are currently 900,000 Muslim immigrants in Spain, most of them Moroccans. Among others, there are three historical elements to Spanish–Muslim relations:

- Spain possesses "Spanish" territories (not colonies) in Africa

- *Hispania* was occupied by Muslims for eight centuries until 1492 (when Granada was captured by the Catholic Kings); nevertheless, 'Orientalists' created a myth, which is still widely held by many writers today, of an idyllic *al-Ándalus* – a time of splendour and harmony between the three civilizations, Christian, Jewish and Muslim.

- A movement of Spanish people who had converted to Islam appeared in the 1970s. From the start, they were very much attached to an Andalusian Islamic nationalism, which is to say they wanted Islam to spread again in Andalusia and thus to the whole of Spain.

Spanish Converts to Islam

This movement of 20,000 members may be a minority but it is prominent in the media and likely to influence public opinion.[1] When the dictator Francisco Franco died in 1975, the *Morabitún* established a network of Muslim communities in Spain. The first, set up in Cordoba under the name *Sociedad para el Retorno del Islam a al-Ándalus*; included among its objectives the restoration of the caliphate and setting up a military high command. Many of the most representative present-day Spanish converts emerged from this movement, even though most left to form less radical organizations. Currently, there are several communities of converts, all very different from, and sometimes opposed to, one another; over a million people have looked at Webislam, one group's website.

In parallel, several Islamic nationalist movements emerged in Andalusia in the 1980s. *Yama'a Islámica de al-Ándalus* was founded in 1980, with various local branches. It established the *Universidad Islámica Averroes* in Cordoba. The philosopher, Roger Garaudy, initially a Communist then a convert to Islam who expressed negationist theses on the Holocaust, was part of its executive committee. However, this university no longer exists.

Cooperation Agreements between the Spanish State and the *Comisión Islámica de España*

As they needed legal associations, Spanish converts created the *Federación Española de Entidades Religiosas Islámicas* (FEERI) and the *Unión de Comunidades Islámicas de España* (UCIDE), whose members were mainly Spanish citizens of Syrian origin. Despite their divergences and while they retained their independence, these two organizations became associated within the *Comisión Islámica de España*. The *Comisión* signed cooperation agreements with the Spanish state in 1992: it received part of the religious taxes, the right to have Muslim cemeteries and for imams to receive social security, and recognition for Muslim religious weddings; religious assistance in the army, in prisons and hospitals followed, as well as the right for schools to teach Islam. The document also provided that working hours and days could be modified to allow for prayers and Ramadan, and that the food available should conform to Koranic principles. Not all of this has yet been enforced. However, the government did reject demands to legalize polygamy. The FEERI and the UCIDE receive state subsidies from the Spanish state and the Muslim communities that are legally registered in Spain are affiliated to these two organizations. Recently, Spanish converts have had to leave the FEERI presidency to pro-Saudi and pro-Moroccan groups.

The al-Ándalus Nostalgia

For many converts, *the Arabs never invaded Spain*. According to this interpretation of history, both the Jews and the Arians, who were persecuted by a declining Church that wanted to impose the dogma of the Holy Trinity, welcomed the Muslims as true liberators who would restore their political, social and religious rights. The dominant view among converts is that the splendour of al-Ándalus contrasted with the darkness of Europe and that only one of the greatest tragedies in the history of humanity – the genocide of Muslims, Jews and Arians – managed to destroy this civilization. This highly ideological and Manichean vision has logical consequences: the return of Islam in al-Ándalus is presented as an authentic and necessary reconquest, as the emancipation from Hispanic occupation, itself illegitimate and bloody.

It is time the myth of al-Ándalus was revised along rigorous historical lines. Such an interpretation is a manipulation of history, it denigrates medieval Spain and Europe, and describes the Islam of the time as superior to arid Western culture. Contrary to the thesis of a peaceful Muslim occupation and a Christian territorial reconquest presented as genocide, it should be remembered that both episodes were bellicose and were in fact, wars. It should be remembered that the Muslim occupation of *Hispania* did not mean easy coexistence between Christians and Jews, and that the cultural and scientific enlightenment of the time was mostly the work of Muslim and Jewish wise men, who were often persecuted because of their ideas and who had cultivated the Greek legacy – both commentary and tradition.

This view of the period rejects idealism and takes historical complexity into account: a few Muslim scholars were able to present themselves as the heirs to Greek culture without renouncing their faith. Above all, it makes it possible to tackle the challenge of restoring true dialogue between cultures. Any other position is likely to provide an ideological alibi to those current Islamists who try to use historical al-Ándalus for their own ends.

What is more, the Spanish government has received unofficial information on potential terrorist attacks.[2] Some radical Islamist leaders do not simply advocate the restoration of al-Ándalus: is it not the case that a jihadist website questioned Spanish sovereignty over Ceuta and Melilla, called for the liberation of both cities and declared war on Spain?

Conclusions

Some Spanish converts are fully integrated and modern; they have good relations with many national and international organizations; they want to be acknowledged as the privileged interlocutors of the Spanish state. Others form small spiritual and traditional communities, apart from the rest of society. But even if most of them advocate a moderate form of Islam, Spanish converts demand separate rights. And they represent a distinct movement within the immigrants who come from the Arab and Muslim world.

A recent Interior Ministry survey gives the following data: 74 per cent of Muslim immigrants say they have a very good or fairly good life in Spain; 80 per cent have adapted to the way of life and the customs of the country; 83 per cent say they face no obstacle to practice their religion. It must be underlined that they believe Spanish identity and secularity to be fully compatible with practising their religion and that they consider that it is unacceptable to use violence to proclaim or spread their religion. Most of them actually say they are strongly in favour of secularity, because all religions are then treated equally. This contrasts with separate claims as voiced by the few Muslim organizations and convert associations that call, for example, for the legalization of polygamy, or the naturalization of the descendants of the *Moriscos* who were expelled from Spain in the seventeenth century – this might include around five million foreigners today.

Conversely, the survival of Arab traces – terminology, monuments, customs, cuisine or festivals – should make it easier for Muslim immigrants to integrate in a secular society that respects freedom of religion.

But in any case, if the immigrants themselves do not express any separate claims, their organizations should not oppose a model of integration that preserves the identity of Europe and the values that flow from it: democracy, freedom of thought and expression, equality between the sexes and individual rights within the community.

Rosa María Rodríguez Magda is a philosopher and member of the Consell Valencia de Cultura.

NOTES

1 Cf. Rosa María Rodríguez Magda, *La España convertida al Islam* (Barcelona: altera, 2006); ibid., 'El Renacer del Islam en España', in *Débats* No. 93, *Pensar el Islam desde Europa*, (2006), pp. 8–15. 2006.
2 *El País*, 5 November 2006.

The Integration Strategies of Different Immigrant Groups in Spain

ROSA APARICIO GÓMEZ

This article deals with different behaviour patterns or strategies of integration adopted by the most numerous groups of immigrants to Spain. This is an issue which does not appear frequently in the Spanish literature on migration. Indeed, no stable and well-defined terminology exists on the subject. It follows that no widely shared approaches and debates on the topic can be found. In these circumstances, the following discussion must necessarily be fragmentary and somewhat subjective: somewhat subjective because I only base my observartions on what I have personally studied in the field, and fragmentary, because my fieldwork refers only to a few of the immigrant groups which in 2000 were the most numerous in Spain. I will later refer to what was empirically observed at the time but, first, I believe it will be useful to make a few clarifications about the terminology I use and to give the reasons behind my approaches.

I will therefore start by explaining how I unintentionally stumbled on the topic and then go on to clarify the meaning which I find convenient to give to the term 'strategies' when connected to the question of the integration of immigrants. Only then will I present some of my findings.

My Access to the Topic of Immigrants' Strategies of Integration

My study of this question came about unintentionally. In 1997, I was involved in a research project on the main obstacles to integration experienced by the most numerous immigrant groups present in the territory of the Autonomous Community of Madrid. For this research, I had taken for granted that these obstacles would appear by examining one by one each of the contexts which are considered 'key contexts' for the integration of immigrants, from their access to the labour market or to housing, to the exercise of their rights or to possible discrimination or xenophobic practices when interacting with others in daily life. But I also took it for granted – as was common in most of Europe

– that if specific obstacles for any of the groups existed, the causes were to be found only in the inadequacy of the laws or in the negative attitudes of the local population. I was thus considering only the role played by the receiving society, without thinking that immigrants themselves also play a role in their integration. In other words, I assumed that everything related to the integration of immigrants would only depend on the actions and omissions of the recipients of the immigration flows.

What the field study showed me did not coincide at all with this view. The Chinese and the Moroccans were differently affected by the rules of access to the labour market or to the health system. The hatred shown by Spanish racist groups had a much stronger impact on people from the Dominican Republic than on people from Sub-Saharan Africa. Peruvians fared worse than Moroccans when dealing with Spanish labour legislation. In sum, it was clear that the laws and attitudes of the receiving society did not decide everything that happens with integration processes; that immigrants had 'something' by which they came to grips with whatever came their way and made them decide what they would be able to obtain in their process of accommodation. This 'something', which was different for the different groups but which was widely shared by its members, is what we came to call 'strategies' inspired by the use which C. Camilleri gives to the word in his book entitled *Stratégies identitaires*,[1] although without strictly adhering to this use.

The Meaning of Immigrants' Integration Strategies

The meaning given by Camilleri to the word 'strategies' refers to ways of solving the problems of identity with which immigrants are confronted, when they find that their identity is not fully understood, valued, or accepted on arrival to the new society. It is then obvious that the 'strategies' of which Camilleri speaks point to different types of psychological options and operations, normally not premeditated, by which immigrants seek to solve the psychological tensions produced by their different identity.

However, in our case, what we have called 'strategies' does not refer to the elaboration of psychological tensions, but to options for specific behaviour patterns which materialize in different forms of action, although undoubtedly these will also relate to psychological processes of evaluation and even of perception of facts and events.

Looking more in general, it is striking that the original and more proper use of the term 'strategies' belongs to the sphere of military language in most of the more widely used European languages, with reference in this case to the planning or direction of war operations, although it has come to be applied to any set of activities consciously planned to attain a predetermined goal. Thus in this sense one can refer to commercial strategies, electoral strategies, and so on.

But of course one cannot think of it in such a way when speaking about immigrants' integration strategies, particularly because no such conscious planning exists on the part of immigrants in order to achieve their integration. None the less, something similar does occur.

To begin with, immigrants who nowadays arrive in a country always speaks with fellow compatriots who already have experienced immigration, either before they start on their journey or – and in greater depth – in the first stages after their arrival. These conversations are typically about what the newcomers have found on their way. This part would be the equivalent to the study of the ground which military strategists do, and it allows the immigrant to map in their mind the itinerary they will need to follow and the obstacles they will have to deal with on their way.

It is from this that the immigrant's strategies will emerge. The veteran, while remarking on the itinerary and the difficulties which accompany the first settlement, will tell what went well with him or her and how they managed it, as well as what did not go so well for them or for other acquaintances, and what is said about this amongst co-nationals. With these conversations, those recently arrived get in touch with the collective memory of their group, and this memory, in its constant reformulation, with time acquires more weight and content. It offers more than what the newcomer already knows and with more authority than just first impressions. And this version not only includes facts but also, more importantly, includes ways of interpreting those facts. By virtue of the latter, some ways of looking at things will be advanced and others will be discarded, so that the field of options which the immigrant will finally consider will become more reduced. For instance, a Chinese immigrant, in order to solve one of the first problems they may encounter, will not go to the social services in the city of arrival, while in contrast, someone from the Dominican Republic will do precisely that because this is what fellow compatriots will have suggested it.

This illustrates a trait which we must include in the concept of immigrants' integration strategies: the strategies always represent an option or an alternative which the immigrant adopts in preference to others in order to solve their difficulties. In the example just given, the difficulties are more or less the same, but at least three choices are open to someone who has to face them: look for help only amongst members of their own ethnic group, go to the social services, or try both. The Chinese will generally choose the first option, people from the Dominican Republic will choose the second, and Peruvians the third. Already at this stage the strategies of each group will begin to differ.

The strategies of the different groups will therefore vary depending on the different solutions which they adopt when they have to face the same or similar difficulties. These would appear empirically in the different practices carried out, but these practices will undoubtedly have been structured in the process of normal interaction within the groups with a common origin, in conversations dealing with the advantages and disadvantages of focusing and interpreting in one way or in another the difficulties which commonly arise when trying to find one's place in the receiving societies.

Having dealt with these conceptual clarifications, I will now go on to explain what the findings were in the study on the integration strategies of the immigrant groups with more presence in Spain at the end of the 1990s, that is, those from Morocco, the Dominican Republic, Peru, Ecuador, China and Sub-Saharan Africa.

Differences Found in the Groups Studied

For more clarity I will take the groups we studied in pairs, based on the similar difficulties they have to face. Sub-Saharan Africans and people from the Dominican Republic quickly realize that their skin colour makes them appear distant and different from the local Spanish population. The same is true of Chinese and Moroccans. Peruvians and Ecuadorians also stand out as typical looking Latin Americans, as of course do Colombians and Bolivians.

Integration Strategies of Dominicans and Sub-Saharan Africans

As has been said, these two groups will have to integrate starting from a similar circumstance: that of being perceived as different because of their skin colour. But already when travelling to their destination their strategies are different. The difference does not lie in paying for the trip, which in both cases they do by getting into debt, although amongst Sub-Saharan Africans, the practice is more often to get into debt with traffickers. The difference lies more in the way they act in connection with the legal restrictions for entry. Sub-Saharans will more often openly infringe these laws through illegal access to the Spanish coasts by boat, while Dominicans find more elaborate ways of avoiding the law without directly confronting it, for instance, entering the country with a tourist visa and overstaying it, or entering by road after landing in other Schengen countries which place less controls on tourists. The prevalence of one or the other option can stamp a mark on the image and consequently on the insertion in the labour market and in daily life of one group or the other, usually going against those who arrived in direct infringement of the laws.

If we now go on to look at the 'type-difficulties'[2] which Dominicans and

Sub-Saharans must face during the first stage of their arrival, we observe that in their initial anxious search for means of survival, Sub-Saharan Africans will seek the help of the social services less than Dominicans, and will instead make efforts to find any type of employment through private channels, almost always of the lowest kind and duration, often alternating with street occupations (street-peddling, truck-loading, parking cars, and so on). But amongst Dominicans a difference must be made: for women there is the possibility of domestic employment in homes and often they rely on this type of work during the first stage of their migratory itinerary.[3] Most Dominican men arrive years after their partners by way of family reunion, and hence, during the first stage of insertion, they do not suffer the pressure to find work and housing that affects other immigrants. We have no data on what happens with those who do not come in this way.

During the next migratory stage, the differences between the two groups become even greater. In the face of the threat of feeling emotionally isolated, Sub-Saharan Africans do not have recourse only to members of their own group in order to reinforce their relationships. They also do not seek support from immigrant organizations or the like. Instead, they open themselves to all sorts of friendships with other immigrants and with natives. By contrast, Dominicans tend to draw inwards towards members of their own group and do not appear to establish friendships easily with others who are not fellow countrymen. On the other hand, amongst Dominicans, some form of family reunion is already far advanced, while amongst Sub-Saharan Africans it is only beginning to develop.

During the same stage, Sub-Saharan Africans seek to extend their informal information, cooperation and friendship networks in order to avoid becoming stagnated in the structurally distant relationships which at the beginning condition their insertion. Dominicans, or rather, Dominican women, have tried instead to help themselves through political activity, originally put into practice through their organization VOMADE and as a reaction to the assassination of their fellow countrywoman Lucrecia Pérez in November 1992; but because of this, their networks have not gone beyond the limits of their fellow compatriots.

As far as discrimination is concerned, the strategies used in response differ, although both experience the same attitudes towards them on the part of Spaniards. Dominicans, on the one hand, react very sensitively to the negative attitudes directed towards them because of their 'blackness', and they often suspect that these attitudes exist even when this is not the case. Sub-Saharan Africans, on the contrary, tend to digest racial discrimination without nervousness and can even joke about it — unless it goes beyond certain limits of indignity.

Finally, in this second stage there are difficulties connected with access to permanent housing. In Madrid, Dominican women have been putting off finding a more permanent home for longer than other groups.[4] This delay stems from the fact that most Dominicans want to save enough money in Spain to build themselves a house in the Dominican Republic, and to return there once they have attained their goal. In fact, however, the return keeps being postponed and few put it into practice.

So much for the strategies adopted by Dominican women. As to those followed by Dominican men and by Sub-Saharan Africans we lack information.

How do Dominicans and Sub-Saharan Africans proceed in the third stage of their migratory process, once they have been able to set the foundations for a normalized life? We know that Dominican women endeavour to give up domestic service. They direct their efforts towards finding employment in commerce – frequently starting their own business. But not all will achieve this as they rarely endeavour to acquire any specific training beyond that which they possessed on arrival. We know little about Sub-Saharan Africans in this stage.

Integration Strategies of Chinese and Moroccans

Immigrants from these nationalities, as was previously observed, have in common the attitude of maintaining a strong adherence to their national cultures. At first sight, it would seem that this would make them adopt similar ways of insertion.

However, although the information offered by their narratives and comments avoids going into detail as to the preferred ways of financing the trip and crossing the borders, it seems that, from the moment they set off for Spain, they rely on their families borrowing money to pay for the trip and they try to play the system in order to cross into Spain without directly infringing the law. The media inaccurately claims that the majority of Moroccans arrive illegally by sea.

The different strategies in the later stages of their itinerary have, however, been well documented. It soon becomes obvious that the Chinese, almost uniquely among immigrants, are able to survive the initial lack of resources thanks to the help of their fellow Chinese. They hardly ever ask for state support. By contrast, even though Moroccans have the addresses of acquaintances and fellow Moroccans from whom they can receive support on arrival, they none the less quickly find out about the benefits from social services for which they are eligible and do not fail to apply for them if the case arises. Many register as unemployed as soon as they are able to do so, and they are the majority amongst immigrants receiving unemployment subsidies.

As to their first housing, the main difference lies in that the Chinese, who will from the beginning find employment from by fellow countrymen, often sleep in the workplace, as is common in China, although if they have relatives amongst already settled immigrant families they would be able to find accommodation with them. Sleeping in the workplace is unthinkable for Moroccans and staying with relatives occurs only in exceptional circumstances. Moroccan immigrants tend to share rooms or a flat with single fellow countrymen until such time as they can consider marriage or bring their wives from Morocco. This happens only in the last stage of their itinerary.

In the next stage, emotional isolation is the 'type-difficulty' which appears in our scheme. But this does not seem to affect the two groups which we are considering, perhaps because their respective cultures provide them with efficient mechanisms for emotional control or perhaps because of the strong connection with members of their own group during the first period of their insertion. Whatever the case, this leads the Chinese to withdraw within their own group, something which does not happen among Moroccans, who are more easily drawn towards the establishment of relationships with their workmates, both Spaniards and those of other nationalities.

Immigrants also must face the threat of stagnation in the precarious employment, housing and relational situations of the first stage. The Chinese seem to sort it well, because they tend to be enterprising and determined, but their self-promotion occurs mainly within the Chinese community. When it comes to Moroccans, a considerable difference tends to emerge between those who started in agricultural jobs and those who started in the construction industry, which are the first jobs more easily accessible to them. Those who go for agriculture are at greater risk of getting stuck in the bad employment and housing conditions in which they found themselves during the first weeks of their migratory itinerary, something which occurs less frequently with other groups. Those who remain longer in rural milieux also seem to make no progress in their relationships with the native population.

But it is during this second stage that these rural workers, as well as those who live in towns, feel most affected by discrimination. They react quite differently compared to the Chinese. This is clearly shown in the way each narrates their experiences. Chinese never speak about discrimination; Moroccans never stop talking about it. However, Moroccans both reaffirm their difference as Moroccans and also object to being treated as if they were different. Undoubtedly the slogan 'equal but different' expresses a noble social ideal but those who insist on being different are usually treated as such, while if someone is treated as an equal then his difference will not be respected.

In a study on the information and support networks which immigrants contact, it appears that friendship networks and those integrated by co-nationals are of more use to all immigrants in order to gain access to housing than social services and NGOs.⁵ Only in the case of the Chinese does it transpire that they also search for housing by looking at the notices posted on buildings in the streets of Spanish cities. However, the study did not consider when they undertake this search, which is what interests us here in order to find out whether or not the Chinese or the Moroccans tend to put off looking for a more stable home, a choice which has important consequences for their insertion. Only other considerations allow us to surmise that both get around the difficulties relatively well, with the exception of Moroccans stuck in agriculture.

Integration Strategies of Ecuadorians and Peruvians

When seeking to finance their trip, Ecuadorians tend to get into debt outside their circle of relatives more often than Peruvians. This increases the pressure on them to earn money immediately upon arrival in Spain. For the rest, about two-thirds in both groups enter Spain circumventing the law rather than breaking it outright, although Peruvians are more happy to live as 'irregulars' by overstaying their permitted period of residence.

In order to mitigate their initial lack of resources, both Ecuadorians and Peruvians resort to private and public aid, whatever the source. Ecuadorians, a large number of whom come from rural areas, show less skill in dealing with the Spanish bureaucracy than Peruvians. In comparison with other groups, both Ecuadorians and Peruvians resort more often to the possibilities offered by Caritas, the large aid network attached to the Catholic Church.

The greater indebtedness of Ecuadorians is perhaps the reason why their first lodgings are more deficient than those of other groups. Cases have been documented of them renting beds by the hour, which never happens among Peruvians. The Peruvians' networks of compatriots are more efficient than those of the Ecuadorians in helping them to get their first provisional jobs.

It is in the next stage that the differences between these groups begin to appear more clearly. Ecuadorians have a strong tendency to compensate for their feelings of emotional isolation by withdrawing into their own group and reducing their relational world to that of their compatriots. Peruvians, on the contrary, are the most active immigrant group in making friends with both Spaniards and immigrants of other nationalities.

This allows them to overcome more quickly the structural stagnation of 'newcomers'. They do this by negotiating individually, rather than collectively or politically, with their employers and landlords. Organizations of Peruvian immigrants, in contrast to those of the Ecuadorians, do not make efforts to appear in the media with the political claims of their compatriots.

Both groups face discrimination. When narrating their migratory experience, Peruvians show that they clearly perceive the occasions in which they are discriminated against, whether because they are immigrants, because they are Latin Americans, or because they are Peruvians. But they also show that this does not surprise them and does not deter them from pursuing their goals; it does not make them feel inferior to those who discriminate them. In contrast, the Ecuadorians' narrations speak about discrimination as if it wounded and hurt them very much, and it seems to make them feel and act as if they were in fact 'inferior'.

Both groups attempt to find stable housing relatively soon. The data suggest that Peruvians tend to be more demanding in the type of housing they want, and will be willing to make greater sacrifices to get what they want.

Given that the conditions in Spain are the same for all immigrants, insertion therefore depends on three variables: a) the educational background of the groups concerned; b) the different sensitivity of the groups in relation to discrimination, and c) the structure of the ethnic network of information and support to which each group is connected.

Up to now we have not put much emphasis on the first of these variables. In fact, however, it is a key factor. The insertion strategy of a Sub-Saharan university graduate is likely to be more similar to that of a Latin American university graduate than to that of a Sub-Saharan who has only attended primary school.

Sensitivity towards real or suspected discrimination is very important. If a Chinese person is treated as someone who is different, they will take it with indifference; a Peruvian will avoid showing how much it annoys them; but to a Dominican or a Moroccan, the fact of being seen as different wounds them deeply and they will have more trouble in negotiating their identity when they meet strangers in daily life.

Finally, another important factor is the structure of the networks from which immigrants seek support within their own ethnic group. The closed character of the networks amongst the Chinese, and also to a certain extent amongst Ecuadorians, will make them seek different channels of insertion from those used by Peruvians or Colombians whose support networks have a marked open structure.

Conclusion

When we were researching the specific obstacles that immigrants in Spain had to face in their process of integration, in Europe it was usual to think about integration in terms of the tasks to be undertaken by the receiving society. The European Council of Tampere of 1999, often quoted at the time, summarized these with the aim of offering 'fair treatment' to all those who came to Europe. In this context, it was not clear what was the point of looking at the strategies used by immigrants. At most, it was considered useful for the purpose of approaching the problems faced by individual persons or groups in isolated actions undertaken by NGOs and local administrations.

However, in only a few years, the context has changed. A noteworthy expression of this change was the European Council in 2004 under Dutch presidency, which began by stressing that integration is a *two-way process*, to which both the immigrants and the receiving society must contribute. The former are now explicitly asked to respect European values and to have frequent interaction with native members of the society. Consideration is being given to the practices shared by the different immigrant groups in their process of integration, in order to prevent any perverse developments or effects of even the most well-meaning political endeavours. If integration is a two-way process, it will be the outcome of the encounter or confluence of the lines of action which each part will tend to follow, an encounter which by definition cannot be deduced from preconceived ideas or unilaterally devised programmes.

Professor Rosa Aparicio Gómez is an academician of the Instituto Universitario de Estudios sobre Migraciones, at the Universidad Pontificia Comillas, in Madrid.

NOTES

1   C. Camilleri, *Stratégies identitaires* (Paris: Presses Universitaires de France, 1990).
2   By 'type difficulties' we understand the most common difficulties experienced by immigrants in the different stages of insertion. During the first stage – that of 'landing' – the main difficulties are those related to the cost of travelling and to the restrictions of entry. In the second stage – 'getting settled'– the most basic problems immigrants would seem to have to face are those related to housing and to means of subsistence. Finally, in the third and last stage –that of 'normalization' or of more permanent settlement – the difficulties are to do with isolation (scarcity of relationships), with bettering their conditions of work and housing, and with experiences of ethnic discrimination.
3   According to a common stereotype most Dominican women would be employed in domestic service. But the data obtained in the study *El Capital Humano de la Inmigración* by R. Aparicio, A. Tornos and M. Fernández (Madrid: Ministerio de Trabajo y Asuntos Sociales, 2003) indicate that this is no longer the case after the women's third year in Spain. From domestic service, they have moved on to other employment.
4   Information collected for a research project on the access to housing of immigrants living in Madrid, partially published in J. Labrador and A. Merino, 'Características y usos del habitat que predominan entre los inmigrantes de la Comunidad Autónoma de Madrid', *Migraciones*, 11 (2002), pp. 173–222.
5   R. Aparicio and A. Tornos, *Las redes sociales de los inmigrantes extranjeros en España* (Madrid: Ministerio de Trabajo y Asuntos Sociales, Observatorio permanente de la Inmigración, 2005), pp. 95–9.

Immigration, Self-Government and Management of Identity: The Catalan Case[1]

RICARD ZAPATA-BARRERO

To link immigration and self-government is a political necessity in confronting the different questions that immigration raises in the context of a stateless nation such as Catalonia. Furthermore, after an initial period of assessment and recognition of the phenomenon, we perhaps find ourselves at a more mature and propitious moment for political innovation in the management of immigration in Catalonia.

When wondering what ought to be the basic elements for shedding light on the issue, we must accept three basic premises (Zapata-Barrero, 2004c). The first is the institutional recognition that the arrival of immigrants forms part of an irreversible historical process, and that in Catalonia it is already a consolidated demographic reality. The second premise is that this new reality has some particular characteristics that makes it very different from previous migrations (even if we accept that we are not looking at a new phenomenon, but one which forms part of the history of Catalonia itself and of its citizenry, the fact that it is a society receptive of immigration). In third and final place exists the premise that the management of immigration in Catalonia has characteristics peculiar to a minority nation.

In this context, for a minority nation like Catalonia to speak about immigration would mean holding a discussion on rights and non-discrimination, on democratic and ethical principles – just as any other state would do. But it would, furthermore, particularly oblige one to discuss the language of identity, to use particular tools, certain political and administrative instruments coherent with self-government and to put in place a worthwhile language policy.

If we agree that, in dealing with immigration, we must accept that we cannot avoid how or from what viewpoint this reality might be interpreted, nor can we disconnect that reality from those values that it provokes, there is clearly a necessity to reflect on what should be the particular context of

Catalonia. This supposes accepting that the descriptive and diagnostic dimension of the situation is inextricably linked to the prescribed and normative character of the constitution of self-government.

For the time being, in the matter of immigration, the Administration of Catalonia (*la Generalitat*) has an executive and administrative function, but no legislative capacity. From the point of view of self-government, it depends totally on what is said and decided by the Central Administration in Madrid, whether or not it is in agreement with that decision. This situation creates an incoherent scenario whereby the government of the *Generalitat* must administer undesirable situations. These situations are a clear indication of the absence of self-government. Let us emphasize three concrete cases.

The first example is the situation of 'those without documents' (whom I call 'those without rights') and the administrative difficulties that this situation causes. This is a typical case in which the *Generalitat* must administer a reality provoked not by itself, but by the central government. In this field, on the issue of immigration, we are faced with a situation more suited to a centralist state than a federal one and we are much more like a French region than Quebec or the Flemish community, which are our most common points of reference.[2]

A second example: Catalonia, in its search for reasons legitimizing its demands for self-government with regards to the question of immigration would not need to limit this claim to political and legislative matters, but should also open a debate on the very identity of Catalonia. The framework of the discussion for reforming the Statute was a missed opportunity for linking the themes of immigration with those of identity. It was concentrated instead uniquely in linking them to responsibility. A debate that links immigration, self-government and identity is essential. The chief minority nations (for example, Quebec and Flanders) directly link their nationalist politics to their policies on immigration. In Catalonia, this link has yet to be made politically and socially.

The third example of a lack of self-government linked to immigration is based on basic democratic reasoning. Catalan society is undergoing an important process of change, but it does not have the tools to demand responsibilities of the *Generalitat* since it is well aware that the *Generalitat* is not in its hands. A clear indicator is the demands made by pro-immigrant groups, directed basically to the central government and not to the *Generalitat*. This fact shows us, moreover, a lack of democratic normality, while in a situation of normal self-government the majority of claims would be directed to the government by the *Generalitat*.

The reasons for debate are more than justified. The result would constitute a combination of arguments particular to Catalonia as a separate entity to Spain.[3] These arguments could be mobilized by Catalonia for the creation of its *public philosophy* of immigration.

In order to identify the phase in which Catalonia currently finds itself, the objective of this article is to supply the terms of the debate from which an *appropriate philosophy* might be constructed. But first let me introduce some specific demographic facts.

Catalonia: A Tradition of Immigration – From a Phase of *Internal* Immigration to the Current Phase of *External* Immigration

Catalonia can be considered as an EU country with a tradition of immigration. One can even affirm that Catalonia is a nation created by immigration. Let us look at the principal phases of immigration in order to define the period when *internal immigration* (immigration from other parts of the Spanish state) changed into *external immigration* (from countries outside the EU).

As one can see in Table 1, immigration has been a central component of Catalan life for the past century. Between 1901 and 1970 (until the last years of Franco's reign), the migratory settlement of Catalonia totalled over 2 million people, which signifies a substantially higher number of immigrants.

Table 1. Estimated Migratory Settlement by the Decade

Period	Net immigration
1901–10	33,669
1911–20	224,300
1921–30	322,079
1931–40	109,823
1941–50	256,718
1951–60	439,874
1961–70	720,442
1901–70	2,106,905
1971–80	256,674

Source: Author's figures on the basis of Cabré and Pujadas (1989).

This high volume of immigration totally changed the composition of the Catalan population. By the calculations of Anna Cabré (1999), by the end of the last decade more than 60 per cent of the inhabitants of Catalonia were there as a direct or indirect consequence of immigration. This percentage would be even higher if we took into account the marriages between Catalans and immigrants. 'The descendants of those who were in Catalonia in 1900 (none of whom, in turn, were Catalan)', affirms Cabré, 'are in a clear minority of one in four' (Cabré, 1999; p. 164). Her conclusion is that immigration will form part of 'Catalonia's contemporary system of reproduction'.

The economic, social, political and cultural impact of this immigration from other parts of the Spanish state is fundamental to the recent history of Catalonia. With the end of Francoism and the beginning of the current democratic period, immigration from the rest of Spain has tended to slow down, but from the middle of the 1990s and above all since 2000, we in Catalonia now expect a new period of immigration due to the growing and accelerated arrival of immigrants from outside the Spanish state. As can be seen in Figure 1 [Figure 1 near here], in the five years between 2001 and 2005, Catalonia has received more than half a million immigrants at a higher pace in absolute terms than during the 1970s. This new situation, differing from previous ones in the source of its immigration, irreversibly converts Catalan society into a multicultural society.

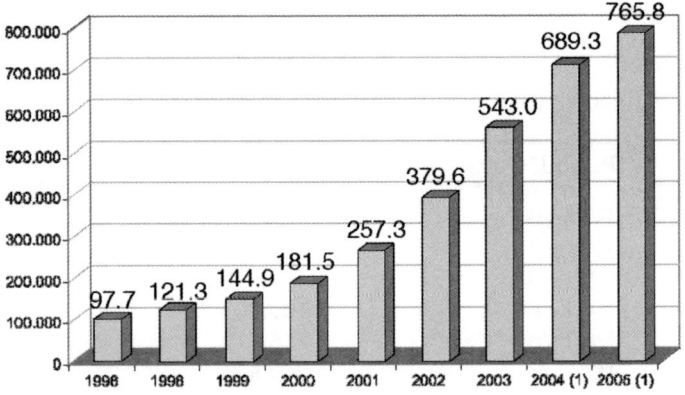

Figure 1 The Evolution of Immigration, 1996–2005

Source: *Pla de Ciutadania i Immigració*. Secretaria per a la Immigració.

In Figure 1 we can also see that this migratory process had two particularly intense periods: between 1911 and 1930, especially during the 1920s; and the 1960s. So we can trace the history of immigration in Catalonia during the past century via three especially significant periods:

The First Period: From the End of the Nineteenth Century to 1930 (the Beginning of the Second Republic in Spain)

At the end of the nineteenth century and the beginning of the twentieth, Catalonia was still a country of emigrants (Cabre and Pujadas, 1999). According to Termes (1984), the percentage of residents not born in Catalonia stood at 3.3 per cent in 1897, rising a fraction to 5.44 per cent in 1910. The situation changed radically in 1916, when, as we can see in Table 2, migratory growth shot up, exceeding natural growth for the first time in contemporary history.

Table 2 The Evolution of the Catalan Population, 1911–30

Years	Total growth	Natural growth	Growth due to immigration
1911–15	55,181	33,489	21,692
1916–20	204,670	2,062	202,608
1921–25	166,922	59,569	107,353
1926–30	279,651	64,925	214,726
1911–30	706,424	160,045	546,379

Source: Cabré and Pujadas (1989).

Economic expansion in Spain during its period of neutrality in the First World War and the public works driven by the dictatorship of Primo de Rivera, caused the arrival in Catalonia of a significant group of immigrants from Aragón, Valencia, the Balearic Isles and, increasingly, from Murcia and Andalucía. Of the increase of 706,424 in Catalonia's population between the censuses of 1910 and 1930, 546,379 – that is, around three-quarters – were due to immigration. This new scenario had a great impact on Catalan social thinkers, not only because of its demographic consequences but also for the economic, political, trade union and cultural changes that it produced. During the 1920s and 1930s, the theme of immigration was at the centre of a social, political and ideological debate revolving around its repercussions on the language, culture and nation of Catalonia and the possibilities and difficulties arising from the integration of immigrants (Calvo and Vega, 1978).

The Second Period: The Franco Era in the 1950s and 1960s

Between the beginning of the 1950s and 1975, 1,400,000 immigrants settled in Catalonia. But as this method does not show us the real flux of immigrants of both types, the real number of immigrants must have been higher, above the 1.5 million mark (Cabré and Pujadas, 1989). The period of greatest immigration was the 1960s with a net average of 72,000 immigrants per year. Profound territorial imbalances in Spain caused a massive migration from underdeveloped rural areas to the industrial centres which were in their most intense stage of development during the period of autocracy. This huge percentage of immigrants into Catalonia came from areas of Spain which were both geographically and culturally distant from Catalonia, thereby increasing the tendency begun in the 1920s (Pujadas, 1983). The magnitude of this phenomenon in terms of following the language and culture of Catalonia, the non-existence of democratic public institutions and the absence of the slightest capacity for self-government, generated an increasing preoccupation with the theme of immigration amongst Catalan thinkers, writers, politicians and academics, who took various positions on the subject of what it meant in these new circumstances to be Catalan and how the nation and national identity should be understood. It was during this period that the reality of the 'other Catalans' was popularly recognized, meaning those Spanish immigrants who had arrived in Catalonia during the 1950s and 1960s.[4]

The Third Period: The End of the Twentieth Century and the First Decade of the Twenty-First Century. The New Reality of External Immigration, Largely from Outside the European Community

During the first few years of the new democratic period, the theme of *external immigration* (from outside the Spanish state), given its scant demographic importance, provoked scant interest. The beginning of the 1990s marked a turning-point. While the central government began for the first time to put into motion an extraordinary amount of regulation and to adopt contingency policies, in Catalonia the first steps were made towards developing immigration policies. Thus, in 1992, the *Generalitat* created the Interdepartmental Commission for the Monitoring and the Coordination of Action with Regard to Immigration and in 1993, it approved the First Interdepartmental Plan for Immigration (1993–2000). During the second half of the decade, with the constant increase in the flow of immigrants, political and social debate on the matter increased (Zapata-Barrero, 2004a). But it was from the year 2000 that *external immigration* really became a consolidated demographic reality, a central element in the dynamic of Catalan demographics: nearly 90 per cent of the population growth between 2001 and 2005 was due to foreign immigration (Pla de Ciutadania i Immigració, 2005–08). In parallel with this process, institutional structures and government acts were formed (Zapata-Barrero, 2004a). During this process of institutional recognition the Secretariat for Immigration was created in August 2000, and in 2001 the Second Interdepartmental Plan for Immigration was approved (2001–04), which would be followed in 2005 by the Citizenship and Immigration Plan (2005–08). In this new context of important transformations in Catalan society as a result of immigration, publications, debates, the declaration and adoption of positions on the subject and the number of initiatives undertaken with regard to it increased (Zapata-Barrero, 2004a).

In Search of the Basis for a Suitable Public Philosophy: First Strategic Actions

If we concentrate our argument on the analysis of the actions and institutional output of the Administration of the *Generalitat* with regard to immigration, and we try to identify those aspects related to self-government, it will not surprise us to discover nothing. This indicates that Catalonia really is at a stage where it has not even begun to construct a suitable institutional framework for its identity.

We can consider the new *Estatut,* recently voted for in a referendum by the citizens of Catalonia (16 June 2006) and by 90 per cent of the Catalan Parliament,[5] as the first legal framework to speak clearly about immigration in terms of necessities and of the claim to certain powers, and the recently approved Citizenship and Immigration Plan as one of the first institutional frameworks to recognize fully that the management of immigration implies a management of identities, since it clearly focuses on citizenship. These two political acts in fact form the basis for a discussion, using empirical arguments, of a *philosophy* for Catalonia.

Inside the process of realizing the *Estatut* and linking it to new realities, the issue of immigration was intensely debated up to the point where Article 138 was created as a distinct field. This field focused principally on two issues: the initial acceptance of immigrants (understood as social and economic integration) and authorization for work permits for those immigrants choosing to work in Catalonia. We must state that Article 138 was thought out more in terms of powers than in terms of the management of identity, establishing both exclusive powers with regard to the acceptance and integration of immigrants and work permits (though subordinate to the concession of residents' permits) under the control of the central government. In other words, Catalonia has no power to say who can enter its territory and in what numbers.

Seen from the perspective of other autonomous communities, Catalonia is a reference point because it already has a tradition of managing immigration. While its activity to date has been administrative, in the third programme (the Plan for Citizenship and Immigration 2005–08), it is taking a qualitative step and will have a political and radical character.[6] The language of citizenship constitutes its principal conceptual innovation. It is right to base this new language on the European Union: the separation of citizenship from nationality is already a discursive reality of the European institutions. This European language has a radical character both towards those states represented in the European Council and towards the Spanish state.

A definition of citizenship linked to residency creates difficulties for the monopoly the Spanish state holds over the definition of citizenship which it links in an almost sacred way to nationality. We can see here two models perhaps forming two philosophies: that of the Spanish state, based on the indissoluble link between nationality and citizenship; and that of Catalonia, focused on residency in place of nationality. This argument is not a privilege of Catalonia's but rather constitutes a European language which Catalonia is adopting as a political strategy.

This new plan should also be seen as the third phase of an historical trajectory. The first programme (the Interdepartmental Plan for Immigration 1993–2000) constituted the first step, with the creation of an interdepartmental network implementing the belief that immigration is a subject that affects all government departments. The second phase came from the second programme (the Interdepartmental Plan for Immigration 2001–04), when it created and consolidated a network of people involved, from the immigrant community, NGOs, trades unions, charities, employers, and so on. During this phase the logic of mutual accusation would give way to a culture of co-responsibility. In those moments we can say that, in Catalonia, the network of people involved is complicit with the Administration and in a joint search for solutions for its management.

Institutional Bases: From the 'Catalan Way' to a Focus on Citizenship and from an Administrative to a Social and Political Interpretation

If we look for the focus of identity, even though the Second Plan for Immigration (2001–04) rightly established the bases of what was called the 'Catalan way', it was only in the recently approved Third Plan that the question of identity was explicitly incorporated as a direction and a criterion in policy design. Although we cannot say that intuition has disappeared from the design of the Plans, their strategic direction becomes more clearly defined each time and we can really speak about a suitable philosophy in strategic political terms.

a) The Interdepartmental Immigration Plan (2001–04)

This plan introduces new variables, particularly the chapter entitled 'A Catalan Way for Immigration', which defines an initial model for the integration of foreign immigrants to Catalonia. This focus on integration implicitly incorporates the variable of 'self-government' in the management of immigration into Catalan territory and highlights three situations which show the complexity of the migratory process. These situations should generate institutional responses adequate to growing immigration, in a search for the suitable management of immigration in accordance with its national specificity.

The *first* situation underlines the fact that immigration increases cultural diversity in the country of arrival and brings to our attention the need for equality with respect to difference as a basic objective of the welcoming country. The *second* situation confirms that immigration is one of the greatest challenges facing European societies and that Catalonia is no exception. It is necessary that proposals should be contributed by the institutions in order to manage this new context correctly, bearing in mind Catalonia's particular culture, society and politics. Finally, the *third* situation (in parallel with the positive substratum created by immigration, not only from an economic but also a cultural and social point of view) affirms that a phenomenon such as immigration, because of its very complexity, can bring about social conflict generated by incomprehension, fear, ignorance and economic inequality. It is necessary to work to avoid these conflicts and, should they arise, to promote the means needed to resolve them, by opening a dialogue and encouraging mutual understanding. In this way it will be possible to maintain a cohesive and socially fair country.

The Catalan Way for Immigration tries to be the starting-point from which the proposals included in the Interdepartmental Plan will follow a precise course: *the fact that immigration holds great hope for the future society and economy of Catalonia, but also great doubts for a Catalonia keen to maintain and increase its political and cultural uniqueness.* This Catalan Way was one of the first contributions of the first Immigration Secretary, Angel Miret. However, its development had a short trajectory. Its second secretary, Salvador Obiols (2002–04), developing certain measures of the Catalan Way, carried out two basic acts: a socializing campaign using audiovisual media, and the provision of Catalan courses as well as the opening of agencies abroad (fro example, Poland, Morocco). These are first steps, but they respond largely to measures directed by intuition rather than to strategic action in the true sense of the term, an understandable fact given that we are in the first stages of the process.

b) The Citizenship and Immigration Plan (2005–08)

In the Third Plan, the Citizenship and Immigration Plan (2005–08), reference to interdepartmentality disappears from the title and a relationship is established between immigrants and citizenship. Thus, from the administrative phase of immigration (interdepartmentality), we pass to a clear focus on social issues and politics.

The new plan proposes a focus on citizenship which follows two approaches. The first consists in perceiving policies in terms of the management of a process of inclusion or equality. The basic idea is to manage the change in status from immigrant to citizen. This implies that, in order to manage this interaction, the basic ideas of equal rights and social and economic equality must be upheld. Also forming part of this dimension of inclusion and equality are issues related to the equality of political rights and the theme of representation (little discussed at present but symbolically important when one considers the presence of people of immigrant origin working inside the organizational structures of the public administration).

Representation and the proportional presence of people in the Administration is a key theme. We refer to representation in terms of proportionality or equality of access to public affairs. Although this question does not form a large part of the plan, it would be wrong to ignore it, as sooner or later it will form part of the social and political discourse.

The second approach deals with the politics of accommodation, that is the management of public zones of interaction between citizens and immigrants (Zapata-Barrero, 2004b). In the political sphere, it is essential to manage this conceptual innovation well, as the language of citizenship might easily turn into a new rhetoric that will eventually cease to function, since it recognizes too many disparate realities. In a discursive context where the whole world talks of 'citizenship', including the designation of immigrants as new citizens, exoticism and paternalism might confuse intentions. Immigrants are not 'new citizens'. To speak of immigrants as 'new citizens' is to use rhetoric to hide the truth: that immigrants are not new citizens. This rhetoric of citizenship is the kind of perverse effect the plan can generate. It is important to avoid these misunderstandings, which popularize a concept to the point where it begins to gain meaning for lack of a clear reference point.

As well as posing the risk of generating certain types of discussion, the 2005–08 Plan has virtues on a conceptual level which establish the bases for a public philosophy. In fact, the conceptual innovation it presents forms one of the basic differences between it and the other plans.

To begin with, the 2005–08 Plan has two important structural novelties: it presupposed its strategic action and field of activity, and it committed to offering a series of indicators to evaluate its social impact and functioning.

But, following the line of our argument, we wish to stress mainly its management of citizenship and its declaration of guiding principles.

In the first place, the management of citizenship-residency implies that residency should become the only criterion in defining policy. From an administrative point of view, this means that registering is a basic criterion that grants the immigrant the right to be included in the public political sphere. Registering expresses the immigrant's desire to establish themselves in Catalonia and to share public spaces with existing residents irrespective of who they are. In this way, the concept of citizenship detaches itself from state-based nationality and attaches itself more narrowly to its relationship with the city and local community.

In the second place, the focus of the 2005–08 Plan ends with the definition of a plural and civic citizenship. It is worthwhile introducing this concept, which integrates the nucleus of the new perception. The focus is on three basic ideas: the value of pluralism, the principal of equality and the behavioural norms based on civic-mindedness. The acceptance of pluralism is key, as it is the condition without which public recognition is impossible. It means that every person who lives in Catalonia, regardless of creed or colour, must recognize pluralism as a basic value in all its manifestations (pluralism of values as well as cultural pluralism; pluralism in its individual as well as its group expression). The value of pluralism includes that of secularism, which means a strict separation between political power and religious beliefs.

Equally, recognizing the fact that the arrival of immigrants from abroad is a new factor generating certain inequalities, the management of citizenship also assumes, as a second keystone, the principle of equality with regard to certain rights and responsibilities necessary for guiding social activity. Later we will see that the principle of equality is as much about equality of treatment as it is about equal opportunities. In fact, in combining these two keystones, one aspires to recognize the reality of pluralism and to construct a vehicle that can identify proposals which minimize those inequalities generated by cultural, linguistic and religious differences coexisting on Catalan territory.

As a basic norm for the guidance of relationships between people (regardless of their mother tongue, culture, religion or skin colour), civic-mindedness forms the third keystone. Civic-mindedness can be defined in practical terms as an 'attitude and norm of public behaviour that allows for the cohabitation of people and assures long-term cohesion'.

Another important idea is embedded in this context of the definition of plural and civic citizenship: that of 'communal public culture'. This concept has the character of an objective-to-achieve, with the management of citizenship with its three keystones (pluralism, equality and civic-mindedness) as its principal tool. In terms of public policy, the approach has a clear strategic intention.

The idea of communal public culture also implies the rejection of what we might call an 'institutional approach to pluralism', in the sense that every culture and differentiated reality inside Catalonia has a form of institutional expression different to the other.[7]

From the viewpoint of premises for action, I believe the best thing is to view them as a mark of the interpretation of and a basic component of the suitable philosophy. They are concepts which help to interpret conflicts and to direct political action. The plan establishes the following for recognition of cultural diversity:

- Human rights and respect for human dignity
- The universality of public policy and the respect for individual rights
- Ensuring the stability and cohesion of Catalan society
- Defence of the Catalan language as the language of resident citizens
- Coordination, cooperation and co-responsibility
- The policy of integration with a European base.

In the first place, we recognize that Catalonia is culturally diverse. Secondly, the respect for human rights and human dignity as a basic principle implies that ahead of any economic, social, or other argument, human rights must prevail. The third principle is the universality of public policy and the respect of individual rights. A link between two important dimensions is created: equal public policy for everyone, and an end to thinking of people in cultural and ethnic terms.[8] Each person can feel they belong to whichever community they choose, but no administration can tell them what they should feel. The recognition of individual rights is important here. In the political field and in Europe we can speak about Holland's experience. This country's great problem was with its focus on 'pillarization' – the administration saw belonging to a community as being the key element in its perception of immigrants, without regard for individual rights. To perceive an immigrant from the viewpoint of their community has very different social and political implications to seeing them as an individual. In the latter instance, one gives the immigrant the freedom to feel part of whichever community they choose (their own will), rather than that community imposed by the administration. Between a communitarian and a liberal approach, the plan has clearly opted for the liberal approach.

Equally important is the fourth principle, which guarantees the stability and cohesion of Catalan society. But for the objectives of this section let us look at the fifth principle: the defence of the language and identity of Catalonia. This aspect is also important on the level of criteria. It is a fact that the immigrant's linguistic reality is different whether one is dealing with society, workplace, relationship with the central administration with regard to immigration papers, the immigrant's shops, markets, or life where they live. Furthermore, while knowledge of the Castilian language may form part of the central government's criteria for entry into our frontiers, it is not a criterion that is pertinent for Catalonia. It is a policy that flies in the face of the process of self-government. I believe it is a common and acceptable to feel that the nationalist process can be affected in a very conclusive way by the immigration of the Castilian language (principally coming from Central and South America) without strategic political action. This fifth principle also implies a recognition that immigration policy is linguistic policy. The two must be closely linked.

Finally, we can speak of coordination (between different administrations), cooperation (between different people) and co-responsibility (in the basic sense of the term, not the responsibility of the various people involved). As the sixth and last premise, we must reassert the importance of seeing integration policy in a European context. We have already seen that a European language is used as much for defining a framework for action as for claiming responsibilities for the central government. This language reinforces the 2005–08 Plan and places it in a discursive reality within Europe. In this framework of European argumentation, the plan makes those basic communal principles its own which were approved by the Council of Europe on 19 November 2004, namely, those relating to a policy of integration of immigrants in the EU and especially the idea that 'integration is a bi-directional process which is dynamic and which involves constant adjustment on the part of both immigrant and existing citizen. Governments must clearly communicate to both sides their rights and responsibilities.'

The Identity Debate: Defining who is an Immigrant in Catalonia, its Historical Traditions, Dual Language and Culture

It is a fact that, in Catalonia, the link between immigration and self-government has an academic, rather than a political or social, character. It is also a fact that academe can supply arguments which make political elites acknowledge the problem. We are now in a situation where the political sphere is beginning to identify the theme in terms of concern, but this is still due to intuition rather than solid arguments.

What is becoming ever clearer is that politics must anticipate the social effects of immigration upon the very definition of a political community. It is a fact that in both Quebec and Flanders immigration reinforces the need to define oneself as a nation-state. While Catalan society has acknowledged that it is facing a new process of immigration from underdeveloped countries, a process that has begun to alter the Catalan urban social landscape, it has not yet fully acknowledged the effect that immigration might have on nation building.

From this point of view, transformations based on immigration also affect minority nations. In this sense, immigration policy is conceived as a policy of reconceptualization and redefinition of the community.[9] Within this framework we must decide which elements can give shape to a Catalan *public* philosophy, bearing in mind characteristics particular to that community.

While Quebec and, to some extent, Flanders, already have a certain tradition of managing the link between immigration and self-government (even if they do not think of this link in explicit terms, as we shall see later), Catalonia is at the key stage for beginning *its* reflections on the subject. It is true that Catalonia has always been a country constructed from immigration and that the arrival of immigrants forms part of its history and identity. But this fact cannot justify the existence of its own form of management for this new phase of immigration and it acquires a radical and political character when we concentrate on its relationship with central government.

To develop this public philosophy is a necessity: we have not yet reached a situation of fracture or Catalan-Spanish polarization with regard to the integration of immigrants. In the other two cases, this division is one of the basic restrictions borne in mind when immigration is managed. In Belgium, for example, the management of immigration is based on social differences which are based on identity,[10] while in Quebec this danger is what prompted the framework of the Programme of 1990 (*Énoncé de politique en matière d'immigration et d'intégration*). This basic danger must accompany any reflection on the theme of immigration.

With a view to establishing the bases of a suitable philosophy and in this way completing important and non-existent aspects of the new plan, I will focus my argument on aspects of identity.

In these initial phases in which Catalonia finds itself, using identity to enter into the debate on immigration and self-government implies having to construct arguments which acquire the characteristics of programmes of investigation. What we can do now is define the principle axes of the debate in the form of its main categories and arguments.

First, if up until now the option has been to focus on citizenship, the argument I will defend is that that focus is incomplete unless one links it to the question of who is an immigrant in Catalonia. This new form of looking at the theme will also generate new dimensions to complete the identity debate. The whole argument tells us that, until we have proper criteria for the definition of an immigrant , the debate on citizenship fuels rhetorical discussion rather than fulfilling a practical political function from the self-government point of view. It is also a fact that Catalonia has a political tradition of self-government, even if it is not a tradition that has traced the evolution of the concept of an immigrant. This double debate, in spite of raising political questions, must not forget to reach society itself. The same political sphere cannot answer its own questions without being aware of Catalonia's social composition.

In order to give this philosophy shape, this approach is a concrete debate which discourages rhetoric and focuses on citizenship. Neither will it tend to fall into nominalism, although it might be in danger of falling into solipsism (that is, a debate about itself), so that henceforward constant reference to the real composition of Catalonia's population will be essential. It is certain that, just as with the debate on fields of responsibility, this debate on identity will focus on Catalonia's (in this case conceptual) shortfalls in managing immigration and will thus constantly produce arguments expressing more desiderata to cover. But this is a general characteristic of political debate in Catalonia. The link between the description and the desiderata of political arguments is a constant in Catalonia, especially where the debate centres on Catalonia's own identity.

If we follow this route, we need to combine two paths of analysis that can be viewed individually but which give a more complete picture when viewed together. The first is comparative logic, especially in reference to other contexts found in similar situations. Using this logic, one would look at the ways in which Catalonia might become a *differentiated society* (just as one had investigated Quebec's model). The answers found will supply the arguments to create a *public philosophy of immigration* and to define who is an immigrant. These differentiating elements require the protection of a state. Let us think, for example, of the Catalan language as an element in differentiating identity.

The second path is that of historical logic. This shows us that Catalonia has a traceable history of immigration. The premise it follows is the recognition that Catalonia is a society of immigration. This fact means that Catalonia was built upon immigration, that its biography is inextricably linked to the arrival of immigrants. Therefore this new wave of immigrants is not exceptional but rather forms part of its historical tradition.

By linking these two strands of logic we create a principal source of arguments in order to begin a debate on defining who in Catalonia is an immigrant. There are two possible batteries of responses depending on whether our point of reference is the Spanish state, or Catalonia. We must also accept that all potential definitions inevitably have a political nature in the sense of expressing, ultimately, a predetermined way of interpreting the power relationships they establish. For example, definitions of Catalonia as a minority nation, a national minority or a 'minoratized' nation all mean different things.

This problem of conceptualizing who in Catalonia is an immigrant is not a theoretical debate but rather has clear practical effects. For example, statistics change dramatically, depending on which definition is used. Who do you count as an immigrant? Do all the cases studied have the same reference point? Is everyone born outside of Catalonia an immigrant? Are those immigrants who came from other parts of Spain during the 1950s and '60s immigrants? Do we need to look at the statistics to compare the numbers of immigrants from inside and outside the EU? Obviously the replies to these questions will vary depending on whether the Spanish state or Catalonia is taken as a reference point. In this same sense, which territorial marking points should one use when statistically analysing Catalonia: the actual political boundary, or the geographical or historical ones? Should northern Catalonia be included? Should southern Catalonia and Valencia be perceived as part of Catalonia? These questions are not mere sub-themes to be explored but are in fact the very nucleus of the debate at hand.

Bearing in mind all these dimensions and looking from Catalonia's point of view, a conceptual difference between immigrants from other parts of Spain and those from outside Spain can be established. The response needs criteria and the discussion on criteria might elicit at least three responses:

- *Place of birth:* 'Born outside Catalonia'. Does this imply that being born inside Catalonia of foreign-born parents means one is not an immigrant? Does it mean that someone born outside Catalonia but resident in Catalonia is an immigrant?
- *Language:* 'Non-Catalan speaking'. This was one of the ways in which Quebec categorized its population.
- *Culture:* Relationship with the indigenous population? Ethnic minority? This is the path chosen by Flanders, namely, categorizing immigrants as an ethnic minority inside its principal area of influence, Holland.

What is clear is that that the selection of a criterion for defining a population made up of people of differing origins is a political option that ends up requiring a definition of community that is national. There are nations that opt for language (Quebec) and those that opt for cultural criteria (Flanders). Following Catalonia's traditions, the criterion of residency in the new Citizenship and Immigration Plan 2005–08 might be the best option. Conceptually, it would be possible to combine this with looking at people's origins in order to establish a dual taxonomy for categorizing the resident population.

The first would have as reference the link between current residency and birthplace. Thus we have the following categorization of the population:

- *Catalan residents*: residents population born in Catalonia, regardless of parents' origins

- *Catalan residents of Spanish origin or belonging to an historic community (for example, the Basques, Galicia*: resident population born in Spain or in a recognized historic community (Basque or Galician citizens)

- *Catalan residents of European origin*: resident population born in an EU country (here we would include also those formally outside the EU, for example, Poles, Hungarians, Czechs, and so on)

- *Catalan residents of non-European origin*: resident population born outside the EU

- *New residents*: members of the population who have recently arrived and who have finished the process of registering themselves, finding accommodation and including themselves in Catalan society.

We cannot ignore the fact that areas of conflict will arise when questions of identity are being debated. We know that in this sense the theme will not be resolved between statists and nationalists. The crux of this conflict can produce very different arguments relating to understanding the tradition of immigration characterized by Catalonia.

To understand this question we must start from the premise that any philosophy of immigration seeks recourse to tradition in order to find answers to those questions raised by the handling of immigration.[11] The problem is that Catalonia has not one but two ways of understanding its tradition: it can understand the presence of recent immigrants (principally from non-EU, developing countries) in two ways, either as a second wave of immigration (the first having been in the 1960s in a non-democratic context and coming mostly from southern Spain and Aragon – that is, internal, Castilian-speaking immigration), or as the first wave of immigration into Catalonia. In the first instance, one assumes that the two waves are comparable (one can even establish a debate to determine the criteria for comparison); in the second interpretation, this immigration is not comparable to that of the 1960s (in general, the immigrants of the 1960s do not admit comparison between themselves and Catalonia's latest new residents). To sum up, rather than having one tradition and various forms of administering it in the future, we are faced with a 'dispute between traditions'. This dispute is fundamental because, depending on the interpretation it produces, we will have different bases on which to produce the arguments necessary for the construction of a suitable philosophy. It would be important first to analyse previous waves of immigration and to identify the elements used to manage them, identifying the main elements of evaluation in terms of the management of identity.

What is also clear is that the question of defining the quality of a member of the Catalan community/society is key. Seeing the issue in terms of a sense of belonging, or in terms of residency, has very different implications, especially when there is no state structure to organize the stability and cohesion of a society in the way that a central government can. Residency as a criterion is more objective than the sense of belonging, which would need to be defined in terms of identity and tradition.

Bearing in mind the new elements we have introduced here, the question of how tradition is defined is fundamental, as it can have two effects which not only alter the stability, but might even damage the cohesion, of society. I am referring to two basic dangers: on the one hand, the danger that society might be fractured to such an extent that a dual or divided society will arise (we can exemplify this symbolically by speaking of a conflict between the 'culture of the bull' and the 'culture of the ass').[12] On the other hand, and connected to the former, it might produce a loss of national hegemony and regression in the cultural public sphere, in other words, a 'Castilianization' of that very same public space that has been recuperated over the past few years. It is a fact that, without public intervention, we must recognize that immigrants are not allied to, complicit with, or participators in self-government.

From a theoretical point of view, including the search for indicators to sustain arguments, we can say that immigration highlights the deficiencies of self-government. This debate must be held in Catalonia and must make reference to the fact that a dual culture exists between the majority and the minority – and that the role of immigrants thus becomes vital. A poorly managed immigration policy could increase the Spanish–Catalan divide. These difficulties are of a practical nature: how can immigrants be integrated into the Catalan linguistic reality when their principle 'references to authority' are expressed in Castilian? (For example, the administration of their papers or their workplace.) In this sense and if we link together responsibilities, self-government and identity, it is clear that the criterion of knowledge of Spanish used by the central government in its management of immigration is not valid for Catalonia. It is, moreover, a negative criterion in terms of Catalonia's self-government. A suitable philosophy of immigration must be able to evaluate those who can or cannot fit in linguistically. It is part of the responsibility of Catalan institutions to ensure they speak to immigrants in Catalan in the majority of public spheres, especially in areas where Castilian speakers are in the majority.

Linking the three dimensions articulated in this argument – the necessity of defining who is an immigrant in Catalonia, the debate over a tradition of immigration in Catalonia, and what we will end up doing about dual language and culture – we create the necessary analytical basis for constructing a suitable philosophy of immigration. Immigration policy must have clear answers to these questions. Resolving these issues of identity and tradition will be the *sine qua non* of a public philosophy of immigration in Catalonia.

The terms of the debate we are establishing have certain outstanding epistemological implications. What we are saying is that in order to have a philosophy of immigration we do not need to follow theoretical and foreign models but must form our own by looking at and analysing our practices and based on our own experience. We do not need to adopt concepts from other realities nor to export concepts. Instead, we need to construct our own distinctive discourse (with those connotations relevant to our traditions). If we speak in terms of the conditions necessary to construct an appropriate philosophy, we can stress the following elements: first, that we must follow the principle of reality and incentive to guide debates. That is to say, all arguments have an empirical reference point that forms a substantial part of tradition. An appropriate philosophy would follow the logical structure of the country.

If we wish to name the three main policies reinforcing the link between self-government and immigration, they would be linguistic, educational and employment policy. These three elements need to be linked when forming policy relating to the accommodation of immigrants in Catalan society. It is certain that in relation to similar cases like Quebec and Flanders, Catalonia is at a disadvantage, because both Quebec and Flanders use languages which in other countries allow them to be part of a state. This fact reinforces the importance of the need for a strong policy on language, including in relation to civil society in Catalonia.

In effect, a public philosophy which links immigration and self-government from the aspect of its identity in relation to an associative world would have to be as necessary as a state would be, especially one that is directly engaged with the fate of immigrants, such as NGOs and other associations of immigrants. A demand that must be made and that must be understood and accepted is that associations must conduct their affairs and requests in Catalan. This is a key policy for self-government. I am not forming a particular defence but rather a question: How can one manage, in terms of identity, the fact that the NGOS and other associations are in favour of the Spanish state rather than Catalonia?[13]

The Final Balance: Strands of Reflection and Future Action

The two languages of self-government (those of responsibilities and those of identity) can be combined to produce arguments contributing to the definition of a public philosophy of immigration in Catalonia, but it is important to perceive the link between the two in terms of priorities. Using only the language of responsibility to legitimize policies might distort the true identity of a community. In order to construct an appropriate philosophy, the language of responsibility must have a communitarian significance and must therefore be based upon a language of identity.

We have seen how the link between immigration and self-government generates a whole series of normative questions that have barely begun to be tackled in Catalonia, either theoretically or institutionally. It is clear that immigration needs to redefine self-government, particularly in order to differentiate between arguments relating to the cultural demands of immigrants on society and those projected towards the Spanish state from Catalonia.

We have also argued that Catalonia must avoid foreign models and/or a-contextual theories when forming a suitable philosophy: the main sources for administering the link between immigration and self-government need to be based on an analysis of Catalonia's own experience as seen, as it were, in the rear-view mirror: using tradition to influence future policy. In this sense, I believe that the experiences of others like Flanders and Quebec show us that the link between immigration and nationalism has different forms in different contexts. In the case of Flanders, its form is negative: immigration poses a threat to national identity. The existence of extreme-right parties (like Vlaams Belang, formed after the Vlaams Blok was banned), whose discourse is clearly anti-immigrant, is an example of the kind of extreme identitary language that should be avoided. In Quebec, on the other hand, immigration is seen in a positive light, as an opportunity. One is nationalized via the main identitary channel — that of the French language. This, in turn, swells the ranks of French culture in a society where English culture currently dominates. This initial perception is fundamental, as it forms the basic premise for helping us to understand different approaches. How should Catalonia interpret this link?

From the very beginning, we see that the immigrant has a linguistic impact: that is the basic fact that directs most action. An immigration policy in a national framework but without state tools is basically articulated as linguistic policy. This does not mean that the other policy aspects are unimportant (such as the level of access and coexistence), but only that they are subordinate to the linguistic dimension.

From a comparative point of view, there are those who advocate an approach based on citizenship (Quebec) for defining policy, while others are focused on an instrumental logic which allows them greater competitive capacity in their relations with other nations like Flanders. What the experience of these other countries also shows us is that the processes of institutional definition also imply a process of conceptual innovation,[14] especially in identifying immigrants, who are at times designated as foreigners (Quebec), and at others as ethno-cultural minorities (Flanders). The criterion for differentiation can be either linguistic or cultural. These two approaches to policy keep a certain distance which bears a relationship to the construction of tradition. This fact must be present as a basic theme in the construction of a philosophy of immigration in Catalonia.

From the point of view of the institutional framework, we can make three proposals: that a document-framework be created to set out a suitable philosophy for Catalonia, that policy direction of a ministerial nature should be created, and that an institute should be established to manage, produce and spread information that will direct government actions in this key issue for its communitary and identitary character.

In Quebec, as in Flanders, document-frameworks have been created which might serve as a reference point for limiting potential policies and reorientating them. These documents, the 1990 *Énoncé* in Quebec and the 1998 *Décret* in Flanders, establish a philosophy relevant to the area in question. In Catalonia, we need to create a document that has been accepted by a political consensus that would fulfil the same function.

From an institutional point of view, there are two ways of managing immigration: either you have a transversal vision as in Flanders, with each ministry dealing with the issue or creating a structure to deal with it, or you deal with it in more specific terms, creating a new ministry to manage the issue, as in Quebec.

Presently, in Catalonia the control of immigration policy has passed from Welfare to the Presidency (where the Secretariat for Immigration, SIM, was created in 2000) and from the Presidency to the Council of Welfare and the Family to coincide with the new legislation and the tripartite government (PSC – Socialist Party of Catalonia, ERC – Republican Left of Catalonia, and IC – Initiative for a Green Catalonia). This displacement of the ministry of the Secretariat for Immigration (from Presidency to Welfare and Family) clearly has political connotations.

If we take the arguments expounded above seriously and if we wish them to form part of an immigration philosophy for Catalonia, it is clear that, whatever that philosophy is, it can only act strategically if it has the institutional force of a ministry.[15] Without this ministerial character, there will be few opportunities for piloting the policy of identity so vital to Catalonia's self-government. This is vital not only in order that Catalonia might negotiate with external factors such as the Spanish central government, but also for its internal politics. It is key that Catalonia should have an adequate political framework to consolidate its interdepartmental network and that it should build an institutional, territorial framework and a network of associations involved in the accommodation of immigrants.

Finally, bearing in mind Catalonia's traditions of immigration management and that it is in its third phase of immigration (if we use the plans as reference), it is incomprehensible that there should be no central institution to manage information, produce arguments and guidance, and also concern itself with technical and professional, as well as academic, training. As examples of the kind of studies still to be carried out and following the guidance of this section, it is imperative that there should exist a quantitative source telling us what people think about immigration according to the variables of identity. For example, apart from the usual social and economic variables, this source could hold information that would indicate whether the sense of belonging (that is, feeling oneself to be 'more Catalan than Spanish' or 'as Catalan as Spanish' – to mention two categories) is a factor influencing attitudes towards immigration and, if so, in what way? Equally, it would be interesting to analyse what resident immigrants' perceptions of Catalonia are and whether they have a sense of belonging.

Catalonia will face great challenges in the future if it takes its task of creating its own philosophy of immigration seriously. The links between self-government and identity require, when dealing with the definition of a new field like immigration, an eminently generational dimension. The one condition necessary for this is that politicians need to make policy; the country's logic must prevail over any partisan interest. In the next few years, we will see if those who govern us have understood that Catalonia faces a new historical opportunity in the shape of immigration.

References

Aubarell, G., Nicolau, A. and Ros, A. (eds) (2004), *Immigració i Qüestió Nacional* (Barcelona: Editorial Mediterrània).

Cabré, A. (1989), *Les Migracions a Catalunya* (Barcelona: Centre d'estudis Demogràfics (UAB)).

Cabré, A. (1999), *El Sistema Català de Reproducció* (Barcelona: Proa-Institut Català de la Mediterrània).

Cabré, A. and Pujadas, I. (1989), 'La Població: Immigració i Explosió Demogràfica', in J. Nadal, C. Sudrià, and F. Cabana (eds), *Història Econòmica de la Cataluña Contemporània*. Vol. V (Barcelona: Enciclopèdia Catalana).

Calvo, A. and Vega, E (1978), 'Generalitat, partits polítics i immigració durant la Segona República', *Quaderns d'alliberament*, 2, 3, pp. 17–27.

Candel, F. (1966), *Els altres catalans* (Barcelona: Edicions 62).

Favell, A. (2001), *Philosophies of integration* (London: Palgrave [1st edn 1998]).

Fontaine, L. and Shiose, Y. (1991), 'Ni citoyens ni Autres: La categorie politique communautés culturelles', in D. Colas, C. Emeri and J. Sylberberg (eds), *Citoyeneté et nationalité. Perspectives en France et au Quebec* (Paris: Presses Universitaires de France), pp. 435–43.

Jacobs, D. (2001), 'Immigrants in a Multinational Political Sphere: The Case of Brussels', in A. Rogers and J. Tillie (eds), *Multicultural Policies and Modes of Citizenship in European Cities* (Aldershot: Ashgate), pp. 107–22.

Jacobs, D. (2004), 'Alive and Kicking? Multiculturalism in Flanders', Paper presented at the conference 'Europeanists' in Chicago, IL (manuscript).

Porta, J. (2004), 'Immigració i Identitat Nacional', Ponència presentada al *2n. Congrés de l'immigració a Cataluña* (17/18 abril).

Pujadas, I. (1983), *La població de Catalunya: Anàlisi especial de les interrelacions entre els moviments migratoris i les estructures demogràfiques*, Tesi doctoral, Facultad de Geografia i Història de la Universitat de Barcelona.

Termes, J. (1984), 'La immigració a Catalunya: política i cultura', reprinted in VV.AA., *Immigració, autonomia i integració* (Departament de Presidencia, Generalitat de Cataluña, 2002).

VV.AA. (2002), *Immigració, autonomia i integració*, Departament de Presidencia, Generalitat de Cataluña.

Zapata-Barrero, R. (2004a), *Inmigración, innovación política y cultura de acomodación en España: Un análisis comparativo entre Andalucía, Cataluña, la Comunidad de Madrid y el Gobierno Central* (Barcelona: Ed. Cidob).

Zapata-Barrero, R. (2004b), *Multiculturalidad e inmigración* (Madrid: Editorial Síntesis).

Zapata-Barrero, R (2004c), 'Davantal: Un marc interpretatiu per gestionar l'immigració des de Cataluña', *Revista Idees*, 22, pp. 76–9.

Zapata-Barrero, R. (2005a), 'The Muslim community and Spanish tradition: Maurophobia as a fact and impartiality as a desideratum', in T. Modood, R. Zapata-Barrero and A. Triandafyllidou (eds), *Multiculturalism, Muslims and Citizenship: A European Approach* (London: Routledge).

Zapata-Barrero, R. (2005b), 'Multinacionalidad y la inmigración: premisas para un debate en España', Prólogo de D. Juteau, *Inmigración, ciudadanía y autogobierno: Québec en perspectiva*, Documents de Treball CIDOB (Serie Migraciones), 6 <www.cidob.org>.

Zapata-Barrero, R. (ed.) (2007a), *Immigració i autogovern* (Barcelona: Editorial Proa).

Zapata-Barrero, R. (2007b), *Filosofías de la inmigración en naciones minoritarias: Flandes, el Québec, y Cataluña en perspectiva* (Barcelona. Icaria) (una primera versión en catalán se encuentra en *Immigració i Govern en nacions minoritàries:*

Flandes, el Québec i Catalunya en perspectiva (Barcelona: Quaderns del Pensament 21, Fundació Ramon Trias Fargas) <http://www.fdtriasfargas.cat/ca/Publicacions/Quaderns.php>.

Ricard Zapata-Barrero is Professor of Political Science at the Pompeu Fabra University, Barcelona.

NOTES

1. Although new arguments have been introduced, particularly the first demographic section, there is a version of this article in Castilian ('Construyendo una *filosofía pública de la inmigración* en Cataluña: los términos del debate', *Revista de Derecho Migratorio y Extranjería*, 10, noviembre 2005, pp. 9–38), and in Catalán (published at the time of the new Statute) ('Construint una *filosofia pública de la immigració* a Catalunya: els termes del debat', a R. Zapata-Barrero (ed.), *Immigració i autogovern* (Barcelona: Proa, 2006).
2. A comparison between policies (philosophies) of immigration in Quebec, Flanders and Catalonia can be found in Zapata-Barrero (2007a, 2007b). The Statute of Catalan Autonomy is the basic institutional norm. It defines the rights and responsibilities of Catalan citizens, the political institutions of the Catalan nation, its competencies and relationship with the Spanish state and the funding of the *Generalitat* of Catalonia. This law was voted in on 18 June 2006 and replaces the 1979 Statute.
3. We cannot say that a debate on this exists in Catalonia, if only at an academic level. Illustrative, though not exhaustive works that have reflected on these issues using Catalonia as reference, include: VV.AA. (2002), G. Aubarell et al. (2004) and Porta (2004).
4. This idea comes from the work of Francesc Candel, *Els altres catalans* (1964), who magnificently described the everyday reality of people migrating into Catalonia.
5. <http://www.gencat.cat/generalitat/cas/estatut/> (September 2006).
6. <http://www.gencat.net/benestar/societat/convivencia/immigracio/>.
7. Something similar to the Lebanese model, where public and everyday life is institutionalized according to religion (schools, hospitals, religion, and so on).
8. We might think of this as explicitly positioning oneself against a communitarian view of immigration and in favour of a Quebec-style *'communauté culturelle'* (Fontaine and Shiose, 1991), which has precisely stood for a focus on citizenship.
9. I am here adapting the idea of the 'politics of reconceptualising community' of Favell (2001), p. 24).
10. Started in 1990 according to Jacobs (2001, 2004).
11. See the first version of this argument in Zapata-Barrero (2005b) and its application seen in a Spanish context in Zapata-Barrero (2005a).
12. In the past few years a (peaceful) war of symbols has broken out in Catalonia. This is represented by Spanish sympathizers bearing on their clothes or vehicles the image of a bull, while Catalan sympathizers bear the image of an ass. Thus we can speak, as I propose, of a 'culture of the bull' and a 'culture of the ass'.
13. It is important to reflect on language. As we have seen, in Quebec the immigration policy is linguistic policy. A study of the relationship between policies, linguistic rights and immigration in minority nations has yet to be undertaken.
14. On the relationship between processes of change and conceptual innovation, see Zapata-Barrero (2007a). For the concrete example under discussion, see Zapata-Barrero (2007b).
15. In fact, if we take Article 138 of the new Statute seriously, only an administration with the Conselleria's range could develop its key tenets.

PART III
Europe's Southern Frontier

PART III

Immigration and the Canary Islands

JAVIER MORALES FEBLES

The Canary Islands have become an entry point into Europe for a large and growing number of irregular immigrants. Two different types of irregular immigration have been observed: the first are those immigrants who arrive in large numbers at ports and airports, generally with tourist visas, and who remain in European territory once their visas have expired; the second comprise a more dramatic category of immigrants who arrive at our shores aboard unsafe boats that put their lives at real risk and which, in many cases, have caused fatalities.

Not all such arrivals are detected. Once they land, it is not possible to keep track of everyone, even those who arrived with visas. Many melt away and so we don't know how many immigrants are not included in the statistics. I would like to stress this at the outset, in order to clarify that the statistical data about migratory movements given below will not reflect the real, inevitably higher, figures. They refer instead only to a verifiable minimum.

The Canary Islands are on the front line of the crisis being played out daily in the Atlantic. In 2006 alone, almost 38,000 irregular immigrants have landed, more than a thousand of whom were unaccompanied children. The unexpected arrival, for example, of more than a thousand immigrants on our shores *in a single day*, meant that our health and emergency services, marine rescue teams, and social and legal services have been stretched beyond their capacities.

To these figures must be added the numbers of people that die at sea, in their attempt to reach the islands: about three thousand people in 2006, according to the numbers provided by the EU Commissioner for Justice, Freedom and Security, Franco Frattini.

According to the testimonies of the non-governmental organizations that work in the field, one in every ten immigrants who starts this suicidal voyage dies before reaching their destination. This situation is made even more dramatic as the Mafias in the countries of origin act with impunity. In several countries, people simply gather at the departure points 'that everyone knows', from whence small boats depart; there, would-be migrants negotiate the price of the voyage and queue up for boarding.

It is almost impossible to know for certain how many people in the Sub-Saharan countries aspire to reach Europe in the next few years, but in September 2006, there were more than ten thousand people in reception centres for immigrants, and this in an island region of only 7,500 square km.

After numerous warnings throughout the last few years, the government of the Canary Islands has succeeded in convincing the Spanish state and the European Union that this is a phenomenon which we must confront jointly. Our regional President, Adán Martín Menis, has had meetings and work sessions with the European Commissioners for External Affairs (Benita Ferrero-Waldner), Justice, Freedom and Security (Franco Frattini), and Cooperation (Louis Michel), in order to raise awareness of the issue. He has already made presentations to the European Parliament on our region's proposed policies.

Notwithstanding this, it must be emphasized that regular and organized immigration is an opportunity for the future of Europe, while irregular and disorganized immigration is a problem for the present. Crucially, border states, and especially border islands, cannot alone assume a problem that is a problem for the whole of Europe: the final destination of the immigrant is, really, the European dream.

Regions like ours, which are on the front line, that is, the first recipients of irregular immigration, must contribute ideas to channel, for the benefit of all, the migratory flows. The fight against irregular immigration must be part of a wider plan of action, including the different phases of migration. Cooperation with the countries of origin and transit, safety and inspection measures, and measures to regulate legal immigration are essential parts of a control plan for such irregular immigration.

Private investment is an effective engine to stimulate the economic development of these countries. The European Union and its member states must support the establishment of a stable and safe legal framework that creates a favourable atmosphere for the development of the private sector in these countries, as well as the creation of tax incentives for investment in sectors and activities that can favour sustainable economic development and the creation of employment.

Cooperation with the countries of origin and transit must reinforce the links between cooperation aid and migration, stimulating good governance and the benefit of those governments that collaborate effectively in the fight against irregular immigration. We will have no influence on the governments of the countries of origin if we do not give them the necessary financial and technical resources to control the migratory flows from the thousands of kilometres of their own coastlines.

The credibility of the immigration policy will be put into question if measures to fight irregular immigration are not adopted. Vigilance and control of external borders, police and judicial cooperation, and readmission and return agreements must go hand in hand with the measures intended to stop the causes of the immigration as well as measures intended to encourage legal immigration.

The first mission of FRONTEX (the European Agency for the Management of Operational Cooperation at the External Borders of the Member States of the EU) is under way on the coasts of Senegal, Mauritania and Cape Verde. Patrol boats, planes and helicopters from Spain, Italy, Portugal and Finland work together in this common operation, with the aim of controlling a maritime space of about two million square km. However, as far as we know, the resources that FRONTEX has at the moment are not able to control the current situation in the Canary Islands. It is necessary to increase those resources considerably to accomplish successfully the mission for which they were created. Vigilance at sea is not only a safety matter; it is also a humanitarian matter. Good management of these resources will avoid the death of thousands of immigrants.

Another very important issue is the readmission and return agreements that must be tackled, together with the establishment of a credible policy for legal immigration, with the application of quotas, like those for which African countries have been asking. In this context, it would be necessary to ensure the real implementation of Article 13 of the Cotonu Agreement, which offers the legal base the EU needs to request the readmission of nationals of signatory countries who are in an irregular situation in EU territory.

Finally, I would like to mention one of the most dramatic situations that has occurred on our islands: the arrival of unaccompanied minors on board small and overcrowded boats. More than a thousand have already arrived during 2006. As a result, the islands' regional authorities have had to manage this humanitarian crisis without the necessary economic resources which could offer these children the opportunities of a developed society. However, over the past few years, we have had an enormous increase in the number of arrivals of such unaccompanied minors. This situation is untenable.

The situation of these children forces us not to be passive, and to take short-term measures. We have data on their families and their homes, and we must encourage family reunification immediately. In accordance with the international agreements on childhood, the interest of the minor is that, whenever possible, she or he should be raised in their family environment. To that end, we must provide them with help in their own countries, and give them the necessary education and training that enables them to find work in their places of origin, near their families. In fact, international agreements on the treatment of children impose on states the duty to alleviate the illegal transfer of children to other countries.

In conclusion, it is urgent that the EU member states jointly assume the responsibility for the hosting and protection of children for whom family reunification is not possible, in order to assure their development, maturity and their better integration into society.

We all know that the migratory problem is very complex, but Europe can confront this situation with greater effectiveness. To that end, right now, we need to act: we need to arrive at a consensus between the member states on the urgent measures to take, and we need the necessary resources, including financial resources, to be funded by the member states.

Javier Morales Febles is the Regional Minister for Foreign Affairs, Government of the Canary Islands.

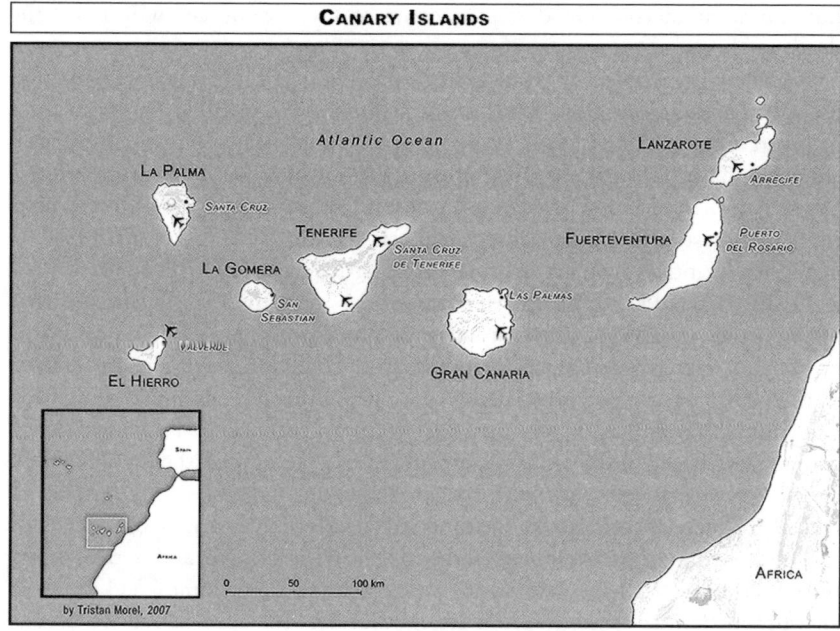

Lampedusa: European Landfall

BRUNO SIRAGUSA

For a few years now, the island of Lampedusa, at the extreme of southern Europe, has been in the media spotlight because of the continuous arrival of rickety boats overflowing with immigrants. I do not intend to burden you with statistics or recall specific events, which are well known, but to present instead some food for thought that has resulted from this phenomenon.

Migration issues are normally dealt with at the macro level: to what extent, for example, can a region, a country, or a continent withstand successive migration flows? But what happens when such a phenomenon hits a specific micro social reality? When it affects a community of 5,000 inhabitants on a small island of 20 square km situated far away, in the middle of the Mediterranean Sea? I will deal with the three most significant aspects of such a situation.

The Arrival of Immigrants

The arrival of illegal immigrants occurs almost always in dramatic circumstances: the rescue at sea, the presence of women and children, if not also of corpses. This phase is dealt with by Italian state authorities, though our fishermen are also at times involved. The authorities administer first aid to the migrants in a temporary detention centre before shipping them to Sicily or back to their countries of origin.

Leaving aside the most dramatic incidents, the migrants are like ghosts for the islanders: they arrive, they are immediately dealt with and they leave again. It is not very different from what occurs elsewhere in Europe.

The (Non)-Presence of Immigrants

In the whole of Europe, migratory issues are not only concerned with immigration control. Millions of immigrants – legal or illegal – have for years been increasingly relevant for the social reality of the countries hosting them.

This has serious implications on the future of European societies which are on the one hand open to multiculturalism (because of political liberalism or economic convenience), but on the other also worried lest they lose their own identity, to the point that we even speak of clashes of civilizations and identities.

In the case of Lampedusa, by contrast, immigrants do not exist. The reason for it is very simple: immigrants are immediately detected and soon sent elsewhere. Therefore, the Lampedusa community does not have to deal with the difficulties of a multicultural society.

The Indirect but Grave Consequences on the Local Community

From what we have described (that is, that immigrants are to us like ghosts and that we have no resident immigrants in Lampedusa), one might infer that we are only marginally affected. Unfortunately this is not the case, because of the indirect consequences which, normally manageable in a normal (macro) context, become a huge weight in the micro social system of a small island. Here are a few telling examples.

The ever more numerous small boats utilized by immigrants, confiscated by judicial authorities, have remained for a long time in the harbour of Lampedusa where they constitute a navigational peril, especially for our fishermen's boats. During a storm, the immigrants' boats were flung all around, producing significant damage to other ships and to the port itself, so much so that the authorities declared an emergency to cope with the resulting situation.

The harbour emergency became a health hazard when the remains of the boats were brought ashore. Disposing of the waste became another difficult problem for such a small island to solve on its own.

The temporary presence of illegal immigrants in the Reception Centre – 18,000 so far, since January 2006– constitutes a considerable burden to public services, such as the water and refuse systems, as well as to our small hospital facilities.

Also, the continuous comings and goings of public officials dealing with migration management clogs the lines of communication between Lampedusa and Sicily. The airline that guarantees a continuous connection is often partially occupied by police agents. Another irony: it is often easier for an illegal immigrant to leave Lampedusa via military aircraft than for a resident of the island, possibly with an emergency, via civilian aircraft.

Lampedusa has also become a symbol in the national political debates on immigration, a situation not without practical consequences and difficulties. Recently, there was recently a public demonstration with thousands of people protesting against the detention centres. Regardless of their merits, such events provoke quite a disruption on our small community.

Finally, the daily and obsessive unwelcome publicity which presents Lampedusa as the island of immigrants causes great damage also to our tourism prospects.

For a New European Immigration Policy: Local Crises

The management of migration involves (or should involve) not only macroeconomic issues, such as the control of migration flows, the agreements with their countries of origin, the coordination of organized crime prevention and repression, or the improvement of the conditions in which immigrants are first received. It should also take into account specific situations, such as those in which the global conditions impact on the most vulnerable local realities which can hardly sustain their impact.

Not much is called for. Maybe a small task force could be set up, with adequate resources, entrusted with monitoring such specific situations and provide the particular assistance needed. It is a simple matter of justice.

Our small community has always responded with civility, in spite of all the discomfort it has had to bear: not a single racist reaction, not a public expression of protest was uttered; instead, great human solidarity was shown towards the people we were called upon to assist. European institutions should support us.

Bruno Siragusa is the Mayor of Lampedusa.

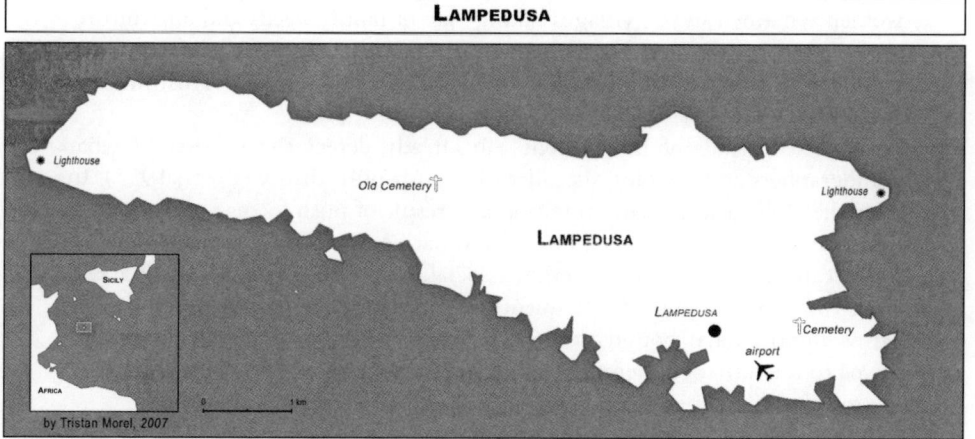

From an Emigrant to an 'Immigrant' Country: A Malta Case Study

JOE INGUANEZ

Introduction

Migration has existed since time immemorial and it has always had geopolitical associations. In the Scriptures, the Abrahamic tribe was essentially a nomadic tribe. What the Italians called *'transumanza'* (sheep migrations), the Spanish *'vaquero'* (migration cattle and horses) and the American 'cowboy' (cattle and horse herders) are also very old forms of nomadic and migratory movements. Similar sheep migrations were also common in Spain and the Balkans.

Very often these seasonal movements, which often lasted for months on end, took the form of a 'to and fro' migratory pattern in search of, literally, fresh and green pastures. These grazing sites gave rise to other economic activities, such as open-air markets where cheese and dried meat were produced and sold, as well as sheepskins and wool. This also had an effect on the social structure since these activities were carried out by men while women remained in the villages taking care of family needs and agriculture. Quite often it had harmful effects on entire economies: the land was cleared for sheep grazing to the detriment of agriculture and forestry. This, in turn, had an impact on the lives of the shepherds.

In these periods of history, one can already detect the validity of what demographers and sociologists called the 'push-pull' theory (Rossi, 1955). In a nutshell, this theory sees migration as a result of push forces at the place of origin and pull forces at the place of destination. Both the push and the pull forces can be of an economic, ethnic, social, or political nature. Having said that, the term 'migration', frequently defined as 'the movement of people across international boundaries' (Latin: *migratio*, removal, change of home) seems to me rather euphemistic. The term 'uprooting' is in my opinion more appropriate than 'migration'.

This 'movement' of people also took place during the Industrial Revolution when whole families moved from rural areas to cities in search of better economic conditions. Later this was followed by shifts of populations between the Old Continent and the New World and Australasia. This has been very well illustrated in William Thomas and Florian Znaniecki's classic *The Polish Peasant in Europe and America* (1996). It is this movement which gave rise to the phenomenon of urbanization and the devastation of the countryside aptly described by the novels of Thomas Hardy and in Wordsworth's most famous poems. This has also taken place in Sub-Saharan Africa. The predominant cause has been economic.

Ravenstein, writing at the end of the nineteenth century, sounds, *mutatis mutandis*, refreshingly relevant to what is taking place at the beginning of the third millenium:

> Bad or oppressive laws, heavy taxation, an unattractive climate, uncongenial social surroundings, and even compulsion (slave trade, transportation), all have produced and are still producing currents of migration, but none of these currents can be compare in volume with that which arises from the desire inherent in most men to 'better' themselves in material respects. [Ravenstein, 1889: 286]

Malta in Time and Space

Malta is a micro-island sovereign state (a territory of 316 sq. km), now a member of the European Union; its history is punctuated with migratory – very often, emigratory – movements. In my view, Malta's history can be properly understood only against the backdrop of these migratory movements.

Malta attained its independence in 1964. Ten years later, Malta changed its Constitution and became a republic and, in May 2004, a member of the European Union. Since about the mid-1980s, the island has transformed itself into a freight transhipment point, a financial centre and a tourist destination.

Geographically, Malta lies in the centre of the Mediterranean Sea at the southernmost tip of Europe, about 93 km south of Sicily. It has no mineral resources, lakes, rivers, or mountains. When in the sixteenth century Charles V of Spain wanted to give Malta in fiefdom to the Knights of Malta, they quite intelligently sent an exploratory mission to see what was being offered to them. As a result of this mission's report, the knights wanted to decline the Emperor's offer, on the grounds that Malta has a meagre supply of water, poor agriculture and no defence – problems which have persisted for centuries. However, since beggars can never be choosers, and since they were offered no

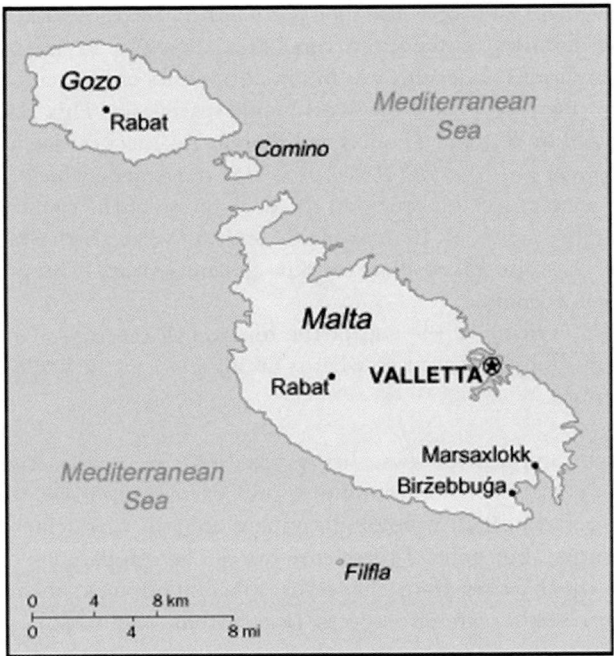

Map of Malta

other alternative, they ended up accepting the offer. Malta's only natural assets were always its strategic geographical position, its natural harbours and the temperate weather.

According to the 2005 population censuses, Malta's population was around 400,000 inhabitants, with an estimated growth rate of 0.42 per cent, making Malta one of the most densely populated countries in the world. Net migration stands at 2.05 migrant(s)/1,000 population.

Politically, it is a democracy administered by a central government, with local councils having few delegated administrative competences. The population is well educated and the literacy rate stands at 93 per cent. Both Maltese and English are official languages, according to the Constitution.

Given that limestone is the only natural resource, and the fact the physical size of the island puts land at a premium, one of Malta's main industries is the construction industry. Though agriculture has made great advances in recent years, especially since Malta joined the EU, it remains a small industry (representing 3 per cent of GDP). In fact, only 20 per cent of the country's food needs are locally produced. The economy is heavily dependent on foreign trade,[1] manufacturing (especially electronics and textiles) and tourism. The latter is, however, going through a sluggish period. GDP at purchasing power parity rate stands at $7,861 billion, with a growth rate of 1 per cent. To date, the highest productive sectors are the services (74 per cent) followed by industry (23 per cent). The labour force stands at about 160,000, with unemployment at 7.8 per cent. However, there is practically no one living below the poverty line.

Given that they live on a small island, the Maltese have developed industries related to the sea. Hence, entrepot trade (goods shipped from one country to another via a third country for further processing or assembly) has always been a major economic activity in Malta, and the Maltese are very well skilled in the practice. Their attitude to the economy is more mercantile than industrial. Mercantile activities even preceded the arrival of the fleet of the Knights of St. John. Luttrell (1975: 53) maintains that, in the fifteenth century, the population of Malta amounted to about ten thousand and he believed that since the island was not self-sufficient it became dependent on piracy and cotton exports to pay for its food imports. In the first century after the arrival of the Knights of St. John in 1530, employment on corsair ships amounted to about a thousand, that is 10 per cent of the population. The peak of this activity was reached in the seventeenth century with the number of those employed in corsairing amount to about four thousand men (Fontenay, 1986: 116ff; Vassallo, 1977: 2).

This points to the fact that, while herdsmen moved in search of pastures for their cattle, the Maltese were in a worse position because they had to cross the seas to other countries to secure their own and their families' 'pastures' and survival, leaving their women at home in Malta! Trade meant not only movement of merchandise but also of men – if not as slaves, then as workers. This was the destiny of most central European Mediterranean countries but, even more so, of Mediterranean islands. Braudel (1982: 78) has noted that emigration was the most common way in which Mediterranean islands entered the life of the outside world. By the end of the eighteenth century, 15 per cent of the male Maltese population was living in a diaspora, trading in Spain, Portugal, Sicily or serving in foreign navies (Vassallo, 1997: 252). During the nineteenth century, the Maltese considered the Mediterranean to be their divinely destined second home. Hence one finds various Maltese settlements in the southern Mediterranean littoral, in such countries as Morocco, Algeria, Tunisia and Egypt, and, in the northern littoral, in Spain, especially Andalusia, Marseilles, the Lipari Islands, Greece, Corfu, Cyrenaica, Cyprus, Constantinople and Smyrna.

However, Maltese emigration reached its peak in the twentieth century, when the country saw a veritable exodus. As in earlier centuries, the main motive for Maltese emigration was the 'push' factor, resulting from a distorted and underdeveloped Maltese economy. It was not political persecution, as in the case of the Berber incursion in search of slaves, of which, in fairness, it must be said that the Maltese reciprocated in kind. Still less was it religious persecution. Instead, Malta was for imperial Britain not only a Chatham Dockyard in the Mediterranean but a strategic post in between two other strategically important British colonies, Gibraltar and Cyprus. No wonder Churchill referred to Malta as the 'unsinkable aircraft carrier'. That is why, ironically, Malta prospered during and immediately after wars: the war machine which has always provided work for the Maltese.

Hence Maltese migration in the twentieth century could easily be explained in terms of dependency theory: the exploitation of its geographically strategic position in terms of Britain's imperial needs in the Mediterranean and, after the opening of the Suez Canal, outside it also. As a matter of fact, Britain never encouraged the development of any industry on the island apart from HM Imperial Dockyard which served Her Majesty's Admiralty and Air Force requirements. This was taking place in a country with the oldest university in the Commonwealth outside Britain and a highly technically skilled labour force.

When military technology started to change radically, and the British allies, the United States and Nato, acquired a strong foothold in southern Italy and Sicily, the Royal Navy declared the British base in Malta to be technologically obsolete. The British government decided to dismantle the base, in the shortest possible time, and keep it simply as a forwarding base. The result was that tens of thousands of Maltese became unemployed and similar number emigrated to Australia, the UK and Canada, with another quota-restricted number to the US. As a result of this, the population of Maltese origin in Australia alone is larger than the present population of Malta.

Maltese Migration 1946–96

Country	To	From	Net Migration	Return %
Australia	86,787	17,847	68,940	21.56
Canada	19,792	4,798	14,997	24.24
UK	31,489	12,659	18,830	40.20
US	11,601	2,580	9,021	22.24
Other	1,647	907	740	55.07
Total	155,060	39,087	115,973	25.21

Source: Attard (1989).

However, at the end of the last century and the beginning of the present one, emigration is turning from brawn to brains. This time, it is being caused by both push and pull factors, and is already resulting in an initial brain drain. The 'push' factor is clear very much among the medical profession. Specialized medics are seeing that if they remain in Malta they will be underemployed in terms of their high-calibre qualifications and clinical experience, which are obtained mostly in the UK and the US. The 'pull' factor, especially after Malta joined the EU, can be also in other professions such as lawyers, accountants and diplomats, and among bureaucrats of all types, including interpreters and translators.

Does Malta receive any benefits from migration? It is believed that the answer to this query is in the affirmative, though it is very hard to quantify it (Cauchi, 1999). Given the tight-knit character of the Maltese family, the more obvious economic benefits of migration are the remittances which these migrants sent to their families, their parish churches and local clubs. Besides this, there is the financial benefit resulting from migrants returning to the mother country for a holiday. Professor Cauchi estimated these holiday-makers to amount to no less than 50,000 tourist weeks per year: 'The wealth in terms of cash as well as expertise that ex-migrants have brought with them is another unknown quantity which one would have thought was of benefit to the old country' (Cauchi, 1999).

For many countries, the cost of emigration is usually in terms of a brain-drain, rather than as a 'brawn-drain', which was more the case with Maltese migrants in the past. The cost of migration to Malta can also be measured more in terms of human suffering caused by the trauma of separation, and the resulting psychological impact, than in economic terms, which, as mentioned above, were positive rather than negative.

In the Way of Others

What about 'immigration' to Malta? Malta is not a country of immigration: that why I put this word between quotes. However, under this heading I am including returned migrants, refugees and third-country nationals entering the country without regular documentation.

Since the 1980s, when Malta started to develop, hundreds of Maltese emigrants returned from their adoptive country to their mother country and found work in Malta. Others simply retired in Malta without taking a job. Others came fleeing persecution, hunger, or unemployment in African countries. Malta has become a *de facto* destination for undocumented migrants. In the last decade of the last century, 2,822 refugees arrived in Malta from Iraq and the former Yugoslavia and 1,968 of them were resettled in Canada, the US and Australia. In this century, most asylum seekers make their way by boat from Libya but do not want to remain in Malta. They arrive in Malta from various nations, mostly from Somalia, Eritrea, Egypt and Sudan. However, Malta was not the intended destination of these migrants (Thomson, 2006). Malta was only in their way as they headed to an Italian territory, whether Lampedusa, Sicily, or Messina. It was only rough seas or wrong bearings that brought these people to Maltese shores.

According to the Office of the Refugee Commissioner, between January 2002 and October 2005 Malta has granted refugee status to 155 persons, humanitarian status to 1,422 persons, while 223 applications were pending. During the same period 1,126 applications were rejected and 54 withdrawn. This increase of refugee population stands out in comparison with the general tendency in the EU. To a lesser extent, the same thing has happened in Cyprus.

This influx has proved to be a huge burden on Malta's financial resources. Besides being fed, these migrants are offered education for their children and free medical services. Some have raised the problem of security and public order. In my view, this position is inflated and ideological. However, in an attempt to show future third-country nationals that Malta is not the ideal place to go to, undocumented asylum seekers were kept in closed detention for eighteen months. Now Malta has regularized its position with the EU regulations, and these migrants are released after one year in detention.

The Empire Strikes Back!

Both emigration from and immigration to Malta is more the result of push rather than pull factors. In the main, this is the result of underdevelopment, very often caused by foreign powers, transnational corporations and corrupt national governments. Underdeveloped Malta sent its migrants to whichever country would receive them. Now the underdeveloped countries which 'belonged' to former European empires or nations are forced to flee – by economic underdevelopment, political persecution and internation or tribal conflicts – to these former colonies, now members of the G8.[2] It is a question of survival. In this geopolitical situation, Malta is only an accidental stepping-stone. However, the price that it is being made to pay is much higher than its economy can afford.

In my view, the solution should be sought in a new economic world order. Trade must substitute aid, which very often has only fuelled corruption in Third World countries and/or re-siphoned back aid money to the country of origin.

References

Attard, L.E. (1989), *The Great Exodus* (Marsa, Malta: P.E.G. Ltd).

Braudel, F. (1982), *Civilisation and Capitalism. 15th–18th Century, Vol I, The Wheels of Commerce* (London: Collins).

Cauchi, M.N. (1999), *The Maltese Migrant Experience* (Gozo, Malta: Gozo Press).

CIA (2006), *The World Factbook – Malta* (Washington, DC).

Dench, G. (1975), *Maltese in London, A Case-Study in the Erosion of Ethnic Consciousness* (London: RKP).

Fontenay, M. (1896), 'Los Fenomenos Corseiros en la "Periferizacion" del Mediterraneo en el Siglo VII', in *Disigualidad y Dependencia. Areas* (Murcia).

Luttrell, A. (ed.) (1975), *Medieval Malta, Studies On Malta before the Knights* (London: The British School in Rome).

Price, C.A. (1954) *Malta and the Maltese: A Study in Nineteenth Century Migration* (Melbourne, Georgian Press).

Ravenstein, G. (1889), 'The Laws of Migration', *Journal of the Royal Statistical Society*, LII: 286.

Rossi, P.H. (1955), *Why Families Move: A Study in the Social Psychology of Urban Residential Mobility* (Glencoe, IL: Free Press).

Thomas, William I, and Znaniecki, Florian (1996), *The Polish Peasant in Europe and America: A Classic Work in Immigration History* (Eli Zaretsky, ed.) (Urbana: University of Illinois Press).

Thomson, M. (2006), 'Migrants on the Edge of Europe. Perspectives from Malta, Cyprus and Slovenia', *Sussex Migration Working Paper*, University of Sussex.

Vassallo, C. (1997), *Corsairing to Commerce. Maltese Merchants in XVIII Century Spain* (Mdisa, Malta: Malta University Publishers).

Dr Joe Inguanez PhD is Executive Director of DISCERN, the Institute for Research on the Signs of the Times.

NOTES

1. The Maltese have during centuries acquired fine skills in entrepot trade. Cfr. C. Vassallo, *Corsairing to Commerce. Maltese Merchants in XVIII Century Spain* (Msida, Malta: Malta University Publishers, 1997).
2. It is not necessary to elaborate on the neo-colonial practices of several types exercised by these countries.

Immigration in Cyprus: New Phenomenon or Delayed Responsiveness?

ANNA PAPASAVVA

In 2005, according to the Cyprus Statistical Service, Cyprus had the highest population growth in the European Union: 2.6 per cent.[1] This, however, is not due to an increase in the birth rate – the fertility index in Cyprus is not more than 1.4 – but to increased immigration into the country. Estimates of the numbers of current immigrants vary between 110,000 and 116,000 non-Cypriots, which corresponds to 13–14 per cent of the total population (854,300 in 2005)[2] and to 14 per cent of the active population. Compare this to the figures for 1992, when immigration into Cyprus had barely begun: there were then 25,506 immigrants or 4 per cent of the population.[3] Therefore whereas Cyprus was a country of emigration from the 1960s on, sending emigrants all over the world, these figures show that it has now become a country of immigration, receiving immigrants from Asia, Africa, the Middle East and the former Soviet Union.

The rapid economic growth maintained in Cyprus during the years following the Turkish invasion in 1974 until the beginning of 1990, and its stabilization to a growth rate of 2–4 per cent, has caused the increased demand in human labour. The indigenous labour supply was no longer sufficient to satisfy the demand for unskilled, labour-intensive jobs in sectors crucial for development and economic progress. As a result, the state decided to change the migration policy, to abandon its previously restrictive labour policies, to grant entry to immigrants and to allow them to work for a restricted period (a maximum of six years) in specific sectors. Foreign workers with limited period residence permits therefore consist the first group of immigrants in Cyprus.

At the same time, the collapse of the Soviet Union has caused the massive influx of citizens and their families from countries of the former USSR, mainly from regions around the Black Sea, who were considered by the Greek government to have a Greek origin and therefore to have the right to a Greek passport. These constitute the second category of immigrants, the Pontians, who come to Cyprus and enjoy the same rights of indefinite residence which are granted to Greek citizens.

Furthermore, international factors, such as the Gulf War and the successive crises in the region, as well as the continuing Israel–Palestine conflict, have caused economic migrants to come to Cyprus, as well as political refugees and asylum seekers.

There is one last and particular category of immigrants: large numbers of women, especially from the former Eastern bloc states, who work as dancers in nightclubs. At the same time, however, many of these women – 'artists' as they are called – are apparently related to the trafficking and even to prostitution for which Cyprus has been often internationally criticized.[4]

Nowadays, immigrants are employed as domestic workers, in the service industry (especially restaurants, hotels and trade), in the construction industry (both in manufacturing and repair) and in agriculture.[5] The overwhelming majority of East Europeans are concentrated in construction, services, agriculture and 'artistic activities'[6]. In these same areas (with the exception of 'artistic activities') there are also immigrants from Syria and Egypt, while workers from Sri Lanka and the Philippines mainly become domestic workers. Moreover, the labour market is largely defined by gender: women are mainly working as domestic workers and cleaners, while men are concentrated in construction industry.

In general, the state's policy from 1990 until today treats the immigrant as a temporary participant in the economic life of the country who will return to their country of origin as soon as their residence permit expires. None the less, the number of immigrants continues to rise rapidly, reaching the levels mentioned above.

Because of the high demand for labour, illegal immigration has also increased. According to the estimates of the Cyprus authorities, most illegal immigrants come from the North of the island, particularly since 2003 when restrictions on the freedom of movement between the two parts of the island were lifted. According to data from the Cyprus police, 5,191 illegal immigrants came to Cyprus in 2005, out of which only sixteen came through the South. Meanwhile, Cyprus has proportionately the highest number of asylum applications in the EU: 9,675 in 2004 and 7,745 in 2005.[7] This is due to the fact that after Cyprus joined the EU in May 2004, Cyprus tightened its visa policy. As a result, a large number of immigrants, mainly from Bangladesh and Pakistan,[8] not necessarily in need of international protection, applied for asylum in order to be able to work, but also in order to be able to benefit from the allowances given by the social welfare services.[9] The government has turned down most of these applications on the basis that they are bogus; it has also taken a hard line towards asylum applicants, forcing them to work in agriculture – since the labour demand in this sector is high – while their cases are being examined. Asylum seekers whose applications are rejected receive no social benefits. Many of these people, who are literally homeless, reside in the Centre for Refugees in Kofinou, set up by the government, where however living conditions are hard.[10] This has driven many asylum seekers to acts of despair, such as hunger strikes, the threat of suicide, even marriages of convenience, which are given very wide publicity by the media.[11]

This last phenomenon, widespread among immigrants, is of particular interest: there are male immigrants who pay Cypriot women to marry them so that they can obtain a work permit: the women typically receive between 1,500–2,000 Cyprus pounds ($3,400–4,500) for marrying foreigners they do not even know.[12] There are also women immigrants married to Cypriot men, sometimes many years older than them, seeking to acquire both a legal means of staying in Cyprus and obviously also a home.

We would therefore say that, Cyprus, until today, has regarded immigrants as temporary residents and it has not pursued policies of social integration. The Minister of the Interior, Neoklis Sylikiotis, has said, 'The state's priority was to satisfy the demand for labour in several productive sectors of the economy, to manage the different consequences of this influx, and to tighten controls against illegal immigration.'[13] Cyprus considered that it had no immigrants but instead only foreign workers: as a result, no systematic and integrated migration policy has been developed which aims either at integrating the immigrant into society or encouraging them to participate in it. The General Director of the Ministry of the Interior, Lazaros Savvides, has said, 'The reason is that no one foresaw the situation we face today. As a result, all the measures taken were short term and all the medium and long-term estimates turned out to be wrong.'[14]

It seems that having decided to take in foreign workers, the state not only had not predicted how many of them would come but also showed that it was not ready to accept, assimilate, or integrate them into society. This policy can only be described as short-sighted and opportunistic. It is also significant that immigration matters are handled by four different ministries and a number of other services with little coordination between them. The result is confusion and ineffectiveness in handling these questions. Furthermore, people in key positions, with no special knowledge and no appropriate training in issues related to human rights, take arbitrary decisions which may seem to be based on xenophobia. An example of this was the refusal by the Immigration Office of the Police to extend the resident permits of a group of Chinese students because they decided that their qualifications were not equivalent to their studies.[15]

Furthermore, the obsolete legislation, which was bequeathed to the

Republic of Cyprus by the colonial regime, is also a factor which leads to much arbitrariness, since it is inadequate to respond to the new demands. The anachronistic legal framework does not allow the authorities to deal with special cases or to use humane criteria. In consequence, immigrants living for years in Cyprus are deported without taking into consideration the consequent separation of families, foreigners are detained in prison *sine die*, or are even arrested as suspects for terrorist attacks.[16] One particularly egregious example was that of a Serbian mother of two children, aged 16 and 17, who applied for an extension to her residence permit so that her children could finish school. Her demand was rejected and she and her two children were deported.[17]

We can say, therefore, that the absence of a carefully planned immigration policy, which would regulate the entry, stay and employment terms of immigrants and provide adequate reception structures and mechanisms for integration or assimilation, has resulted in the failure of immigrants to integrate and the failure of Cypriots to accept immigrants as members of society.

It is no accident, we believe, that these people, being literally marginalized by Cypriot society, know very little of the culture and – more significantly – of the language of the country in which they live. Even people who have lived in Cyprus for more than five years can hardly communicate in Greek, choosing instead to use English. The immigrants, for their part, say that they have not learned Greek because they did not consider it necessary. English is sufficient for them to live and survive. As far as communication and contact with the Greek Cypriots are concerned, these do not seem to go beyond the professional framework. Having thus created their own different framework within society, they do not seem to aspire to anything more than that. The state, for its part, does not oblige these people to learn Greek and nor does it provide them with systematic language lessons.

In view of this isolation and marginalization of immigrants, it is fair to conclude that neither citizens nor the state were prepared to accept such a large number of immigrants. In a society, which since 1974 and the violent division of the island by the Turkish Army, is relatively homogenous, foreign elements become obvious and frightening. Cypriots are sceptical that immigrants of a different culture can integrate, and they consider immigrants to be a threat to their own relatively homogenous society. They seem to prefer, I believe, to isolate immigrants or to show a certain indifference towards them, instead of trying to incorporate them into society.

We can therefore conclude that immigration in Cyprus is not considered a social fact which has to be faced, but instead a social problem, bringing many troubles into today's society. Foreigners quickly become a scapegoat for contemporary phenomena such as unemployment, criminality, the increase in divorces and family breakdown, and even demographic changes.

The media have played an important role in this by publicizing figures without analysing them, in order to create a stir, and painting a picture of Cypriot society as generally xenophobic.

It can even be claimed that the state's policies have sustained and even encouraged the xenophobia of Greek Cypriots. Its incapacity to realize in time this phenomenon and to develop a systematic integration policy has left free space for impulsive or even sometimes arbitrary deliberations from the several organs of the society. This becomes even more obvious by the way that the whole legal and administrative system works. There is a certain institutional racism reflected in the slogan, 'We use immigrants as and when we need them.'

Hence, it goes without saying that Cyprus, as an EU member state and a modern and democratic country, as it would like to be considered, needs a more integrated and long-term immigration policy with a more humane approach. This must be based on an acceptance of the immigrant as an integral part of contemporary society. It is at least encouraging that there have been considerable debate on this recently, and the authorities seem to be starting to take the issue seriously. The EU's Directives and international conventions, already signed by Cyprus (the Directive on long-term residence, the various conventions against racism, discrimination and human rights violations) are adopted and need to be implemented. The recent provision of language lessons for foreigners and the intensive language courses being introduced in schools for non-Cypriot students, are some of the signs that a new policy is being put into place. Let us hope that this leads to a change in the perception of immigration in Cyprus.

Reports and Articles

European Commission against Racism and Intolerance, Second Report on Cyprus, Adopted on 15 December 2000, Council of Europe.

ENAR (European Network Against Racism), 'Η αντιμετώπιση του του Ρατσισμού στην Κύπρου' [Responding to Racism in Cyprus] (Cyprus: 2006).

KISA, 'Asylum law, policies and practices in Cyprus – An overview based on asylum cases', June 2005.

KISA, Press Release, 'Migrant's charges of police corruption', 6 September 2006.

KISA, 'Παρανομούν Εξωθώντας Γυναίκες και Ανηλίκους στην Εξαθλίωση' [They are breaking the law by driving women and children in despair], 17 January 2007.

Ministry of the Interior, 'BCs: Issue and renewal of temporary residence and employment permits for third country and nationals', Nicosia, 17 December 2004.

Ministry of Labour and Social Insurance of the Republic of Cyprus, Department of Labour, Κριτήρια και Διαδικασία για την Παραχώρηση Αδειών Εργασίας σε Αλλοδαπούς / Αμοιβή και Όροι Εργοδότησης [Criteria and Procedure for Work Permits to Foreigners/Wage and Employment terms], 2 December 1991.

Ministry of Labour and Social Insurance of the Republic of Cyprus, 2005 Annual Report, Chapter 5, Department of Labour

Polykarpou, Doros, ENAR Shadow Report 2005, *Racism in Cyprus* (Cyprus: 2005).

Report on the Republic of Cyprus against Discrimination in the Fields of the European Union Acquis, Policy and Measures Against Discrimination, Nicosia, June 2003.

Report of the Republic of Cyprus on the Implementation of the Conclusions of the European and World Conferences Against Racism, Nicosia, Cyprus, June, 2003.

Στατική Υπηρεσία Κύπρου, Δημογραφία 1990–2005 [Statistical Service of the Republic of Cyprus, Demography, 1990–2005].

Statistical Service of the Republic of Cyprus, Population Projections 2002–52, 21 January 2004.

Στατική Υπηρεσία Κύπρου, Έρευνα Εργατικού Δυναμικού, 2005–06 [Statistical Service of the Republic of Cyprus, Research on Human Labor, 2005–006].

Statistical Service of the Republic of Cyprus, 'Cyprus in the EU scale', 2006.

Στατική Υπηρεσία Κύπρου, Πορτραίτο της Ε.Ε. – στατιστική «πρωτιά» κάθε χώρας, Ιανουάριος 2006 [Statistical Service of the Republic of Cyprus, 'The EU portrait', January 2006].

Statistical Service of the Republic of Cyprus, Populations Summary Data, 3 August 2006.

Στατική Υπηρεσία Κύπρου, Εγγεγραμμένοι Άνεργοι, 2007 [Statistical Service of the Republic of Cyprus, Registered unemployed, 2007].

Τμήμα Ερευνών Ελληνική Τράπεζα Λτδ, Κεφ.5 ΔΗΜΟΓΡΑΦΙΚΑ [Hellenic Bank Research Department Ltd, *Annual Review 2006*, Chapter 5 'Demography' (Nicosia: 2006)].

Trimikliniotis, Nicos, Report on Measures to Combat Discrimination in the thirteen candidate countries (VT/2002/47), Country Report Cyprus, Second Draft, May 2003.

Trimikliniotis, Nicos, 'Guide to Cyprus Legislation on Anti-Discrimination' (Nicosia: INEK, June 2005).

Trimikliniotis, Nicos and Corina Demetriou, 'Active Civic Participation of Immigrants in Cyprus', , Country Report prepared for the European research project POLITIS, Oldenburg, 2005 <www.uni-oldenburg.de/politis-europe>.

Έκθεση των ΗΠΑ αξιολογεί τις προσπάθειες στον αγώνα κατά της εμπορίας ανθρώπων [The USA Report , Published on SETimes <http://www.setimes.com>, 7 June 2006.

UNHCR, Background Note on the Protection of Asylum Seekers and Refugees in Cyprus, June 2003.

UNHCR, 2006 Country Operations Plan for Cyprus.

UNHCR Statistical Yearbook 2004, Cyprus

Anna Papasavva is Project Director of the Daedalos Institute of Geopolitics in Cyprus.

NOTES

1. 'Cyprus in the EU scale', Statistical Service of the Republic of Cyprus, 2006.
2. The figures refer to the residents of regions who are under the control of the Republic of Cyprus, with the estimated number of Turkish Cypriots.
3. Τμήμα Ερευνών Ελληνική Τράπεζα Λτδ 'Κεφ.5 ΔΗΜΟΓΡΑΦΙΚΑ', [Hellenic Bank Research Department Ltd, *Annual Review 2006*, 'Chaper 5 Demography' (Nicosia: 2006)].
4. The US government in a recent report has included Cyprus in the list of countries under monitoring, because it 'has not provided enough evidence of its effort in solving the crucial questions of human trafficking for sexual exploitation.' (*Trafficking in Persons Report*, June 2006 <http://www.setimes.com/cocoon/setimes/xhtml/el/features/setimes/features/2006/06/07/feature-01>
5. See Annual Report of the Labour Department for the year 2005.
6. We use this term to refer to the sector where the last category of immigrants we described is mainly employed.
7. Jacqueline Theodoulou, 'Illegal immigrants have their rights too', *Cyprus Mail*, 28 February 2006.
8. See 2004 UNHCR Statistical Yearbook <http://www.unhcr.org/cgi-bin/texis/vtx/country?iso=cyp>.
9. Emilia Strovolidou, 'Cyprus: The twisted reality behind the statistics', UNHCR News Stories, in Nicosia and Rupert Colville in Geneva, 13 January 2005 <http://www.unhcr.ch/cgibin/texis/vtx/country?iso=cyp>.
10. Fwtini Panayi, 'Βοηθούν πολιτικούς πρόσφυγες' [They offer help to political refugees], ΣΗΜΕΡΙΝΗ [*Simerini*], 12 October 2005.
11. Stefanos Evripidou, 'Asylum seekers choosing death', *Cyprus Mail,* 24 July 2005.
12. Costas Savva, 'Γαμπροί και νύφες έναντι £4.000' [Grooms and brides for the amount of 4000 CP], ΣΗΜΕΡΙΝΗ [*Simerini*], 16 April 2006.
13. Neoklis Sylikiotis, 'Μετανάστευση και Μεταναστευτική πολιτική' [Migration and migration policy], ΠΟΛΙΤΗΣ [*Politis*], 20/11/2006.

14 Marios Dimitriou, 'Η ένταξη μεταναστών με μακροχρόνια διαμονή' [The integration of long-stay immigrants], ΣΗΜΕΡΙΝΗ [Simerini], 3 December 2006.
15 Gogo Alexandrinou, '"Πόρτα" σε Κινέζους φοιτητές' ['Door' to Chinese students], ΠΟΛΙΤΗΣ [Politis], 13 February 2007.
16 Yiorgos Michailides, 'Σοβαρές καταγγελίες από αιτητές πολιτικού ασύλου' [Heavy accusations by asylum seekers], ΣΗΜΕΡΙΝΗ [Simerini], 12 November 2003.
17 Militsa Polemitou, 'Ελευθερώστε τη Γιασμίν' [Freedom to Jasmin], ΣΗΜΕΡΙΝΗ [Simerini], 24 March 2006.

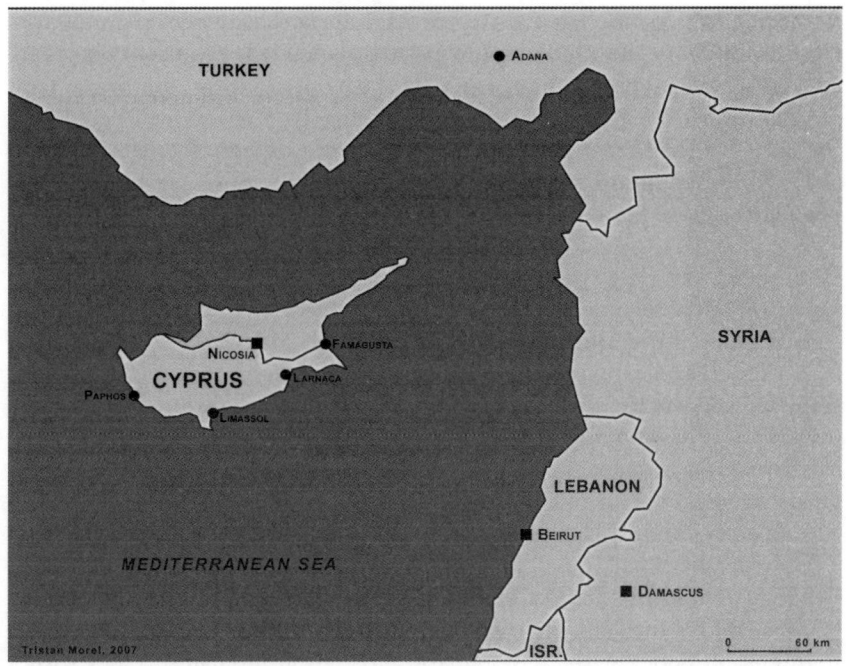

Interview with Neoklis Sylikiotis, Minister of the Interior of the Republic of Cyprus

People say there are between 80,000 and 100,000 non-Cypriots in Cyprus. What is your opinion about the number of the immigrants nowadays in Cyprus and what do they represent as a percentage of the population?
According to statistical data, in Cyprus there are at the moment about 60,000 legal immigrants, 40,000 European citizens, while it is estimated that there are about 50,000 illegal immigrants. This is equivalent to 20% of the total population, with legal immigrants representing 10% of the total population and 14% of the active population. Compared to those of other European countries, where immigrants generally represent some 6–7 per cent of their population, these figures are pretty high.

One could say that the state has suddenly found itself faced with a situation that it cannot control. Why have we not realized the significance of this question, at least not until today?
It is a fact that Cyprus has suddenly gone from being a country which sent emigrants all over the world to a country which now receives immigrants. The peculiarity of Cyprus in comparison to other Mediterranean countries is the fact that it is situated in a politically and economically unstable region. Furthermore, as a modern and peaceful country, despite the political problem, and a member of the EU, Cyprus is a transit point to other European countries.

However, what had caused the current problematic situation is basically the fact that we thought that the foreign workers who were coming to Cyprus to work would fill the gaps in the labour market and then leave. This reflects an immigration policy without systematic integration mechanisms and with a rather short-sighted approach.

This is why we have already started planning a new immigration policy, with united integration programs, a complete application of EU Directives, places for receiving immigrants and asylum seekers, and an improvement in the conditions of detention for illegal immigrants. This is of course for me, as Minister of the Interior, one of my main goals and priorities.

It is well known that on the European level there is a particular mobility towards this question. How and to what extent does this influence the measures that are taken in Cyprus?

At the European level there are different approaches, as there is also diversity of the needs and the possibilities of the several EU countries. That is why no united immigration policy has been yet planned. It is, however, generally accepted that we should look at the causes which oblige people to migrate in search of a better life: for example, the crucial problems of the poverty, unemployment, illiteracy and health care in the migrants' countries of origin. The EU should hence offer financial support to these countries, in order to help them improve their infrastructure. As for the EU Directives, which are aimed at creating a more uniform migration policy, Cyprus has the obligation to adopt them in its legislation and to implement them.

Several ministries and many departments and services are dealing with immigration. In effect this creates a general confusion and ineffectiveness. How should this situation be dealt with?
As a matter of fact, we intend to put in place better coordination, as well as a more humane approach, which would be less adherent to the legal aspect of the question than to the political. We should insist, as a democratic country, on respect for the human rights of immigrants, either they are legal or illegal. The question of the deportation of the illegal immigrants who have families in Cyprus, is also of a particular significance. As a country which has signed the International Conventions and the European Directives, we are obliged to respect them.

There is also a discussion concerning the incompetence of several officials, who, while being in crucial positions, do not have the appropriate knowledge and education. As a result, many arbitrary decisions are taken, such as the deportation of persons living for years in Cyprus, the segregation of families, and so on. What measures could be possibly taken in order to avoid this situation?
The lack of knowledge is evident more in the area of political asylum than in migration. When it has to do with asylum, we must investigate whether the asylum seeker is indeed in need of protection or if he is just using the asylum system to extend his stay in Cyprus. These are the cases where a special knowledge is needed. For this reason, seminars are organized either with UNHCR or the European Union.

As for the immigration issues, the problem lies mainly in how the legislation is interpreted. We often notice that officials use their discretion not in favour but to the disadvantage of the immigrant. An example is the EU's Directive on the status of third-country nationals who are long-term residents, which was implemented in Cyprus in January 2007 after a year's delay, into which the Directive on family reunification has been also incorporated. Apart from the main provisions, there are parts of these directive which leave room for different interpretations. As a result, even though the decision had been taken to extend residence permits until the full implementation of the Directive, many officials decided otherwise. This reflects in reality a sterile legal approach.

The way someone uses their power, greater or smaller, indicates how progressive or conservative they are. That is why we say that a more humane approach is needed for the handling of these questions, which however can not be imposed by the law. Significant in this aspect is also the attitude of the politicians and the media, as the xenophobic trends sometimes have their influence in society.

If therefore we, the politicians, have a more humane attitude towards these questions, the officials could be also affected by this.

Another reason for the inefficient conduct of these questions is obsolete legislation, which was bequeathed to the Republic of Cyprus by the colonial regime. How do we deal with this?
The law for Foreigners and Immigration has been modernized with some modifications. The law for Political Refugees is about to be voted on, and a bill for trafficking will be sent to Parliament.

What are the measures taken in order to control illegal immigration?
Cyprus has dealt with illegal immigration and has almost solved the problem. We have repatriation agreements with a number of countries.

The problem, then, is within the 'internal borders' of the country. Illegal passage from the northern to the southern part of the island, especially since 2003, is a fact. How do we deal with this question? In your speech at the opening ceremony of the Daedalos Institute of Geopolitics conference on migration in December 2006, you mentioned that cooperation with the 'authorities' in the North is needed in order to deal with this problem. How could this happen?
I mentioned this not only at your conference, but also in the European Council of the Ministers of the Interior and Justice. In order to deal with this problem, it is necessary to cooperate with Turkey and to coordinate with the EU. This point was also made by the Italian Minister of the Interior, Giuliano Amato, in the Council of the Ministers of the Interior and Justice in Dresden, in Germany. The Turkish Minister of the Interior has reacted rather evasively, saying that we cannot talk about a political question in a technical discussion.

As for the occupied areas, something peculiar is happening. While on the one hand, the EU continuously encourages the opening of crossing-points between the North and the South, the de-mining of several parts of the Green Line, and so on, on the other hand, it also blames us for an insufficient surveillance of the Green Line.

The case of the British bases lying between the occupied areas and the South is also significant. They [are] in reality a hole for illegal immigration, despite our efforts to coordinate with the British bases.

The question of the asylum seekers is also of a great importance. There is a long-term delay between application for asylum and the final decision. This actually encourages illegal immigrants, especially those who cross from the occupied areas, to apply for asylum even if they do not really need political protection, as is true in the majority of cases. In order to facilitate the procedure and to deal with this problem, we have recently decided that officials of the Asylum Department should conduct interviews immediately and *in situ*, that is, at roadblocks.

Another crucial and important issue for Cyprus is human trafficking. The recent report of the US Congress into human trafficking, and particularly prostitution, puts Cyprus in the unflattering category of countries under surveillance. However, while the authorities are aware of the situation with cabarets and artistic visas, we perceive certain inertia of the state towards this question. Why is that happening?
I would not call it inertia. We acknowledge the gravity of the question and we are taking several measures in order to control the problem.

Within the framework of the Law on Human Trafficking, as well as the relevant EU Directives on the matter, we are currently examining the legal framework in order to establish stricter criteria. At the political level, measures are being taken to tighten up our policy on artistic visas. I would like to specify, as an answer also to the criticism we receive for issuing artistic visas for not solely artistic activities, that these visas are not only given to women coming to Cyprus to work in nightclubs. They are also given to music bands or other artistic groups who come to Cyprus for specific purposes and for a specific period of time. What we can do, however, is to control the place where a person with an artistic visa will work.

Furthermore, I have to underline that today we are putting particular emphasis on informing possible victims, through our embassies in certain countries.

In addition to giving the public a broad range of information on this question – through relevant material, the media and other sources – and the stricter and more systematic control of nightclubs by the police, the creation of a refuge centre with specially trained staff are some of the measures to be taken to deal with this problem.

It is worth mentioning also that we will be gradually applying the policy of communal priority, according to which the residents of EU countries are given a priority for these permits.

We believe that trafficking is another question that has to be dealt with globally, like immigration, not only from the point of view of the receiving countries but also from that of those who prepare and offer these persons for this kind of jobs.

It seems that not only the state, but also society was not prepared to accept such an large number of foreigners. It is also significant that, since 1974, Cypriot society has been in general terms relatively homogenous. Do you believe that the absence of a well-planned immigration policy has contributed to some extend to the phenomena of xenophobia and racism among Greek Cypriots?
We must admit that there is xenophobia in Cyprus and that we are not ready to accept the different and particularly, the immigrant. This xenophobia is unfortunately being reproduced in the speeches and attitude of some politicians, as well as some parts of the media. We should not, for example, exaggerate the facts and arbitrarily accuse foreigners every time there is a crime. It is of course easier to accuse the foreigner than the local.

We can ascertain that Cypriots, in general, do not easily accept the foreign or the different. This position is also reflected in education, where I particularly had the chance to work in the past. Our education still has a monocultural character. However, since we are now a country of receiving immigrants, which means that there is a large number of immigrants in our schools, we should give emphasis to a more multicultural approach and thus give the opportunity to those children to be integrated within the school community.

It is therefore important to stress the principles of equality, solidarity and cohesion through the integration mechanisms, as well as to emphasize the importance of the knowledge of the history and the culture of the others. At the same time, we should help them come closer to our culture. The Cypriots, as people who have lived abroad (at least most of us), should sense more the necessity of the social integration of immigrants.

That is why society needs to have more information about these people. Phenomena, such as, for example, blaming illegal immigrants for unemployment, are a result of insufficient information available to the citizen and should be confronted. Besides, while contacting foreign organizations and organizations for immigrants' support, I have many times insisted on the fact that this ministry is ready to support them and work towards their integration in the society.

There is a unique phenomenon in Cyprus: the vast majority of foreign workers does not speak Greek and, what's more, they do not seem to want to learn it. Why do you think this is happening? What are the state's or even society's responsibilities? What are the measures to take

in order to deal with this situation?
First, we should underline that there is no legal obligation to know the language in order to obtain a residence permit. For example, during the discussion for the implementation of the EU Directive on long-term residents, there was a suggestion that such a language requirement be introduced but it was not accepted. Thus, while the Directive aims to integrate immigrants, in Cyprus we did not consider it necessary to include a provision for the language, even if knowing the language is a strong element of integration. At this point, maybe we should take into account the fact that in Cyprus you can gain Cypriot citizenship without knowing either of the official languages of the state. It is also significant that in Cyprus there is a wide use of English. However, it is necessary to increase the number of programmes available for learning the Greek language and to make them systematic.

Sometimes people express fears of the danger of a demographic change of the island because of the large number of foreigners with respect to the small size of the island and its population. Are these fears legitimate?
We could say indeed that because Cyprus is a small island, it is possible to face such a problem. However, panic and the exaggerated attitudes do not help to deal with it. They generally take the form of a hysterical reaction towards a question that must be treated more globally and humanely.

Do you think that we could talk about having ghettos within the towns, such as, for example, in Nicosia within the walls, or old Limassol?
The low rents in the centre of towns are a very important factor in attracting immigrants. We could not however talk about having ghettos, at least not at the level of other cities as Berlin, London, Paris, and so on. We should, however, confront these phenomena while it is still early. The responsibility of course of the Ministry of the Interior towards this is really important. What we should do, for example within the walls of Nicosia, is to work towards the development of the region in order to give the motives to the locals to return back to the centres and a mixed population be thus created.

Hence, measures should be taken for the cooperation of the responsible authorities in order to deal with this phenomenon, which reinforce, among others, the xenophobic trends and, what's more, marginalize the foreigners.

Geopolitical Affairs interviewed Minister Sylikiotis on 14 March 2007.

Chinese versus Palestinians?

YOHANAN MANOR

This somewhat off-putting title is trying to express in a nutshell the widespread idea that, at some stage, and for security reasons, Israel stopped importing Palestinian workers and replaced them with other migrant workers from countries as far away as China.

Most of Israel's Arab citizens[1] may consider themselves as belonging to the Palestinian people. But this feeling does not change the economic, social and civil fact that they are part and parcel of Israeli society and of its labour market. In other words, they are not included in the category 'Palestinian workers'.

The Palestinian workers come from the Gaza Strip and the West Bank,[2] two regions that were under the control respectively of Egypt and Jordan until they were taken over by Israel in 1967. They are indicated on the map, 'Israel and Palestine 1947–67'.

Chinese workers were never the majority or the most important group among the foreign workers in Israel. But the use of their name was an easy and striking way to point out the apparent absurdity of importing workers from faraway countries, instead of the Palestinian workers so closely at hand and so apparently well adapted to Israel's needs.

For more than fifty years, since the 1950s, the Palestinians resorted massively to migrant work to make a living. Between 1950 and 1960, more than 450,000 Palestinians left the West Bank, which was under Jordanian rule, for the Gulf States, at a rate of 45,000 a year. Between 1967 and 1989, this rate fell to less than 8,500.

It is generally asserted that, because of the security problems caused by the first intifada (1987–93), Israel decided to stop the influx of Palestinian workers and to replace them with foreigners. However, this assertion is not correct. The latter did not *replace* the former but were rather *added* to them. Together, the two influxes reached such a magnitude that their cumulative effect has begun to have devastating consequences. The Israeli government had no alternative but to address urgently the issue and to adopt a policy to curb both of them.

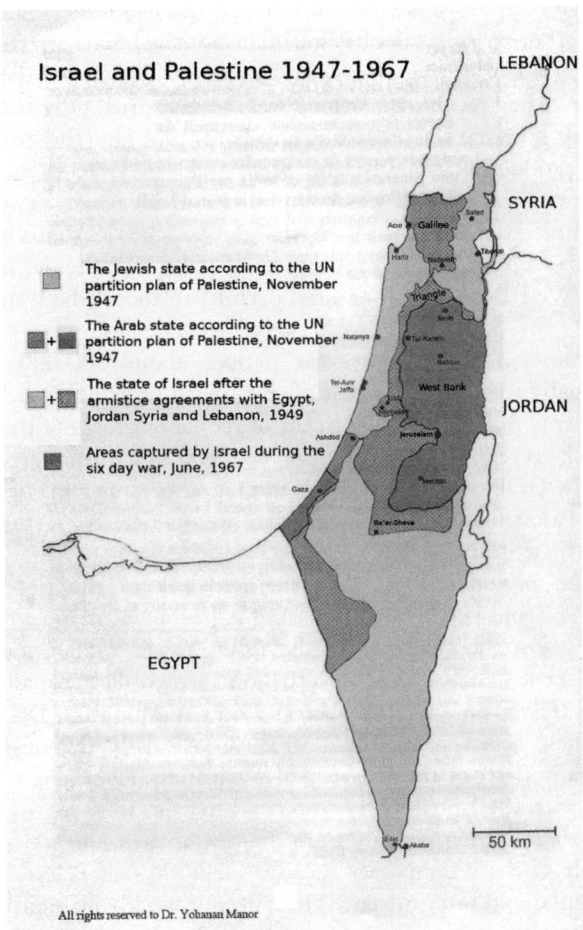

The Flow of Palestinian Workers

The phenomenon of Palestinian workers in Israel started almost immediately after the Six-Day War in June 1967. It stemmed from the disappearance of the demarcation line between Israel and the West Bank, and between Israel and the Gaza Strip.[3] From the moment of its inception to today, this migration has passed through four main stages.

Between 1967 and 1987, the flow of Palestinian workers developed gradually from a sporadic, limited and unregulated phenomenon to a steady, very large and partly regulated one.

In July 1968, the Ministerial Committee in charge of Economic Affairs decided to authorize the work of Palestinians in Israel, in sharp contrast to the conclusions of a committee of experts headed by Professor Michael Bruno.[4] One of their major recommendations was 'to authorize only the free movement of goods between the territories and Israel, but not [the free movement] of factors of production, notably labourers'.[5]

The authorization process was completed and formalized two years later when the Ministerial Committee in charge of Security Affairs gave also its backing. This step led to the opening of official governmental offices in the cities of the West Bank which granted permits to those who wanted to work in Israel.

During most of this twenty-year period, about 200,000 Palestinians worked annually in Israel, only half of them with work permits. This overall figure represents more than 20 per cent of the labour force of the Palestinian territories. The revenues earned by those workers, which amounted close to 30 per cent of the GDP of the West Bank and Gaza, became a major asset for the burgeoning Palestinian economy, and they were a decisive factor behind the dramatic increase in its GDP.[6]

The second period started with the outbreak of the first intifada in December 1987 and ended with the 'Oslo Agreement', namely the exchange of letters between the chairman of the PLO, Yasser Arafat[7] and the Prime Minister of Israel, Itzhak Rabin,[8] and the joint Israeli–Palestinian Declaration of Principles (DOP) on 13 September 1993, in Washington, DC.

During these seven years, there were almost daily demonstrations and strikes in many parts of the Palestinian territories. These often deteriorated into clashes with the Israeli forces who tried to calm them down by isolating and cutting them off temporarily from the rest of the country.

As a result of these temporary measures, the flow of Palestinian workers became irregular and intermittent. Their overall weekly or monthly numbers remained the same, but with huge daily fluctuations. Agriculture and construction, two sectors largely dependent on Palestinian work, were seriously disrupted, and this at a time of an exceptional economic boom due to the massive influx of new immigrants from the former USSR.

Building contractors became both unable to deliver on their undertakings and also extremely worried by the growing number of petty acts of sabotage.[9] They urged the government to authorize the import of foreign workers. Finally, after almost two years of pressure, they were successful. From 1990 onward, there has been a steady and growing flow of foreign workers.

The third period covers the seven years between the beginning of the implementation of the Oslo Agreement in 1994 and the outbreak of the second intifada at the end of September 2000. It started with a stabilization phase marked by a return to a regular daily flow of Palestinian workers

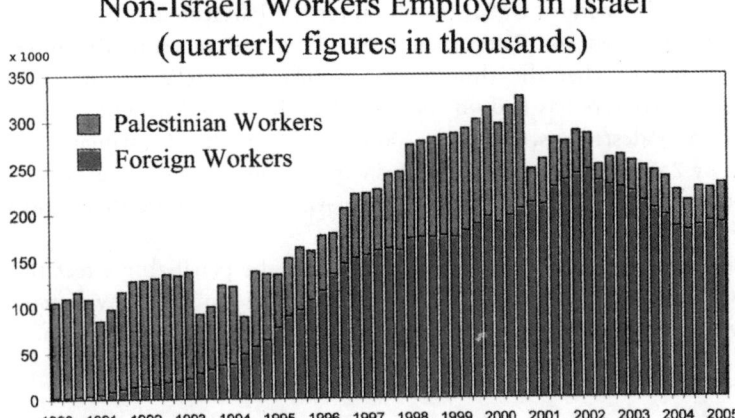

Source: Finance Ministry, Economy and Resreach Division, Economy Information (Hebrew) Jerusalem, June 2005, page 9

between 1994 and 1996, followed by a phase of growth which rapidly reached the highest level ever, namely 120,000 Palestinians workers with permits in 2000.

The last period starts with the second intifada, characterized by the increase of suicide bomb attacks carried out by the whole spectrum of Palestinian organizations against civilian targets in Israel's main cities. It reached its climax at the beginning of March 2002, when there were two or three suicide bomb attacks a day which caused several scores of fatalities and several hundred casualties. The Palestinian terror offensive was stopped thanks to Operation 'Protector Shield' launched by the Israeli forces in the West Bank. As a result of this operation, which lasted several weeks, the number of suicide bomb attacks was reduced to two or three a month.

During these twenty bloody months, the number of Palestinians working in Israel plummeted drastically; for example, in 2001, they numbered only 12,000.[10] Contrary to the past, and in spite of further progress in the neutralization of suicide bomb attacks – notably thanks to the erection of a security barrier built, in general, close to the 1967 demarcation line in the West Bank[11] – their numbers never regained their previous high levels. According to Israeli Central Bureau of Statistics, in 2005 the number of Palestinians working in Israel was 43,300, with only 20,000 holding work permits.[12] In July 2006, the number of declared Palestinian workers in the Israeli economy was 14,700.[13]

According to the World Bank, every 10,000 Palestinians authorized to work in Israel generates US$120 million for the Palestinian economy, that is, about 2.5 per cent of Palestinian GNP. One can measure the huge and dramatic impact that the almost total closure of Israel's labour market to Palestinian workers has had on the rate of both unemployment and poverty among the Palestinians. Between 2000 and 2005, the Palestinian economy shrank by 25 per cent, resulting in an unemployment rate of about 30 per cent and putting half of the Palestinian population in the West Bank and Gaza below the poverty line.

None the less, one should not ignore as well the penalizing effect that the massive work of Palestinians in Israel has had on their own economy, depriving it of its sole relative advantage: low labour costs. As explained by Israeli economic researchers:

> The exposure ... to the high wages paid in Israel has caused a grave distortion in the Palestinian labour market ... The higher salaries in the [Palestinian] territories, by comparison with those in Jordan and Egypt, together with the security problems, have made them less attractive for investments. Hence, the textile industry has shifted from Israel to Jordan and not to the Palestinian territories ... Working in Israel is indeed a kind of addiction which increases the income of individuals and even of the Palestinian Authority, reduces the rate of unemployment in the short term, but strikes a blow at the development process of the Palestinian economy in the middle and long term.[14]

This point of view is also shared by Palestinian policy makers who stress however the conditions needed for reversing this trend:

This phenomenon carried another disadvantage for the Palestinian economy. It created an artificially high level of wages that put the Palestinian employers in a difficult situation of competition over Palestinian workers. Palestinian employers had to compete with Israeli employers, who could afford high wages given Palestinian standards (albeit very low wages by Israeli standards). This reduced incentives for investments in Palestine, and consequently hindered economic growth ... The interests of the Palestinian economy, and the Palestinian welfare more broadly, could require a gradual reduction of the number of Palestinians working in Israel, but only if paralleled by the opening of Palestinian borders with the outside world as necessary to attract investments of the kind that can increase production and exports.[15]

The Flow of Foreign Migrant Workers

One can point to three main stages in the influx of foreign workers in Israel. The first one covers a period of the six years from 1990 to 1996, during which the increase in the influx of foreign workers is quasi-exponential: starting from almost nil in 1990, it jumped to 50,000 in 1994 and almost tripled in two years to 140,000 in 1996 (see the graph entitled 'Non-Israeli Workers Employed in Israel'). During the second period, from 1996 to 2001, there is still a substantial annual increase in the number of foreign workers but far less steeply, passing from 150,000 in 1997 to more than 250,000 in 2001. The third phase, from 2001 to the present, is characterized by a reversal in the pattern of the trend, namely by a slight decrease and then a stabilization in the number of foreign workers: 200,000 in 2002, 189,000 in 2003, 200,000 in 2004 and 178,000 in 2005.

We have noted above how the measures taken by Israel to suppress the first intifada restricted the flow of Palestinian workers, turning it into an intermittent and unreliable one, and how in turn Israeli employers tried to overcome this chronic disruption by urging the government to authorize the import of foreign workers.

No doubt the increasingly irregular and unreliable supply of Palestinian workers explains their partial replacement by foreign workers. But these factors alone cannot explain the staggering increase in the influx of foreign workers. The major reason for this spectacular increase is the 'discovery' by the Israeli entrepreneurs and farmers of the highly lucrative character of importing migrant foreign workers.

In the early 1990s, for an Israeli employer, the cost of a foreign migrant worker was far cheaper than the cost of a Palestinian. The gap between the two was probably not as large as contended by some observers who claimed that 'A Chinese worker costs $10 for 10 hours of work per day, while a Palestinian costs $30 for the same number of hours.'[16] The actual gap was however substantial enough to become the main incentive for importing foreign workers. For example, in 2003 the gross salary for a Palestinian or a foreign worker in the construction industry was NIS 2080 (Israeli New Shekel), but the cost to the employer was respectively NIS 3110.55 for the former and NIS 2098.86 for the latter, that is, a difference of more than 30 per cent.[17]

One should also take into account another subsidiary incentive for the import of foreign workers, that is, the commissions paid by the foreign migrants looking for a work permit: $9,000 in the case of a Chinese worker, $6,000 for a Thai, $5,000 for a Filipino and $2,000 for a Romanian. In theory, these commissions were illegal, but were common practice. They are paid to go-betweens or to recruitment agencies that retrocede part of them to the future employers and even to government, as this has actually been the case with China.[18]

The import of foreign workers has therefore been very lucrative for both the recruiting agencies and the employers. Together they have turned into powerful lobbies putting pressure on the government to ensure that the quota of work permits allocated to them is at least maintained and even increased.

Table 1 Number of work permits attributed by sectors of activity 1996–2002

	1996	1997	1998	1999	2000	2001	2002
Construction	75,000	57,000	43,000	35,000	34,000	45,000	45,000
Agriculture	18,000	18,000	18,000	18,000	18,000	22,000	27,000
Home Nursing*	8,000	12,000	15,000	15,000	22,000	33,000	38,000
Industry	5,000	3,000	4,000	2,000	7,000	2,000	2,000
Total	106,161	90,192	80,632	70,172	81,646	102,886	111,380

* Home nursing refers to care services provided to the elderly and the handicapped.
Source: Israel's Immigration Administration

Table 1 shows clearly that the number of permits grew substantially in 2001 and 2002, whereas during these two years the Israeli economy was in a deep recession, notably as a consequence of the second intifada.

Two-thirds of the foreign workers in the sector of home nursing were Filipino women. In the other sectors, almost all the workers are male. In the agricultural sector, they come from Thailand and Nepal. In the construction and industry sectors, their countries of origin are Romania, the former USSR, China, Turkey, Bulgaria and various countries of South America. In the catering and hotel business, they originate mainly from the Philippines, China, Thailand, Nepal and India.

In 2005, according to the Israel's Central Bureau of Statistics,[19] foreign workers with permits came from more than a hundred countries, but 80 per cent came from five countries, namely, Thailand (29%), the Philippines (23%), the former USSR (12%), Romania (12%) and China (6%). By contrast, those without permits who had entered the country on a tourist visa came mainly from twelve countries: the former USSR (23%), Jordan (14%) Romania (7%), Brazil (5%), Poland (5%), Columbia (5%), Czech Republic (5%), the Philippines (4%), Mexico (3%), Egypt (3%), Turkey (3%) and Hungary (3%).

A Devastating Runaway Success

The import of foreign workers may have started to compensate for the irregularity and unreliability of the supply of Palestinian workers, but it rapidly exceeded this goal and developed at an accelerated pace, because Israeli employers discovered the huge benefits they could draw from it.

This discovery did not put an end to or even curb the employment of Palestinian workers, which continued unabated because it was still very lucrative for Israeli employers, the average salary for a Palestinian worker being at least 55 per cent less than that of an Israeli worker.[20] So what occurred was not at all a compensation for the irregular flow of Palestinian workers and its replacement by a new flow of foreign workers, but instead a parallel increase of these two flows, eventually reaching unprecedented levels.

On the eve of the second intifada, more than 300,000 non-Israeli workers were employed in Israel. The actual figure was probably closer to 400,000, as one Israeli official has hinted. This represents no less than 6 per cent of Israel's total population, a very high ratio, apparently the highest among OECD countries. The phenomenon was even more spectacular when the percentage of these non-Israeli workers in Israel's business sector was considered: in 1999 this reached a peak of 16.5 per cent.[21] (See the graph, 'The share of foreign workers and Palestinians in the business sector, 1995–2005'.)

During the second intifada, the flow of Palestinian workers dwindled drastically. This was not at all the case with the flow of foreign workers, whose numbers continued to grow in spite of the deep recession of the Israeli economy. The import of foreign workers seemed to follow an illogical pattern. It seemed to grow almost independently from the state of Israel's economy, and to be largely disconnected from its needs. This kind of runaway pattern was caused first and foremost by the extremely lucrative character for both go-betweens and employers of importing foreign workers, especially when, in addition to the differential incentive, the law regarding the payment of the guaranteed minimum wage was not respected.

It was also boosted by two other factors: on the one hand, the growing number of foreign workers preferring to work without a permit, and on the other, the bypass of the quota system set up to bring this number under control.

The trend in the growth of foreign workers without permits appears clearly in the graph below. The ratio of those with permits to those without was 3:2 in 1995. In 2001, this ratio was inverted to 1:2 and has remained such till today.

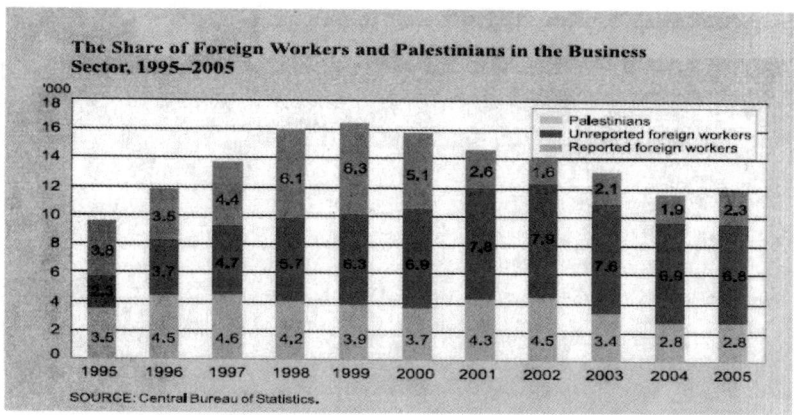

Source: Bank of Israel, Annual Report, 2005, p. 209

A first factor for this reversal was the growing number of foreign workers entering Israel on a tourist visa. In 1995, they numbered 46,000 and in 2001, 139,000.[22]

Another factor was the growing number of foreign workers 'losing' their permits as a result of their employers' decision to end their contracts or transfer them to other employers, or as a result of their own decision to leave their employer and find another one.

This last case is the most intriguing, since it seems to run against the self-interest of the foreign worker. Quite surprisingly, it is when these 'legal' workers, that is, with a permit, go 'illegal' by giving up their permits, that they manage to improve their economic status. According to a study by the Ministry of Industry, Trade and Labour,[23] the main reason for the loss of legal status is not because immigrants overstay their visa period, but instead because they quit their legal employer because they feel they are being exploited. In their new 'illegal' status, foreign workers often improve their bargaining position.

The illegal migrant is then no longer bound to an employer or obliged to work in one sector of activity: they can choose between various jobs and even undertake several small jobs, depending on the wage offered.[24] Pay is higher on the black labour market, as testified by an illegal foreign worker: 'I have been illegal for several years. I work for a company eight hours a day, five days a week. I earn between $800 and $1,000, depending on the overtime worked. I have a private insurance policy for which I pay 185 shekels ($37) a month and I send home $400 each month.'[25]

Sometimes the quotas allocated by the government to the different sectors were not respected by the administrations in charge of their implementation.[26]

However, the major breach in this system stemmed from the lack of quota in home nursing. As explained in a report of the Bank of Israel:

> There is no quota for permits in this sector; the number is determined according to need. The conception is that senior citizens are better off being cared for in their home environment, which also involves less spending from the state budget than hospitalization in a geriatric institution ... One of the problems in this subsector consists of social welfare workers leaving their employers to work without permits at higher wages. This harms the needy person, and leads to the entry of additional foreign workers on the same permit.[27]

This very large influx of foreign workers has been a source of growing concern for Israel. Some have pointed to the very serious risk it represents in the near future for the delicate ethnic balance within Israeli society. Until recently, the legal provisions preventing foreign workers from acquiring permanent residence and afterwards citizenship were sufficient to brush aside this danger. However, recently a breach has been opened in these protective measures, when for humanitarian considerations Israel has had to grant permanent residence and citizenship to scores of the children of foreign workers born in Israel.[28] For the time being, this is a very limited number, but none the less it constitutes a legal precedent with potentially large implications.

Palestinian officials have often emphasized that resorting to Palestinian workers was far more convenient for Israel since 'most of these workers did not have to stay overnight and thus did not become an ethnic community inside Israeli society'.[29] This point of view was also shared by most Israeli officials. One of them stressed that 'Palestinian labour is preferable to other foreign labour because the latter becomes a demographic problem while the Palestinian worker goes home after a day's work.'[30]

So the Palestinian workers have not been perceived as a demographic danger and even less as an economic or security one. Until recently, the Palestinian workers were not considered to jeopardize the employment of unskilled Israeli workers. For many years, the Israeli government considered these Palestinian workers as no more than a leverage of its policy of the stick and the carrot towards the Palestinian Authority, and had ostensibly ignored their negative effect on sectors of the Israeli economy. However, in the last few years there has been a change in this perception.

Contrary to the conventional view, the daily movement of Palestinian workers in and out Israel was never considered as a real security risk, even at the height of the Palestinian terrorist offensive in 2002. Such a threat was indeed often invoked by Israeli officials to justify stopping the flow, more as a way to force the Palestinian Authority to take measures against these terrorist attacks than out of a genuine fear that Palestinian workers could be involved in them. As stressed categorically by former Brigadier Dov Sedaka:

> Palestinian workers are nowhere near being a risk. Throughout all the years of Palestinian labour in Israel there was only one incident in which a licensed worker carried out an act of terrorism. The Palestinian side was very attentive to monitoring workers, and the magnetic card system worked for Gaza. Indeed Gaza shows how a fence can enhance security monitoring of workers.[31]

So for thirty-five years, from 1967 to 2002, these two flows of Palestinian and foreign workers were largely considered by Israeli officials as an economic blessing, with no negative demographic, social, or economic side-effects. But from 2002 onwards there has been a change in this perception. The negative economic effects of these two flows have been at the core of this change.

Three main areas of economic damage have been attributed to the large-scale import of foreign and Palestinian workers. The most manifest damage has been the increase in the unemployment of unqualified Israeli workers. The rate is very high in underdeveloped areas but above all in the Arab sector. The second damage refers to the weakening of the bargaining power of the Israeli worker, the reduction in real salary, the increase in social inequalities and the subsequent increase of social allowances in the national budget.[32]

At the beginning of 2004, Prime Minister Ariel Sharon addressed this issue in a special debate at the Knesset devoted to social inequalities:

> The government does not believe in transferring money to people who can work ... The situation in which we nearly found ourselves, that of a society supported by allowances will never return ... The employment rate in Israel for the employment age groups is 10 per cent lower than the average in the developed countries around the world. Some estimate the damage caused to us as a result of this at $9 billion in our annual product ... We will not be able to continue for ever when citizens of this country live off of unemployment allowances and foreign workers build the country in their stead. A country with 280,000 unemployed citizens cannot allow itself 300,000 foreign workers. We decided to put an end to this situation.[33]

Another additional damage is the slowdown in the modernization and productivity of the areas in which Palestinian and foreign workers are employed, notably agriculture and construction.[34].

Reining in the Runaway Process

The economic and social consequences of this runaway process, exacerbated by the deep recession caused to the Israel's economy by the second intifada – notably the almost complete halt to tourism and the steep reduction in foreign investments – forced the government to adopt a policy able to rein in this process and get control over the influx of foreign workers in Israel.

Up until the end of 2001, the government had limited its involvement in the import of foreign workers to the issue of work permits and to the enactment of laws detailing their economic and social rights. In 1991, a law was adopted on migrant workers' rights to paid holidays (twelve days), public and religious holidays (nine days), maternity allowances, health insurance and redundancy payments. A new law on the employment of foreign workers was passed in 1999 and amended in 2001. It extended to Palestinian and foreign workers the law on the guaranteed minimum wage and its related benefits, made it compulsory for the employer to supply decent accommodation, and authorized a residence of five years and three months for all workers keeping on with the same employer.

From the end of 2001, the government began to implement policies aimed at curbing the growing flow of foreign workers. In the meantime, it started by reducing the number of permits allocated to the various sectors, except to 'home nursing'.

During the summer of 2002, after endorsing the recommendations of the Rachlevsky Committee on 'the reform of the labour market', the government adopted Decisions 2327 (July) and 2469 (August) on 'the estrangement of foreign workers'. By these decisions, the government was for the first time conducting a clear policy with regard to the employment of foreign workers. This policy was based on the following five principles:

1. Ending the anarchy in the allocation of permits by centralizing in one administration the data on them[35] and their attribution
2. Putting a ceiling on the number of permits which in the long term will have to be below 30,000
3. Reducing the number of foreign workers in Israel, with a target of 50,000 by the beginning of 2003
4. Increasing the cost of employing a foreign worker compared to employing an Israeli worker by levying a 8 per cent tax on the employer
5. Strictly enforcing the laws on labour and on entry to Israel.

An 'Office of Migration' was set up to implement this policy. It was put under the supervision of the Ministry of the Interior and was to centralize control over powers previously exercised by different ministries in this field. It was also to be in charge of 'preventing the entry of new foreign workers, making use of the unused permits within the quota ceiling by resorting to the pool of foreign workers already working in Israel'.[36] In 2003, the target number for the reduction of foreign workers in 2004 was set at 100,000. In August 2004, the target for 2005 was cut to 70,000.

The implementation of this policy fell short of achieving these ambitious targets. But it succeeded in both reducing substantially the number of foreign workers and increasing notably the number of unskilled Israeli workers in the agricultural and construction sectors. From September 2002 to the end of 2003, between 65,000 and 75,000 foreign workers were expelled or 'left on their own free will'.[37] According to the chief of the immigration police, 'Some 90,000 foreign workers were expelled in the last 18 months.'[38]

The improvement of the employment trend of Israeli workers in the construction sector is also very clear. In 1999, 120,000 Israeli workers were employed in this sector, that is 49 per cent compared with 59,000 foreign workers (24 per cent) and 64,700 Palestinian workers (27 per cent). In 2004, the figures were, respectively, 128,700 (66 per cent), 48.500 (25 per cent) and 17,000 (8 per cent).

According to a report issued by the Finance Ministry, this was a proof that the combination between:

> ... [t]ight control on the number of foreign workers, the measures taken by the government to increase the cost of employing foreign workers, and the reduction in the amount of social transfers, has led to the replacement of Palestinian workers by Israeli workers (mainly Israeli Arab workers) and not to their replacement by foreign workers ....[39]

These encouraging but insufficient results led the government to intensify its policy by adopting in August 2004 the recommendations of the Endorn Report, which asserted that the best way to reduce the number of foreign workers was to go on making it more expensive for employers to employ them. The government decided that from May 2005, work permits would not be handled any more by employers, but instead by registered special agencies. These would be in charge of paying the wages and related allowances, and would be obliged to respect the guaranteed minimum salary and pay the taxes on the employment of foreign workers. Moreover, this annual tax was to be increased from $900 to $2,600 and the mobility of foreign workers among employers of the same sector was to be authorized. One should also recall that at the same time, in June 2005, the Israeli government decided to withdraw from the Gaza Strip, to dismantle all Jewish settlements, and to reduce to zero the import of Palestinian workers until 2008.

The implementation of all these decisions does not seem to have brought about the planned further reduction in the number of foreign and Palestinian workers. After a reduction of more than 20 per cent in 2003 and 2004, the number of foreign workers rose by 2.7 per cent in 2005. According to the Bank of Israel, this rise was 'attributable to economic growth, which boosted demand for them, combined with a relaxation of the policy aimed at reducing their number'.[40]

It is too early to assess fully the efficiency of Israel's policy to reduce drastically the number of foreign workers. The basic assumption of this policy was that the key factor in the irresistible growth in their numbers was the profit employers were getting from employing them. Therefore this policy sought to reduce the profit margins and to make it more expensive to employ foreign workers, through different means such as divesting the employers from handling the work permits, taxation and a strict implementation of labour laws, notably the minimum wage and its related benefits.

A decisive factor in the success of such a policy is its strict implementation. To say the least, this has not been the case – as was noted by the State Comptroller, who has denounced the regrettable tendency of the authorities in charge to neglect the enforcement of the labour laws on the employers. Any relaxation, failure or hiccoughs in this respect will cause the policy to collapse.

Dr Yohanan Manor is the chairman of the Centre for the Monitoring of the Impact of Peace in Jerusalem.

NOTES

1. Israel's Arab population is concentrated in Galilee, the 'Triangle' and Jerusalem, namely the north and central part of the violet area on Map 1.
2. Since Israel's withdrawal from the Gaza Strip in the summer of 2005, the Palestinian Authority has full control over the area. The West Bank is still divided into zones A, B and C. In zone A, the PA assumes full responsibility for internal security, public order and civilian matters. In zone B, Israel retains supreme authority in matters of security. In zone C, Israel has full civil, security and military responsibility.
3. In 1949, the West Bank was annexed by the Kingdom of Transjordan which thus became the Kingdom of Jordan. The Gaza Strip was not annexed by Egypt but simply put under its administration.
4. Professor Michael Bruno was Governor of the Bank of Israel from 1986 to 1991
5. Finance Ministry, Bureau of the Director General, Economy and Research Division, *Economic Information* (in Hebrew), June 2005, p. 8. Quoting from Dr Shlomo Sversky (2005), *Mekhir haYohara – HaKibush – ha Mekhir shelsrael Meshalemet (The Price of Arrogance. The Occupation- The Price Israel is Paying)* (Tel Aviv: Sifrei Mapa, 2005).
6. Between 1978 and 1988, the GDP of the West Bank and Gaza rose from $0.73 to $1.9 billion (see Government of Japan and the World Bank, *Aid Effectiveness in the West Bank and Gaza*, 2000, Annex I, p. 1).
7. By which the PLO recognized the right of Israel to exist in peace and security and committed itself to a peaceful resolution of the conflict.

8   By which Israel recognized the PLO as the representative of the Palestinians in the peace negotiations.
9   Arik Merovski, 'Palestinian workers return, and so does sabotaged construction', *Haaretz*, 13 April 2005.
10 Daniel Gottlieb, *The Effect of Migrant Workers on Employment, Real Wages and Inequality. The Case of Israel – 1995 to 2000,* July 2002, p. 6 <http://www.boi.gov.il/deptdata/neumin/neum121e.pdf>.
11 Cf: Arnon Soffer 'Nous avons tracé la clôture', in *Israël en Israël, Outre Terre,* 9 (2004), pp 110–30.
12 The figures provided by the World Bank were slightly different, namely 18,800 with permits, 18,600 without permits and 7,400 holding foreign passports.
13 Israeli Central Bureau of Statistics (CBS), October 2006.
14 Finance Ministry et al., *Economic Information*, p.19.
15 Ghassan Khatib, 'Dependency and Exploitation' in 'Palestine Workers in Israel', *Bitter Lemons, Palestine–Israel Crossfire,* 20 June 2005, p. 2. Ghassan Khatib was Minister of Planning in the Palestinian Authority.
16 Michael Ellman and Smain Laacher, *Migrant Workers in Israel: A Contemporary form of Slavery*, Joint publication of the Euro-Mediterranean Human Rights Network and the International Federation of Human Rights, June 2003, p. 9 <http://www.euromedrights.net/90>.
17 Ibid., Appendix C, p. 42.
18 Ibid., p. 25.
19 CBS, 30 July 2006.
20 Kav La'Oved ('Workers' Hotline,' an Israeli NGO), Annual Report 2006, Part II (Tel Aviv), p. 8.
21 Annual Report of the Bank of Israel, April 2006, p. 209.
22 State Comptroller, Report Number 55 B (in Hebrew), 2005, p. 377.
23 See Ministry of Industry, Trade and Labour, Research and Economy Administration, *Factors Leading Migrant Workers to Illegal Employment*, referred to in Kav La'Oved, Annual Report 2006, p. 4.
24 Ellman and Laacher, *Migrant Workers in Israel*, p. 16.
25 Ibid.
26 State Comptroller, Report Number 55, pp. 376, 401.
27 Annual Report of the Bank of Israel, April 2006, p. 216.
28 Their total number is estimated at between 3,000 and 8,000. Kav La'Oved, Annual Report 2006, p. 7.
29 Khatib, 'Dependancy and Exploitation' p. 2.
30 Brigadier General (Res) Dov Sedaka's interview 'Not a Security Risk' in 'Palestine Workers in Israel', *Bitter Lemons, Palestine–Israel Crossfire,* 20 June 2005, p. 7. The reality is somewhat different. Since 1967, more than 70,000 Palestinians from the Gaza Strip and the West Bank have settled in Israel. Through marriage with Arab Israelis, a substantial number have requested and obtained Israeli citizenship. Some were authorized to reunite with their families from which they had been separated during the 1948 war and were also granted Israeli citizenship. In addition, more than 80,000 Palestinian from the West Bank have settled after the Six-Day War in East Jerusalem and got permanent residency. See Soffer, 'Nous avons tracé la clôture', p. 117.
31 Brigadier Dov Sedaka headed for many years headed the civil administration in Gaza and after that in the West Bank, interview 'Not a Security Risk', p. 6.
32 Finance Ministry et al., *Economic Information*, pp. 11–14.
33 January 19, 2004
34 Finance Ministry et al., *Economic Information*, pp. 11–14.
35 There were no less than six data bases on these permits.
36 State Comptroller, Report Number 55 B, p. 377.
37 Ibid., p. 375.
38 Ruth Sinai, *Haaretz*, 6 March 2004.
39 Finance Ministry et al., *Economic Information*, p. 14.
40 Annual Report of the Bank of Israel, April 2006, p. 214.

PART IV
Europe's Southern Marches

PART IV

Migrations and Algerian–French Relations: A Shared Responsibility

SID AHMED GHOZALI

Despite and beyond the demarcations and successive changes of borders which date centuries if not millenniums ago, the phenomenon of migration has historically been one of the most powerful factors of development. Moreover, contrary to the belief prevailing in the North of the Mediterranean, it is for the South that migration has been of vital importance, long before the people of the North perceived migration as a threat to their way of life. The issue of emigration remains closely linked to the global development of our region and therefore to the sub-regional integration processes. It is coincidental that in the case of migration, it is the negative connotations that prevail, as governments favour a reductive, security-focused approach. These two trends follow the same logic, and the persistence of this logic does not facilitate the long-term management of the phenomenon. [When I talk about shared responsibility, I refer to the failures of a very diminished action and mainly to the inaction both of the part that fails and the part of the huge task that remains.]

In the North, the political classes – including the extreme right and left – believe that the long-term solution to the migration problem resides with the development of the countries of emigration. However, in developed countries, it is always the security angle that is highlighted, especially when elections are imminent. In 1995, the Barcelona Declaration marked the starting-point of the Euro-Mediterranean Partnership, a wide framework of political, economic and social relations between the member states of the EU and their partners in the southern Mediterranean countries. However, this partnership remains inadequate: the positive aspects of free exchanges and police cooperation are counterbalanced by, for example, the erection of anti-immigrant 'walls', and the deportation of illegal immigrants. The issue of terrorism is handled with the same incoherent and hypocritical approach. The roots of evil are widely accepted to be of Islamist origin, without accepting the role of the West in being in league with those Arab regimes that comply with western interests to the detriment of their own citizens. The subsequent lack of legitimacy, as well as bad governance and bad feeling in profoundly frustrated populations, are also forgotten.

By a tacit connivance between the imperial superpower on the one hand, and the repressive governments and advocates of violence in the Arab-Muslim world on the other, the 'inexorable' progress of extremism is thus enshrined in Huntington's so-called 'clash of civilizations'. Huntington's theory is widely used to explain the destructive effects of 11 September 2001 and the extreme exploitation thereof by the US Administration. The weight of historical assumptions burdening the relations between the West and Islam, the development of different trends of Islamophobic thinking and a distorted perception of Islam have nourished confusion and misconceived ideas. The thesis of the 'clash of civilizations' prevails today in the media with its corollary, the tantamount importance of security issues when dealing with global issues in general, and the so-called inter-Mediterranean partnership in particular.

The implicit effect of this objective 'agreement of interests', which is at the basis of the unspoken connivance between the respective centres of influence, is this: each party intentionally uses the other's religious issues for its own purposes; first, to sway public perception and to justify the monopolizing of the wealth of others by war, second, to demarcate an 'imperial legitimacy' of their power, third, to appear as the sole power who will defend their populations from those who dare to threaten them.

Adjacent to this political situation lies the fact that the Mediterranean Partnership project suffered from the absence of a utopian vision like that which led to the unification of Europe. There is also the institutional distance between the parties to the North and the South of the Mediterranean: established democracies to the North, and to the South, countries which have not quite resolved the differences between the demands of the functions of the state, and those of justice. The gap grows wider still, since neither party to the partnership plays the game as prescribed by the project's rules, in particular, Section 2, which calls for the respect of human rights and the rule of democracy. Indeed, the founding premise of Barcelona was rooted in the idea that security is linked to development and that only the State of Law could guarantee both, through good governance, both in the Mediterranean and elsewhere. There is an absence of vision therefore, but also of coherence between the different applications of Barcelona as regards the sense of the partnership. 'Short-termism', in the sense of those exceptions and aberrations sought by politicians in democracies every two or three years, is contrary to any partnership project worthy of its name.

Since ancient times and in different situations, Algeria and France have been brought more than once to the world's attention by certain characteristics of the Franco-Algerian case. Because of an extended shared

Algerian–French history and the inadequacies of the Algerian media to comment, international observers continued to blame, as regards their perception of Algeria, the public and private media in France. Forty-five years after independence, France and Algeria's shared colonial past still continues to weigh heavily on the relations between the two countries; the relationship between these two countries and their peoples very often remain marked by a number of traumas and psychological, political and other disputes which date to that period. Certainly, Algeria was the essential link of the French colonial empire,; and of course periodically, insurrections have had an enormous human cost over a century: some 15 per cent of the population were lost in the seven-and-a-half years of war during the first confrontation. However, the Algerians want to have good relations with France. The Algerian 'anti-France' trend and its French counterpart of 'anti-Algeria' are marginal factions that some people manage and exploit on behalf of the one side or the other for the purposes of internal politics. The human dimension of the bilateral relations between Algeria and France has no equivalent in the world. If we add together the numbers of Algerian citizens in France, the French citizens of Algerian origin and the French who formerly lived in Algeria (the *pieds-noirs*), there are more than six million people living in France who are connected with Algeria in one way or another and who feel that whatever is happening there concerns them. And I am not including here the generations who were politically marked by the French adventure in Algeria; or those who remain deeply involved in French–Algerian developments, who represent nearly one-fifth of the French population.

However, despite the presidency of Charles de Gaulle, France has not produced a policy on Algerians in France, and we cannot talk of a French policy on Algeria. De Gaulle thought of Algeria as 'a narrow door through which passed the entire French policy on Africa'. His successors, and in particular Valéry Giscard D'Estaing had trouble thinking of Algeria as an independent state, rather than a continuation of the old colonial empire. It was customary in diplomatic circles to speak of Algeria as France's back yard or private game park. The exchanges that the Evian agreements favoured soon became common in the key sectors, if not degraded, and the Algerians thought that the French partnership tended to abuse its dominant position in these sectors. For example, in 1972, Algerian vineyards were dismantled, when Algerian wine exports plummeted to less than 4 per cent of their 1962 level; land assets, mines and housing were nationalized, as was the banking sector. After a wave of violence against Algerian citizens in France from 1970 to 1973, Algeria stopped sending new emigrants.

The 1990s were marked by a renewed deterioration in the relations between Algeria and France. While difficulty was already being experienced in maintaining good relations between the two countries; a new sense of mutual suspicion arose to complicate relations and to confuse people's minds, to the extent that it affected the internal consistency of the positions presented by both the media observers and the political decision-makers of each country. There was a reluctance to face up to a phenomenon as fundamental as terrorist obscurantism, which many people in France considered to be a defensible cause in the 1990s. The fact that for years France refused to condemn the terrorist barbarism was a means by which to cover it up and ultimately to legitimise it. It is no small paradox and a scandal that this misrepresentation of Algerian reality abroad should occur mainly in countries whose officials, media and civil society both claimed to be friends of Algeria and also resolute enemies of international terrorism.

Today, there is a pressing need to focus on the southern shores of the Mediterranean; the approach to the problematics posed by the issues of freedom, democracy and development must be refined, particularly, as a result of geopolitical factors such as the eruption of fundamentalist violence, the issue of emigration, the Middle East's descent into the hell of war and, finally, the growing status of our region as a 'future major holder of oil reserves'. This factor has been a development that has been taking shape for nearly for a decade and one which may lead to the region 'politely' expelling the European from their 'back yard' in the Mediterranean. To quote historian Joseph Rovan: the Maghreb 'is not only the business of the French but that of Germans and all Europeans, in the same way that Poland cannot be the business of the Germans only but that of France and of all Europe'. Only if it ceases from comparing North Africa to the Middle and Near East, can the European Union join the countries of the South of the Mediterranean in an original process of democratization, long-term development and credible unification.

Nothing seems more hated than the formula used by some people that 'we are condemned to unite' or 'condemned to cooperate', as if said cooperation is to be viewed as a worst-case scenario, instead of a situation that is in all our national interests. The profession of faith in the Euro-Mediterranean cooperative effort is contained in our readiness to make a zone of peace and prosperity in our common area. Peace and prosperity, development and security, good business and friendship: each term cannot be realized without the other, each supplements the other. We must come out of the past, and look to the future: prosperity will increase with the increase of cooperative exchanges.

But the Maghreb countries negotiated partnership agreements with the EU with the most diminished contractual capacity; the negotiations were conducted in an uncoordinated and dispersed manner, with some countries weakened by internal strife. The Arab Maghreb Union was in disrepair and the level of inter-Maghrebi exchanges of goods and services practically non-existent, never mind that the Maghrebi negotiators did not factor in the re-evaluation of their assets regarding elementary security data (that is, the transformation of the region according to its oil potential) in their strategies. But it is by achieving a minimal level of institutional development and decisively registering in a process of good governance, democratization and respect of private freedoms that our region will manage to give the new Euro-Mediterranean relations a perspective of maximum benefit and high potential.

European assistance must live up to this potential within the context of a more long-term vision

The basis of our partnership with Europe must be one of successful transaction, in other words a lasting win–win one. Each partner here measures their interests against those of the other, within a context of mutual self-respect. Neither partner should attempt to impose their will on the other: it is in both their self-interests to agree to a single course of action for both parties.

The laws of economy do not constitute a science so precise as to enable one partner to conclude on a way of functioning that he would impose on the other. Free enterprise is based on a free arbitrator and the ultimate conviction of the businessman. There is nothing more anti-economical and anti-social than the so-called reforms that would prevail over the conviction of the sovereign States concerned. In this context of imposing 'reforms' from outside, yesterday's ideology of nationalization has become today's ideology of privatization. The worst privatization is the one that believes privatization to be a goal in itself or the one that we suffer as a way of thinking imposed by the International Monetary Fund. The only good reform is the one that we put into practice out of conviction, not one that is imposed. The region will benefit from Euro-Mediterranean integration only to the extent that they have this belief. The Europeans must bear the consequences regarding the redeployment of agreed investment, in view of this integration, and also support the conditions for the establishment of good governance and more generally, in regard to the long-term projection of their vision for the future.

In the same manner that the market economy and the processes that lead to it cannot be simplified down to a mathematical formula, the edification of the State of law and fundamental freedoms cannot be imposed from outside. Partners must openly communicate to reinforce their partnership but they must also carefully distinguish between generous support from a loyal partner and unwelcome interference from an overbearing one.

However, the clauses on the respect of democratic freedoms and on good governance in general must not appear in Euro-Mediterranean partnership agreements merely for form's sake. The complacency of partners who do not comply with these clauses is as disturbing as the interference in the internal affairs of another State.

The sharing of universal values between partners must go in both directions: respect for what is received, on the one hand, and the rejection without ambiguity of inadmissible clauses, on the other. This has not been the case, for example, as regards certain ambiguous positions here and there in Europe as far as fundamentalist violence is concerned: when such barbarism is unleashed, the absence of condemnation is equivalent to the legitimization of the barbarians' cause, which human conscience cannot accept. Has September 11 made the West conscious of the reality of the terrorist phenomenon, its real dimensions and the conditions which are necessary to eradicate it? Maybe not, as the West's response to terrorist violence is even more violence, presented at best as a legitimate defence which will not, however, produce lasting solutions. Military supremacy will never overcome the terrorist phenomenon. In the end, terrorism will be overcome by the establishment of good governance, personal and public freedoms, and development. These are the conditions necessary to overcome the foundations of terrorism.

Let's come to today – to the challenge linked to the perspectives of a profound development on the geopolitical level, regarding security. The history of the Middle East in the second half of the last century forms a precedent in terms of the dangers linked to such a development. The United States proclaims its 'supreme' interest in the region because it contains at least the two-thirds of the world's proven oil reserves: a proportion that increases to nearly three-quarters with the inclusion of the Arab Maghreb. This increase is due mainly because of the potential discovered in the Algerian Grand Erg Oriental. Algeria, which was for a long time classified as a natural gas-supplying country only, should soon see itself as one of the major holders of oil reserves. Given that the reserves of non-OPEC countries have increased at a rate six or seven times *less* than global demand, it is likely that 80 per cent of new demand will depend on reserves located in all Mediterranean and Arab-Persian countries. Even without mentioning the important position of the Middle East as far as crucial East–West naval and air routes are concerned, the region is of vital interest from a commercial and military point of view for the United States' position throughout the world. The Strait of Gibraltar, the Sicilian Channel and the Suez Canal are the Mediterranean counterparts to the strategic straits of Bab el-Mandeb and Hormuz, the pressure points for all security forces centred on the Middle East. The countries adjacent to these pressure points are therefore prone to experiencing 'interference' in their security. This situation has two implications for the southern Mediterranean:

- The increasing interest of the United States in the region could bring about a possible increase of the contractual power to the benefit of our countries.

- The increasing interest of the United States in the region could also foster an increasing concern about the possibility of new threats.

The evolution of the European Union fascinated the Maghrebis because it was not constructed by brute force, but by common will, the force of law and democracy providing a precious point of reference to unitary aspirations in North Africa.

The Maghrebis must use this process as a catalyst and a source of inspiration for their own project. The European should also help to make successful the Maghrebi project, for the benefit of the inhabitants of both shores of the Mediterranean. It depends on the Maghrebis to compensate for their delays (institutional, financial, technological, cultural and social) by using the capital of a complex and venerable history, with a territorial and ethnic, linguistic and cultural continuity. Following the cruel war that left Europe in pieces, the founding fathers of the CEE rose to the challenge of bringing together peoples who had ended up regarding each other as 'biological' enemies. Nothing like that happened in the case of the Maghrebi nations; the Maghrebi idea is anchored in a common culture, while leading at the same time to historical nationalism. Our most evil crises or tensions did not break the unity of our peoples. At worst, together with the multiple, futile attempts of unions, a Maghrebi pessimism has been aggravated, which is more attributable to the disputes and fractures between generations, with all resulting disaffection against States, politics and public issues in general.

As in Europe, so too the Maghreb. The high road to integration passes through the modernization of institutions; tidying one's own house; achieving an acceptable level as regards the State of law, democracy and the respect for individuals; marching towards the practice of good governance. Implementing these measures will best release creative energy in a way to form the foundation of a credible Union, in compliance with the aspirations of the peoples as well as the requirements of times. Whether it plays on specific fields or on a global scale, single leadership without cooperation is incapable of producing the safe and peaceful world to which everybody aspires. The superpower, for all its supposed good intentions, exposes itself and exposes us to acts derivative of totalitarianism. The Gulf wars, the tragic failure of Palestine, the conception of an 'axis of evil' and the deviation of the anti-terrorist struggle are enough to show how dangerous is the imperial choice that prevails today.

Until China, Russia, or India manage to claim a more significant role, only Europe might be the necessary counterbalance towards equilibrium and an increasing security in the world. Europe cannot escape from a triple challenge: unite the conditions of political Europe, accompany the countries of the southern Mediterranean in the restoration of the State of Law and the course of real democracy, and contribute to the management of the real risks of North Africa becoming like the Middle East. Prosperity will come along with our exchanges, which while still below the potential of the region, amount to some ¤50 billion; assistance and cooperation will give us the means to make this sum five times larger. Isn't that a stake that deserves to be considered?

Sid Ahmed Ghozali is the former Prime Minister of Algeria.

The Sudanese Population in Egypt

DOREYA AWNY[1]

Egypt and the Sudan enjoy privileged relations at historical, geographic, sociological and strategic levels. Geographically, the Sudan is the prolongation of Egypt; Egyptian Nubia is the northern part of Sudanese Nubia, with the Nile as the obvious link. The Pharaohs unified both countries on several occasions; the Pharaohs from the North occupied the South before the black Pharaohs (the Kingdom of Kush) occupied Egypt in their turn. In fact, Egyptian civilization under the Pharaohs did not stop at Aswan; there are nearly three hundred pyramids in the Sudan, almost more than in Egypt. These relations continued and even deepened with the conquest of the Sudan by Mehemet Ali in 1821, the Turco-Egyptian period until 1881, and the Anglo-Egyptian occupation regime in 1899. These relations have never been broken since independence in 1956. This is partly due to the al-Azhar mosque and the Coptic Church; Sudanese students and intellectuals in Egypt, those at the Egyptian University in Khartoum, and businessmen also play a part.

There has been an intense two-way migration between the two countries for hundreds of years. The Sudanese go to Egypt to buy cereals but also to escape drought. The Egyptians buy wood, incense, gold and ivory from the Sudan. The borders, which were artificially fixed in the nineteenth century (at the 22nd parallel), remain porous, particularly in Wadi Halfa and on the Red Sea. Caravans move freely between the two countries along a 1273-km-long border (the Sudanese littoral on the Red Sea is 853 km long). Christianity and then Islam spread to Nubia and beyond from Egypt.

In 1976, independent Sudan signed the Wadi El Nil (Nile Valley) treaty with Egypt, thus ratifying the established situation: in Egypt, the Sudanese enjoy, if only theoretically, the same rights as the Egyptians. In 1995, relations between Cairo and Khartoum flared up after President Hosni Mubarak was the target of a terrorist attack, believed to have been carried out by Sudanese Islamists. The Wadi El Nil treaty was immediately abrogated but it did not stop hundreds of Sudanese from crossing over the border every day.

As the Sudanese regimes – whether military or Islamist – began a harsh policy of repression against political opponents, migrations increased and thousands of Sudanese were practically forced to leave for Egypt. Most were

Muslims and came from the North. Some were Christians and animists who had already fled to Egypt because of the civil war in the South from 1955 to 1972; the instauration of sharia law in 1983 also contributed to their exile in Egypt. Moreover, war broke out again in January 1983 and lasted twenty-one years, until January 2005: almost four million people were displaced or became homeless and over two million people died.

Before this conflict even ended, another bloody war broke out in Darfur in 2003; it has been raging since then, with tens of thousands of people fleeing the country.

It is believed that between two and five million Sudanese live in Egypt and make up 90 per cent of the refugees there, alongside Palestinians, Ethiopians, Somalians and Eritreans. But not all the Sudanese who live in Egypt are refugees. The word only appeared in 1989 when the Sudanese President Omar al-Bashir came to power.

Egypt, a country of emigration, has become a country of immigration and transit

Asylum seekers who are bound for Western countries go through Cairo, where the Regional Office for the United Nations High Commission for Refugees (UNHCR) is located. Its officials may well claim to have resettled the greatest number of refugees ever in the world: they have only resettled a quarter of Sudanese asylum seekers. The United States, Canada, Australia and Finland cooperate with the UNHCR through private sponsoring programmes.

Egypt has signed the 1951 Convention on the status of refugees, the 1967 Protocol on the status of refugees and the 1969 OAU Convention on the specific problems linked to refugees in Africa. Yet the government has sent to the UNHCR a letter listing major restrictions on education, health and employment, even though the situation did improve in some areas. The country's policy remains all the more ambiguous as Cairo did not adopt the dispositions or the laws which would make it possible to implement the international conventions to which it subscribed.

Sudanese Statistics in Egypt

According to the UNHCR, the number of Sudanese asylum seekers who were registered between 1994 and the end of 2005 was 58,535. In December 2005, 31,990 Sudanese obtained political refugee status. Only 16,675 of these were resettled in host countries (315 were not able to reach these countries because their official papers were incomplete); in other terms, 28.48 per cent were asylum seekers and 52 per cent were blue-card holders, that is, political refugees. This means that 15,000 Sudanese who obtained political refugee status were not resettled. Most of them remained in Egypt, for 'integration on location'. Sixteen thousand Sudanese asylum seekers were rejected and their application files were closed. Over ten thousand were not granted an interview because the UNHCR stopped interviews when a peace agreement was signed in 2004.

The situation in 2005 was as follows: some 71 per cent of Sudanese with political refugee status and who had the right to be resettled in a host country remained in Egypt.

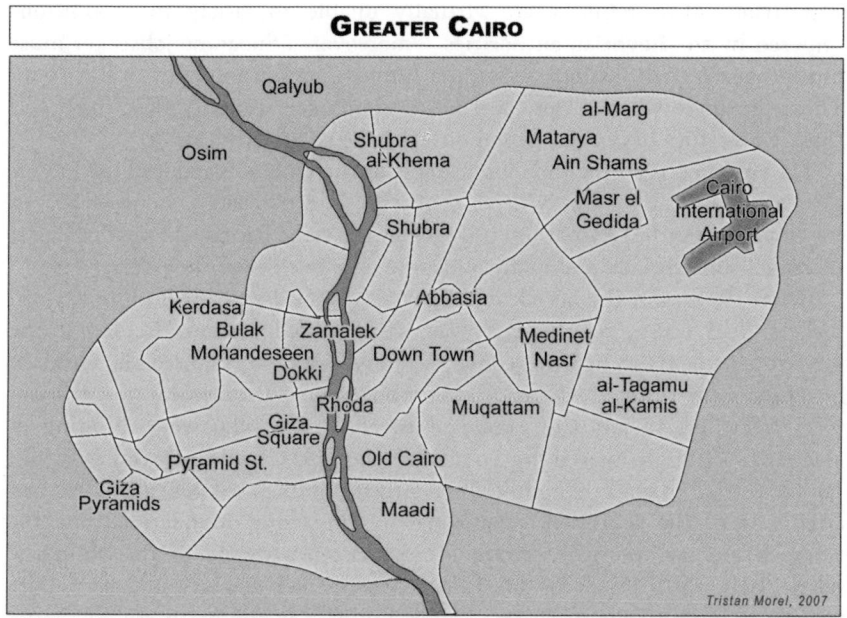

Who are these Sudanese people?

There are around twenty ethnic groups in the Sudan, the major ones being the Arabs, the Beja in the North, the Nuba in the centre, and the Dinka and Nuer in the South. Several surveys have shown that refugees come from all the ethnic groups and tribes.

The Arab-speaking Sudanese, who are generally educated, integrate better and can benefit from the support of the large diaspora in Egypt. As there is no language barrier, it is easier for them to find work. Many of them are political refugees who do not want to obtain political refugee status and do not want to abandon their nationality. For others, it all depends on their level of education and the help that the Sudanese diaspora in the West can provide them. Many were teachers and businesspeople who find life in Egypt mediocre; but those who come from camps for displaced populations enjoy their new environment, despite the overcrowded and poor habitation. The Sudanese who attained political refugee status have a better life than those whose files had beene closed. Around thirty organizations, churches and centres help them, depending on their vocation.

The Sudanese who have obtained political refugee status (blue card) have the right to work in Egypt. But unemployment affects 25 per cent of the population. Most refugees are virtually unable to satisfy the conditions imposed by the Egyptian authorities – moreover, a Sudanese jobseeker must prove that his qualifications do not put him in competition with an Egyptian. Those Sudanese who do not have political refugee status (yellow card) and those whose files have been closed do not have the right to work.

The situation is all the more unbearable as the UNHCR reduced the help it gives to refugees by around 72 per cent, that is, a monthly stipend of $55 in 1998 was reduced to $15.50 in 2002. Over the same period, the cost of living increased substantially and many Sudanese now live below the poverty line.

Before 1995 and the attack on Mubarak, all Sudanese, including asylum seekers, had access to the Egyptian state school system. In 1992, the government in Cairo had overcome its reservations and granted the children of refugees the right to education. Yet schools were already overcrowded and were reluctant. In any case, most of the Sudanese who were thinking of resettling in English-speaking countries (the US, Canada, Australia) would have preferred to send their children to private English schools, if the fees had been lower. The UNHCR established a scholarship foundation, but the refugees are not yet fully aware of its existence. Some of the churches, particularly the Church of England, have opened primary and secondary schools to improve the situation of a few hundred young refugees.

The Egyptian government did not open public hospitals to refugees until 2005. Refugees still cannot be treated for long-term illnesses, receive expensive medical care or have surgery. Many Sudanese complain about the services in public hospitals. There are often clinics in the mosques, the most famous being the one in Mustafa Mahmoud Mosque – precisely where the 2005 events happened. Non-Muslims, who dare not go there, are generally seen to by the clinics of several churches, most of them Church of England.

Sudanese do not have access to low-rent accommodation and most of them find it very difficult to find a place to live. It is virtually an impossible task for those who do not have political refugee status. Generally, the Sudanese are urban refugees and the markets that are managed by their compatriots give them the possibility to retain their food culture. Similarly, they are attached to their African languages and cultures. They are gathered both in the centre and in the periphery of Cairo. The greatest number of them lives in Arba Wa Nuss, the famous 'four and a half' because it is 4.5 km east of Greater Cairo; it is not a camp but an *Ashwaiya*, a shantytown among the hundreds of similar settlements surrounding Cairo where millions of Egyptians live. The Sudanese, most of them from Darfur, live there in an overcrowded and precarious environment; they meet up with Egyptians but do not really mingle. Other refugees, particularly those who are Christians, live around the churches – mostly Anglican, sometimes Orthodox – in the Sakkakini, Abbassiya, Zeitoun and Ain Shams neighbourhoods in the centre and to the north of Cairo. There are very few Sudanese refugees in Alexandria.

All these frustrations have been increasingly felt over the past two years. The Sudanese have lost all confidence in the UNHCR, which is really the only body they depend on; they had never expected anything from the Egyptian authorities. Such is the situation which led to the sit in and the tragedy after a riot began in August 2004 when the UNHCR considered that the May and June peace agreements put an end to all applications for political refugee status – including applications from Darfur where the war was still raging.

The Three-Month Sit-In

The sit in began on 29 September 2005 when a dozen or so Sudanese demonstrated in the park that is located opposite the Mustafa Mahmoud Mosque 'in order to attract the attention of the international community and find a solution to the problems of the Sudanese'. The location was a logical choice, as the UNHCR had been interviewing asylum seekers nearby for months on end. The park is big enough for a very large number of protesters, who had access to the water rooms of the mosque. It is in the middle of a very busy neighbourhood, with businesses and clinics, which means the protesters could be heard by many passers-by. The small demonstration thus began, but the number of Sudanese in the park increased rapidly. It is believed that a month later, there were always between 800 and 2,000 people in the park. On average, 1,500–2,000 people remained in the park during the three-month sit-in. According to estimates, there were 3,000–4,000 Sudanese in late December 2005, just before they were evicted: half were young single individuals; the other half were women, children and elderly people. It seems that news spread by word of mouth. A similar proportion of Muslims and Christians were involved; at least sixteen tribes were represented, from all over the Sudan. All the actors in the protest showed full solidarity, independently from their ethnic group or religion. One must bear in mind that most of the Sudanese in the park had a blue card and were waiting to be resettled in the West.

The Media and Civil Society

'We are the victims of poor management', or 'We refuse to assimilate', or again 'Where is the international community?', or even 'Who will guarantee our rights?' ... these were the slogans on the placards which were on the park gates. Others paid tribute to the Sudanese who were murdered or disappeared in Egypt. Neighbours reacted differently to the protesters: some were hostile, refusing to hear their demands, blamed them for drinking alcohol and 'making love in the open'; others encouraged them and gave them food. A security service for the protesters organized meetings with the media, with lawyers and NGO representatives. The Arab-speaking and English-speaking Egyptian press covered the event; so did the African press and a number of websites. But only one week before the protesters were evicted did the international press begin to seriously focus on the event.

The Protesters' Demands

The protesters demanded:

- the resumption of the interviews which had been stopped in 2004
- no 'racism' and no 'discrimination', no 'ethnic preference' when choosing candidates for an interview
- the re-opening of the application files that had been closed by the HCR
- the end of arbitrary arrests
- the end of pressure put on refugees so that they would go back to the Sudan.

But what the protesters wanted above all was an option that would be final: resettlement in a western country or any other viable solution. The UNHCR then responded with increasingly hostile communiqués, finally representing most asylum seekers as economic or fake refugees. For instance, in the communiqué of 30 October 2005, refugees are accused of being illegal immigrants, their obstinacy the one and only cause of the failure of negotiations, and the sit-in organized on the basis of unfounded rumours. The protesters denied everything as a whole: the demonstration followed as a result of their negative experience with the UNHCR.

The Egyptian Authorities and the Authorities of the Host Countries

From September to December, the authorities looked on kindly; their relations with the protesters remained peaceful. According to the protesters themselves, they could not have remained in the park for three months 'without the protection of the Egyptian police'.

Daily Life During the Sit-In

Although it was uncomfortable, the park became a shelter. Observers noted that the demonstrators felt safe there. They did not have to pay rent. They were gathered together, and the fact that agents surrounded them strengthened solidarity among them. Over time, the situation improved. Areas were reserved for women and children. Plastic covers replaced sheets to protect them against the wind and the rain. Communal meals were prepared on camping stoves. Neighbours and NGOs provided the protesters with food, clothing and blankets. even a makeshift clinic was set up, which a Sudanese doctor visited regularly – even though seven to eleven people died before the protesters were evicted.

Children were taught in Arabic and English. Conferences were held every day so that refugees could keep informed, be advised and guided. Very quickly, the demonstrators organized themselves administratively speaking and were very proud of it: they elected representatives to negotiate with the UNHCR and formed a public relations committee as well as a security service. Five 'leaders' emerged: requests were passed on to them and the refugees trusted them, at least initially. However, after the beginning of December, these 'leaders' began to lose some of their standing in the community.

The Agreement of 17 December 2005

On 17 December, the UNHCR announced that it had come to an agreement with the five 'leaders'. They referred to an 'amicable' settlement, which only concerned the Sudanese whose names were on a final list given by the 'leaders' to the UNHCR. Among some interesting points, were mentioned the possibility that applications would be reopened, that it might be possible for the Darfur Sudanese to obtain a yellow card, and that those who had a yellow card might have their files reviewed.

But there were also threats in the agreement: if applications were rejected, the rejection would be final and the UNHCR would no longer be responsible. No date or place was given for future interviews. Everything was vague. The demonstrators were given a single chance: the sit-in would end and the Sudanese would leave the park in small groups of around twenty people and meet with the UNHCR representatives; the Darfur refugees would be seen first; once the interviews were over and the status of each person was clear, the Sudanese would go home – if they had a home – or they would get a place thanks to some financial help. But under no circumstances could they go back to the park. This process needed to be completed in four days at most, but the Sudanese found it difficult to imagine that 2,000 interviews could be carried out over such a short period.

Mixed Reactions from the Demonstrators

On 19 December, the UNHCR representatives went to the park to explain the terms of the agreement. Yet most of the Sudanese did not agree with it, even though their 'leaders' had signed the document. They were not able to reach a consensus. Confusion and uncertainty prevailed and several people declared they would not leave the park before everyone had been interviewed. A *written* guarantee was given as a condition, while the UNHCR considered that the best guarantees for the Sudanese were public opinion and the international press. But the refugees knew that if they left, they would give up their main asset. Even those who were in favour of the agreement refused to leave the park before all of the applications had been reviewed. Contacts between the refugees' representatives and the UNHCR were suspended. On 22 December 22, the UNHCR sent an official letter to the Egyptian Foreign Affairs Minister to inform him that it could do no more; it thus accepted the eviction of the demonstrators if necessary. Confidence was still high among the demonstrators, who were getting organized for the winter.

Eviction

The eviction was a striking event, both politically and emotionally. Witnesses, demonstrators, NGOs, the UNHCR, the authorities and the media were all in agreement on how it happened. In the afternoon and evening of 29 December, many policemen and tanks were deployed all around the park; the police promised the refugees that it was all about protecting them 'in case the Muslim Brotherhood organized a demonstration'. One thing is certain: neither the police nor the Egyptian government gave the demonstrators the slightest indication that the protesters might be evicted.

Around 1 a.m., witnesses saw four thousand anti-gang police, sixty buses, ten tanks and six ambulances move into position around the park. Senior officers and the demonstrators' representatives met. The government promised the latter that they would be transferred to well-equipped camps where they would be treated humanely, given shelter and fed – it would later be known that they were training camps for the police and the army, which had been set up hastily. If they refused to leave, they would be evicted by force. The Sudanese demanded guarantees that they would be treated well and would be safe; they wanted their representatives to visit the camps first. These requests were rejected. From 2:30 a.m. to 5 a.m., water cannons were used on the protesters. Representatives from the embassy and delegates from the Liberation Movement for the Sudanese People went back and forth, trying to convince the refugees, but to no avail. The anti-riot police began warming up, singing patriotic hymns and running around the park. From 5 to 5:30 a.m., there was an attempt at a last negotiation then the anti-riot police started to evict the refugees: they were pushed inside the buses; women, children and elderly people were trampled; the buses left very quickly towards the camps, some in Cairo, some on the outskirts of the city.

The Egyptian authorities later pretended that the eviction was carried out without violence. On the contrary, it has been proven that the police used disproportionate violence, including tear-gas and electric coshes; officers tried to stop the soldiers who were beating up the Sudanese. As all the exits from the park were blocked, the demonstrators had no chance of dispersing. Witnesses said that the wounded were left unattended and that everyone were forced to cram inside the buses, including wounded children, many of whom died when they reached the camps. In fact, two to ten bodies were seen around the park. The park was deserted but scattered with personal items.

After the Eviction

On 2 January 2006, the UNHCR called a meeting; nearly seven hundred people had received medical care, the bodies of the dead were not buried until 21 March in Egypt. The illegal immigrants were taken to three prisons; blue-card or yellow-card owners were released in small groups. On 30 January, the Egyptian Embassy in Britain announced– without releasing any figures – that all those who were able to prove they had political refugee status (entry visa, asylum application registered by the UNHCR) had been freed; after many contradictory statements, the spokesman for the Foreign Affairs Minister promised that no refugee would be sent back to their country. On 11 February, all those Sudanese who were still in prison were released and their residence permits were renewed for another six months.

The Role of the Sudanese Embassy

The position of the Sudanese Embassy was ambiguous. A human rights organization managed to obtain a list, drawn up by the embassy, of the people that the Sudan intended to repatriate.

The Official Egyptian Reaction

According to the official position, Egypt had acted as a mediator between the demonstrators and the UNHCR, and the eviction had not been violent. Special relations between the two countries were emphasized. Like Adel Imam, the 'goodwill' ambassador for the UNHCR, the Egyptians underlined the intransigence of the Sudanese demonstrators, who were against any form of compromise.

Egyptian Opposition and International Opinion

The Egyptian opposition and civil society condemned the abusive methods of the police in the strongest terms. On 30 December, NGOs published a communiqué against the Interior Minister, 'a criminal who only knows the language of force'; it also criticized the country's media for 'misleading public opinion on the three months of humiliation that the refugees had to endure'. There were calls for the Interior Minister to resign, for concessions from the government and the UNHCR, and for more transparency on the disappearances and deaths. On 5 January 2006, Amnesty International called on the Egyptian government to commission an independent and impartial investigation by international experts, which would strengthen the government's credibility. On 9 January, twelve NGOs wrote to the United Nations High Commissioner for Human Rights, Louise Arbour, to criticize the fact that the UNHCR had voiced no public condemnation. The International Federation for Human Rights even condemned the violence, asked for an investigation into the eviction operation and advised Egypt to adopt a new law to give refugees better protection.

One Year Later

There is a consensus among the officials of the refugee aid organizations whom I interviewed (the UNHCR refused to cooperate and referred back to its website): there has been no improvement in the situation of refugees since December 2005. The same uncertainty pervades all levels; in fact, the situation has deteriorated since potential employers will not choose Sudanese for fear of 'trouble'. The UNHCR's policy and mentality have not changed; just under three hundred refugees have been resettled in the West. The Egyptian government has not commissioned any investigation; even the results from the post-sit in autopsies remain secret. But to our knowledge, no one has been repatriated. The Sudanese are generally very reluctant to go back to their country; organizations gave assistance to some five hundred who wanted to go back to Khartoum, but Sudanese still come to Cairo in huge numbers.

Conclusion

The tragic end of the Cairo sit-in is the result of a series of failures involving: the international community, the UNHCR, the Sudanese regime, the Egyptian authorities, NGOs and finally the 'leaders' of the protesters.

The international community cannot prevent civil or interstate wars and cannot put an end to them either. It is unable to force dictatorships to show a minimum of respect for human rights and people prefer to leave for the unknown and risk their lives rather than carry on living under repressive regimes. Tragic situations become extended and more numerous, particularly on the African continent – and the Sudan is a prime example of this. But Palestine, Iraq and Afghanistan have also become victims, because the new international 'order' refuses to look at the roots and the real causes of these problems and prefers to 'manage' conflicts rather than to put an end to them.

The UNHCR office in Cairo bears the greatest responsibility for the events of December 2005. They are in charge of sorting out the true refugees from the economic refugees. One must acknowledge that it becomes an impossible task since their budget is shrinking while the number of refugees increases. The UNHCR never admitted that they alone were responsible for granting political refugee status and laid the blame on the Egyptian authorities. According to several researchers and journalists, they put 'amateurs', not qualified professionals, in charge of interviewing people; the interviews were held in public, in the street, rather than in an office. The UNHCR's blatant hostility against the Sudanese and this besieged fortress mentality means that the Sudanese have lost all confidence in the organization.

The Egyptian authorities also bear a heavy responsibility. For twenty-five years, they had imposed an exception law which forbade all gatherings, but then let a sit-in of 3,000 people last for three months in the very centre of Cairo. The excessive violence that followed remains inexplicable, even if the disproportionate use of force among the Egyptian police is famous.

NGOs, the press, the 'leaders', all had the duty to inform and convince the demonstrators, who could only rely on rumours about possible resettlement in the West.

Recommendations

The following are recommendations emerging from the aftermath of the 2005 sit-in:

- Considerably increase the means of the UNHCR and the NGOs that deal with displaced people.

- Improve communication between these organizations and refugees.
- Set up pedagogic information websites.
- Redefine clearly the responsibilities and the mandate of international organizations.
- Restore the refugees' confidence at all cost.
- Help poor countries like Egypt that become immigration and transit countries to offer refugees better living conditions.
- Use part of the funds given by the European Union to protect its borders in that sense.
- Urge the Egyptian government to pass the necessary laws to enforce international conventions.
- Demand an investigation into the December 2005 events.
- Prosecute and punish the mafias who 'smuggle' asylum seekers.

Doreya Awny is a journalist for *Agence France Presse*.

NOTE

1   This study is based on the research carried out by various centres such as the *Forced Migration & Refugee Studies Program* (FMRSP) at the American University in Cairo; on news bulletins and articles from the Arab and international press. I went to see the Sudanese and tried to interview them; to no avail, because they refuse to talk. I met with the leaders of the main churches and institutions who look after these refugees; and with actor Adel Imam, a 'goodwill' ambassador for the UNHCR and an official at the Egyptian Foreign Affairs Ministry.

PART V
South of the South

PART V

South of the South

The Geopolitics of African Migrations to Europe: International Hypocrisy about the Right and Need for Africans to Emigrate

GEORGES TADONKI[1]

Modern African patterns of migration are gradually becoming an important issue in world affairs. Although not often publicized, as was the case for the South East Asian boat people, the desperate movements of Sub-Saharan African to Europe constitute a global challenge which is currently neglected but which has the potential to become more acute in decades to come. This article does not intend to address the issue of migrations into Europe comprehensively; its focus is Sub-Saharan Africa, its people, current migration trends and the impact of these on Europe and also on their homeland. African modern global migrations are at a turning-point. They represent a massive intellectual and human capital loss for the continent but they also constitute a crucial source of investment in local communities, as well as a massive contribution to inward capital flows from abroad which are needed in African countries. Money transfers to families at home (for example, via Western Union) are an interesting characteristic of these trends.

European Arrogance and Security Posture vs Partnerships

The cultural shock of 9/11 and the associated security concerns which now affect western civilization, including the European citadel, have resulted in a much-distorted perception of migrants, their motives, goals and objectives. Historically, African migrants are neither terrorists nor beggars. While the majority of African immigration to Europe is for economic reasons, almost all economic problems in Africa have their roots in political failure, ethnic rivalries and in a pathetic inability of African leaders for decades to turn their countries into business-friendly nations, where their own citizens can dream, build and prosper securely.

Most scholars and researchers now recognize that migrating to western countries is an important investment for most migrants, their families and communities. Each migrant pays a very high financial and psychological cost for his adventure, without any insurance. Many lose everything, including their lives. During the last ten years, the total travel cost of African migrants to western shores can be estimated be more than $6 billion. This cost is still increasing but people prefer to take the risk instead of investing that money at home. It is important for research in foreign affairs to help build a better global understanding of African contributions to the complex movements of human and financial capital that sustain global economies. This is a new fundamental geopolitical challenge to North–South relations. One part of the world cannot indefinitely control the quantity and quality of migration flows. It has never happened in the past and will not happen in the future. However, Africa and the western world need a win-win partnership on migration, which requires intellectual resources beyond migration laws and police. Failure to address this will only result in medium to long-term devastating consequences on global security.

A growing number of western officials and diplomats now openly declare 'We cannot receive all the misery of Africa.' While this may sound trendy and aimed at their electoral constituencies, the consequences of such attitudes are debilitating. Should Europe be opened to all poor Africans? The answer is obviously no; it is an issue of national sovereignty on immigration policies. However, the question is grossly misleading, as Africa is definitely not a 'poor continent', with its extraordinary mineral and energy resources, as well as its human capital. The real issue is the unacceptable level of poverty and social exclusion in Africa. While this was agreed in the millennium goals, very little has been done to help millions of Africans enjoy globally acceptable living standards, as well as peace and the rule of law in their societies. Addressing social vulnerability or poverty and governance in Africa is a challenge to international relations and the global movement of capital and resources. Global peace is deeply linked with a greater and equitable share of global prosperity. What is required is less intervention in Sub-Saharan Africa to protect ex-colonial economic dominions and preferential relations, and more transparency and the mutual respect of civil societies. There is a need for greater international accountability on investments in productive business and industrial partnerships in Africa: these will certainly generate growth, jobs and a better place to live.

While it is important to build a framework for migrations that benefits all parties, the international community has failed both to recognize the right of Africans to migrate and to offer protection to migrants. Despite an abundance of international treaties and agreements, the migrants caught in an 'illegal' situation in the host country are often abused, tortured and treated below humanitarian acceptable standards. We have witnessed shameful 'charters' commissioned to return African migrants home forcibly, often chained and brutally manhandled until they are delivered back to Africa by European security forces at a cost that could have been better used to address the issue. The international community can no longer witness these situations without reacting.

International Context of African Migrations

Growing global opportunities for business and human development, and a wider access to information and communications technology, have set the conditions for greater human mobility. In the case of Africa this is exacerbated by years of poor governance and the failure to plan and respond effectively to the hope of young generations in search for a better life.

The twenty-first century could be seen as one of opportunities, of greater freedom and modernity. However, it is also pathetically a century of sharp global contrasts. Although the international community can be acknowledged for major achievements in the last few years in Sub-Saharan Africa – for example, debt reduction, humanitarian assistance in chronically affected countries, support to fight HIV/AIDS and better governance in Africa – the continent remains far below global average living standards. Abject poverty, unemployment, natural disasters and wars constantly remind the most optimistic Africans at home that it is time to run before it is too late for them to afford it. Thus the pragmatic movement of African hordes, in search for a better shared prosperity, also demonstrates complex patterns of panic. Skilled or not, poor or rich, educated or not, people vote with their feet where ballots either don't exist or have no meaning. They run as far as possible towards places where they can exercise their right to prosper, and earn a decent wage for their strength and skills. The right to prosper is a fundamental human right, which should be protected and promoted as much as democracy. Africans aspire to live in countries where their investments are sustainable and their life pensions guaranteed. Africans migrate away from the impunity of careless leaders who have failed in their mandate to build sustainable societies. As long as there are in Africa four civil servants for every one small business, insecurity will push out those who can still afford the adventure to western shores.

However, historically coming late to the game, the African global economic emigrant too often ends up as a 'criminal', having challenged the migration laws of host countries. Several reports highlight a growing score of abuses by western states exercising their sovereign right to regulate and maintain the rule of national emigration laws. However, the tainted approach of a decadent nation-state concept fails to understand the root causes of modern African emigrations, and to properly manage its potential of shared opportunities. As if looked down on from above, the African is treated with extreme arrogance by the citadel. The question is how long will it last? What will be the geopolitical consequences? A European immigration policy based on win-win partnerships with Africa is long overdue. This policy must go far beyond security-based and project-based approaches that provide only patches to an exploding barrel.

International Hypocrisy

'The trend of increasing numbers of migrants taking ever greater risks to reach the European territory is almost symbolic of the EU's failure so far to effectively address the root causes of migration' (Amnesty International, 2006). The entire western world shares the same hypocrisies, which can be described with a few observations: globally, African migrants are negatively affected by a generally poor observance of international obligations under human rights and refugee laws. The principles of international burden-sharing and protection are still a dream. The result is a constant abuse of basic rights and human dignity suffered by migrants in the name of the laws of states. But no human being is illegal, and the principle of 'illegal immigrant' is potentially dangerous when used without international control.

The persistence of colonial trends in international relations is reflected by the collusion of failing states and global interests driven by the thirst for natural resources. This is exacerbated by the emergence of a new giant, China, that has taken the game to the next level. We are witnessing a new scramble for Africa, or rather for African resources, which is further squeezing the continent and its people into another forty-year period of lost geopolitical dreams. No lessons learned from Congo, Nigeria and Sudan seem to work for the African masses. Instead, the key global players now know how to get the oil but are not interested in changing the lives of those who really own it.

The poor understanding of the real dimensions of forced displacements in Africa is further complicated by European arrogance and the failure to understand and maximize the benefits of migration for all parties. Despite this, it can be argued that in all its forms, African migrations to Europe result in net economic gain for host countries and net loss for poor African countries, which did not anticipate it through sound policies. Labour force and brain-drain losses are obvious, but there are more geopolitical pitfalls to be investigated.

International Obligation to Protect Migrants

It may be arguable, but the classic concept of the nation-state is dead and will not help failing immigration policies in European countries that feel threatened. It is important to understand Africa's late arrival in the historical trend of global economic migrations and its right to a share, as the continent itself has been an amenable host to several hordes of European and Asian migrants. There is an international obligation of accountability in the protection of African migrants, mostly uprooted people that do not intend harm in their quest for freedom and sustainable prosperity.[2]

While freedom of movement is recognized as a civil right within national borders, no state currently allows full freedom of movement across its borders, and international human rights treaties do not confer a general right to enter another state. According to Article 13 of the Universal Declaration of Human Rights, citizens may not be forbidden to leave their country. Curiously, there is no similar provision regarding entry of non-citizens in a sovereign country. It is an important aspect that needs to be addressed urgently. In this regard, an international campaign for the right to 'emigrate' is due.

Host states should be reminded of their responsibilities and accountability for 'uprooted people' at home and in host countries. This should be translated into respect for refugees and migrants' rights in the fight against 'irregular migrations'. It is important to recognize protection as a key right for all emigrants, irrelevant of their legal status in the host country, as they do not have an intention to harm. Equally fundamental is the 1990 UN Convention on the Rights of All Migrant Workers and Members of their Families.

Recently, we have witnessed the EU participating in high-level meetings on the links between migration and development, such as the Rabat Conference in July 2006, the UN High Level Dialogue on migration and development in September 2006 and the Tripoli Conference the same year. However, more needs to be done through an effective engagement with civil society in vulnerable societies and disadvantaged groups, to address the root causes and work towards a better ethic of migration in a more equitable world. Migration is today at the point where international trade was fifty years ago, says Mr Dhananjayan Sriskandarajah of the Institute for Public Policy Research in the US. For many at that time, the current governance system for international trade was unimaginable. 'Those thinking about a new international framework for managing migration face remarkably similar challenges,' he has stated. 'How to design a system that leads to freer and fairer flows of people, skills and remittances?' 'I dread to think of the scenes we may be contemplating in, say, 20 years if we do not make a massive consolidated effort to create jobs and opportunities in West Africa,' said UN Special Representative for West Africa, Ahmedou Ould-Abdallah. 'What is happening now is only a tip of the iceberg, compared to what will occur if urgent solutions are not found.'[3]

References

Amnesty International, *EU/Africa: Migration and development* (20060<www.emhrn.net>.

African Union, Executive Council, *The Migration Policy Framework for Africa* (20060.

Baldwin-Edwards, M., 'Migration into Southern Europe: Non legality and markets in the region', *Insecurity Forum* (2005) <www.insecurityforum.org>.

Guild, E., 'Migration and the Challenge to National Sovereignty', Ninth Session of the Complex Dynamics of International Migration Interdisciplinary Seminar on the Conceptualization of the Migration Phenomenon 2005–2006 (2006)<http://cdim.cerium.ca>.

Human Rights Watch, *Stemming the Flow, Libya, Abuses Against Migrants, Asylum Seekers and Refugees*, Part III (2006) <www.hrw.org/reports/2006/libya0906/libya0906web.pdf>.

Ibrahim, M., 'Securitisation of Migration: a racial discourse', *Insecurity Forum* (2005) <www.insecurityforum.org>.

ILO, Southern Africa Multidisciplinary Advisory Team, 'Labour Migration to South Africa in the 1990s', Policy Paper Series No. 4 (Harare, Zimbabwe: ILO, 1998).

International Organization for Migration, *Identifying International Migrants* (2006)<www.iom.int>.

Koenig, D., 'Toward local development and mitigating impoverishment indevelopment-induced displacement and resettlement', Working Paper No. 8 (London: Refugee Study Centre, 2002).

Larsaeus, N., 'The Use of Diplomatic Assurances in the Prevention of Prohibited Treatment', Working Paper No. 32 (London: Refugee Study Centre. 2006).

Mutume, G., 'African migration: from tensions to solutions', *Africa Renewal*, 19, 4 (2006) <www.un.org/ecosocdev/geninfo/afrec>.

Oliver-Smith, A., 'Displacement, Resistance and the Critique of Development:From the grass-roots to the global', Working Paper No. 8 (London: Refugee Study Centre, 2002).

Russell, S.S. (2002), 'Refugees: Risks and Challenges Worldwide', Center for International Studies, Massachusetts Institute of Technology, Migration

Information Sources (2002) <web.mit.edu/cis/www/migration>.

Turton, D., 'Conceptualising Forced Migration', Working Paper No. 12 (London: Refugee Study Centre, 2003).

Selected Legal Instruments for International Migrations

1. Universal Declaration of Human Rights
2. Convention on Prevention and Punishment of Crime of Genocide – 1948
3. International Labour Organisation Convention No. 97 Migration for Employment Reviszed – 1949
4. Convention Relating Status Refugees – 1951 and ITS Protocol – 1967
5. Convention on Political Rights of Women – 1953
6. International Labour Organisation Convention Discrimination IRO Employment and Occupation – 1958
7. Convention Against Discrimination in Education – 1960
8. International Convention on Reduction of Statelessness – 1961
9. International Labour Organisation Convention No. 118 Equality Treatment of Nationals and Non Nationals in Social Security – 1962
10. International Convention on Elimination of All Forms of Racial Discrimination – 1965
11. International Covenant on Civil and Political Rights – 1966
12. International Covenant on Economic, Social and Cultural Rights – 16 December 1966
13. International Labour Organisation Convention No. 143 Migrations in Abusive Conditions and Promotion of Equality of Opportunity and Treatment of Migrant Workers – 1975
14. Déclaration sur la race et les préjugés raciaux, Paris, UNESCO – 1978
15. Convention on the Elimination of All Forms of Discrimination Against Women – 1979
16. International Labour Organisation Convention No. 157 Concerning Establishment of an International System for Maintenance of Rights in Social Security – 1982
17. Convention Against Torture and Other Cruel Inhuman or Degrading Treatment or Punishment – 1984
18. Déclaration sur les droits de l'homme des personnes qui ne possèdent pas la nationalité du pays dans lequel elles vivent, Assemblée Générale des Nations Unies, 13 Decembre 1985
19. Convention on the Rights of the Child – 1989
20. International Convention on Protection of Rights of All Migrant Workers and Members of Their Families – 1990

21. Protocol to Prevent, Suppress and Punish Trafficking in Persons – Especially Women and Children – Supplementing UN Against Transnational Organized Crime – 2000
22. Protocol Against Smuggling of Migrants by Land, Sea and Air – Supplementing UN Convention Against Transnational Organized Crime – 2000
23. United Nations Convention Against Transnational Organized Crime – 2000

Georges Tadonki is Senior Regional Information Adviser, UN Office for the Coordination of Humanitarian Affairs (OCHA), Johannesburg.

NOTES

1 This article expresses the author's personal view and not that of the United Nations.
2 'Uprooted people', a concept that well defines most African immigrants, was introduced in by the World Council of Churches in 1995.
3 Quotes from Sriskandarajah and Ould-Abdallah are from G. Mutume, 'African Migration: From Tensions to Solutions', *Africa Renewal*, 19, 4 (2006).

Migrations and Nation Building in Black Africa: The Case of Ivory Coast since the Mid-Twentieth Century

PIERRE-AIMÉE KIPRÉ

Introduction

In the deep crisis that Ivory Coast is currently experiencing, an essential part of the debate is focused on the issue of immigrants. Some people argue that all immigrants living on Ivorian soil[1] should automatically be granted Ivorian citizenship by the state. Others think that Ivorian citizenship should not be granted even to those who have been resident in the country for a long period. The debate polarizes those who believe that citizenship is a right granted by birth in a particular state and those who believe that it derives from descent. This antagonism partly explains the magnitude of the country's internal conflict, as all states of West Africa are represented in Ivory Coast by their immigrants. Even though the number of immigrants[2] is currently decreasing, the country still houses the highest percentage of foreigners (26 per cent of the population in 1998) in black Africa.

Born as an independent state in 1960 out of the French empire, a state which overturned all the socio-political and economic structures in place before the twentieth century, and like the rest of the countries of the French empire, the state of Ivory Coast eventually achieved national consciousness by means of associations which thought of themselves as specifically Ivorian, on the pretext that they had been created by communities which had been in place before colonization.[3] But what is the point of arguing about citizenship rights when there have been massive migrations in Africa since the colonial conquest, and when these have still not introduced the same concept of citizenship in all African states?

A crossroads of different peoples even before the twentieth century, Ivory Coast was the French colonial territory in West Africa where there was the largest and most varied migration throughout the colonial period. An analysis of this case can illustrate some aspects of state-building in post-colonial Africa and some of the paradoxes of the geopolitics of contemporary Africa, as well as the socio-political challenges of regional integration.

The purpose of this article is to clarify the issue of migration in Ivory Coast and to show why Ivory Coast has become the principal pole of attraction for West African migration between the mid-1950s and the end of the 1980s. It will discuss the perception that Ivorians and immigrants have of one another and of their role in the construction of the country – a perception which is divided between collaboration and enmity. Finally, it will look at ways of escaping from approaches which generate conflict and above all at how to respond to socio-political challenges, in Ivory Coast as well as in other West African states, in order to give regional integration a chance.

The Scale of Migration in Ivory Coast since the Mid-Twentieth Century

The history of migration in Ivory Coast shows how the colonial period marks the starting-point for massive migrations in this country, flows which have remained large in the post-colonial period, at least until the 1980s.

In the inter-war period, migrants were largely obliged to leave their country of origin to flee forced labour. From the 1940s onwards, migrants were essentially seasonal workers, as forced labour was abolished in 1946 and the 'Code of indigenous labour' adopted in 1950.[4] Whether or not they were organized by the SIAMO trade union (*Syndicat interprofessionel d'acheminement de la main d'oeuvre*), convoys of immigrants from other colonies, or from Savannah regions of the colony itself, were brought to the forest zones. This stimulated coffee and timber production in the 1950s: the 'boom' in population accompanied the economic boom.

Far from falling at independence, the influx of migrants grew (see Figure 1). The wave of *coups d'état* in Africa and the political uncertainty caused by certain regimes meant that there were occasionally politically refugees. But, in general, because of the strong economic growth which Ivory Coast experienced (an average rate of 9 per cent a year from 1959 to 1979), it was above all economic refugees who came into country from 1960 to 1980, with a sharp rise in the period 1970–80 at the time of the climate crisis in the Sahel countries. During the 1970s and 1980s, there was an influx of Fula herdsmen to the northern savannahs. As in Mali and Burkina Faso, their relations with the peasants in this zone became difficult in the 1980s.

Above and beyond these significant economic factors, the demographic explosion which had occurred all over West Africa since the 1950s also explains the magnitude of the movements of population towards those regions where agriculture was expanding. In terms of the jobs and income other than those which such an economic system allow, the economic policy for both the colonists and the post-colonial period did not anticipate the effects of the demographic revolution. Thus the cocoa, coffee, timber and, later, oil sectors – that is, the forest zone of Ivory Coast – imported the bulk of able-bodied workers including migrants. The dynamism of the port of Abidjan, from its opening in 1950 onwards, and the structural effects of economic growth in secondary towns did the rest.

Thus, since the end of the colonial era, Ivory Coast has had a classic geopolitics of migration. There are zones of emigration (the whole savannah and the Sahel) and zones of immigration (the forest regions and their peripheries). The trends thus created cause an insidious depopulation of the so-called 'poor' regions, which are reduced to simple suppliers of labour and which are therefore major areas of emigration. During the colonial period, Upper Volta (now Burkina Faso), French Sudan (now Mali) and Niger found themselves in this position, to the benefit of the peanut plantations in Senegal (the seasonal movement of the so-called '*navétanes*') and then later above all to the benefit of the coffee and cocoa plantations of Ghana and Ivory Coast. In Ivory Coast itself (see Figures 3 and 4), all the savannahs of the North and the centre were depopulated for this reason: in 1998, this part of the country, which represents 60 per cent of Ivorian territory, received only 13 per cent of the immigrants and counted 27 per cent of the whole population, against 40 per cent at the end of the 1950s.

Studied on the basis of biographies or censuses, these population movements changed both the composition and also the geographical layout of the Ivorian population. Over time, Ivory Coast became the main centre of attraction for the migratory flows in West Africa. Already in 1937, migration to the Ivory Coast was larger than migration to Senegal. In 1955, Ivory Coast had more immigrants from Upper Volta and French Sudan than Ghana did. This long-term trend, from which the Ivorian authorities sought to profit at the end of colonization (see the second four-year plan of 1958–1962) was maintained and even increased after 1965 and into the mid-1980s, making Ivory Coast the principal destination for West African migrants, with more migrants than any other African or European country.[5] The percentage of foreigners in the total population, which was estimated at 11 per cent in 1958, was 18 in 1965, 22 in 1975, 28 in 1988 and 26 per cent in 1998 (see Figure 1).

However, from 1980 onwards, Ivory Coast, like the rest of black Africa, suffered a serious economic depression and fell back on plans for structural adjustment. Job opportunities fell. Migratory trends dropped off, although there were no large-scale returns of immigrants to their countries of origin.

Longer-standing immigrants, those who had been in the country for fifteen years or more, and foreigners born in Ivory Coast, soon became at least as numerous as more recent immigrants.[6] However, a close analysis of the statistics and of the accounts given by certain immigrants shows that, at the turn of the 1980s, not only did Ivory Coast start to become a stepping-stone to other destinations, especially western Europe (above all France and Britain) but also Ivorians themselves started to emigrate, especially to Europe.[7]

In other words, the migration issue changed. In the context of the social and economic crisis of the 1980s, the social effects of the demographic increase (the population rose by a rate of 3.5 per cent a year in the 1980s) and of the effective integration of millions of long-standing immigrants began to be keenly felt. Although these latter were still regarded as foreigners, many of them had been born in the country.[8]

Ivorians and their Long-Standing Immigrants: Connivance and Rejection during the State-Building Process

At the end of the 1950s, as was the case with other colonies,[9] Ivory Coast was confronted with the question of immigration and foreigners. Acts of violence were perpetrated against Dahomen and Togolese immigrants in October and November 1958 which the colonial authorities had difficulty containing.

However, in French West Africa, whether through the promotion of pan-African ideas or through the action of federal parties like the *Rassemblement démocratique africain* (African Democratic Alliance) or the African Regroupment Party, and trade unions (for instance, UGTAN, the General Union of Workers of Black Africa), there were numerous friendly exchanges. People fought for common goals, especially since they were all in the same colonial boat. Then, it seemed, the divisions were not between 'foreigners' and 'natives' but between colonizers and colonized.

In fact, there were attempts to exploit the immigrant populations for political ends, since their desire to return to their countries of origin remained permanent since the forced displacements of the inter-war years.

Indeed, as soon as the colonial relationship was set up, the balance of power set up by the colonial system reinforced the ancient feeling among West Africans that there was an irreducible difference between them and the white colonists. This feeling was present in the fight for political independence. At the same time, colonization introduced new perceptions and definitions of the foreigner which eventually superimposed themselves on traditional ideas. First of all, there was the establishment of borders between territories. Then, above all in the inter-war period, a whole pseudo-psychological discourse grew up which classified various 'native' communities in all the colonies according to certain labels ('the courageous and docile Voltaians', 'the recalcitrant Senegalese', and so on). Finally, there were administrative practices founded on 'the control of native populations' and 'the respect of local customs'.[10] These encouraged populations to identify themselves with certain territories. The post-colonial authorities were to keep these policies intact.

There was evidence of this territorial consciousness throughout French West Africa when the empire promoted the autonomy of each colonial territory in the framework law of 1956. Numerous functionaries and assistants of the colonial regime returned to their countries of origin, while seasonal migration continued to be more significant than permanent migration (see Figure 5.)

In sum, when the existing state was created in 1960, it was not the case that the values of hospitality and traditional welcome were the rule everywhere or for all people. They had already been weakened by the ideology and socio-economic policies of the colonial system. It was on the basis of those policies that nation-building was implemented throughout West Africa. When the French colonial empire was moving towards independence for its colonies, the acts of violence committed in October and November 1958 against the Dahomen and the Togolese in Ivory Coast showed how extremely sensitive the question of immigrants had become at the end of the colonial period. It also raised the question of how to integrate immigrants into colonies like Ivory Coast, into which the colonial power had directed the bulk of migrants in French West Africa.

However, from 1961 onwards, the new authorities took a number of new initiatives in this area. They wanted to free up access to the whole of Ivorian territory to immigrants as much as to local Ivorians, including all those who had moved to different parts of the country to work the land. The law on nationality of 14 December 1961 awarded Ivorian citizenship on demand to all immigrants from the former French West Africa who had come to Ivory Coast before 1960. In 1966, a bill allowing dual citizenship was drawn up which would have given the same civil rights as those enjoyed by full citizens to all nationals of the *Conseil de l'Entente* (Council of Understanding: Benin, Burkina Faso, Ivory Coast, Niger and Togo). Even though this bill was withdrawn by the government because of certain concerns,[11] every African immigrant none the less had the right to settle anywhere in Ivory Coast and to work the land. President F. Houphouët-Boigny said, 'The land belongs to him who makes good use of it.' By the same token, all foreigners were allowed to repatriate all their goods to their country of origin, without any restrictions. Finally, from 1980 onwards, circumventing the opposition expressed in 1966 and the restrictions introduced in the citizenship code of 21 December 1972 (there was a long naturalization procedure and a probationary period of ten years before the applicant could enjoy full civil rights), a presidential decree provided that all nationals of West African states could participate in every election in Ivory Coast, just like Ivorian citizens.

Until the 1990s, this set of measures constituted the official framework of Ivory Coast's policy on immigration. President Houphouët-Boigny and the single party which governed Ivory Coast were determined to forge all the inhabitants of the country into one mould of citizenship 'without distinction of age, region of origin or sex, in order to create a coherent nation'.[12] For the new authorities, independence meant integration of immigrants based on the strategic choices made in 1959 when the *Conseil de l'Entente* was set up, including the strengthening of the position of Ivory Coast in African–French relations and in those of the West with black Africa in general. This strategic choice was the opposite of the 'federalist' Senegalese strategy adopted at the end of colonization, although the economic aspects of the Senegalese option, especially the development of plantations from 1950 onwards, were retained. This policy of attraction was a success to the extent that it attracted huge numbers of immigrants from all over West Africa, but it is surprising to note the modest number of naturalizations between 1965 and 1998: in 1998 there were fewer than 100,000 naturalizations (see Figure 2). Perhaps the immigrants were not interested in obtaining Ivorian citizenship. Perhaps naturalizations were in fact only granted very grudgingly, even under President Houphouët-Boigny. Perhaps people simply accepted their lot, whatever it was.[13]

In general, the majority of the Ivorian population adapted to the official position.[14] In villages and towns, the settlement of immigrants or of Ivorians from other parts of the country posed no problems. The recruitment of foreign labourers on plantations or in industry was also problem-free before 1980. It seems that there was a great deal of good will towards immigrants, as peasants were happy to give them the use of parcels of land without being pressurized to do so by the authorities. There was also non-African immigration too: the French colony in Ivory Coast was the largest in black Africa, almost 50,000 even by 1975, and Syrians and Lebanese were also numerous (300,000 in 1985). This hospitality not only corresponded to economic and political needs, it was also presented as being an untouchable African value. At a time of one-party and even one-man rule,[15] the exercise of civil rights seemed to be universally accepted.

Even if some have interpreted this official position as being one of exploitation of immigrants, immigration started to become an issue not so much for political reasons but instead as a result of social and economic concerns. The civil rights accorded *de facto* to the 'foreigners' from West Africa were not criticized so long as their 'quasi-citizenship' was understood as a way of restricting the labour market and business opportunities outside the agricultural sector. For some people by the end of the 1970s, though, it appeared to be accentuating a *de facto* monopoly of certain immigrants in certain sectors: charcoal production, construction materials, car repair, small retail, and so on.

Moreover, until the agricultural crisis became a land crisis and a general social crisis in the mid-1980s, with the return to their villages of numerous unemployed town-dwellers, criticism and hostility were expressed only by the urban classes, usually more so in private than in public. By contrast, in the south of Ivory Coast, with the arrival of the Fula herders after the climatic crises of the 1970s in the Sahel, the countryside became the area where there were an increasing number of violent and even bloody clashes between herdsmen and peasants.

Besides these incidents between Ivorians and foreigners, which were always rapidly contained by the authorities, it is important to emphasize that, on the political level and between Ivorians, the monolithic political system did not prevent a perverse alliance being created between those in favour of integrating the West African region and those in favour of 'an Ivorian nation'.[16] The first movement was supported by a section of the middle class which wanted to attain positions of power within or against the one-party system: it was based on an analysis of local mechanisms of patronage to arrive at an ethnic and trans-national vision of politics. The second movement, described as 'nationalist' by the media at the time, was expressed principally at sports matches against other African countries from the mid-1970s onwards. It emphasized 'Ivorian interests' without ever saying what exactly they were. The first truly serious acts of violence committed against 'foreigners' since independence occurred after a football match between Ivory Coast and Ghana in 1985.[17]

These acts of violence and these expulsions of foreigners, which occurred in a number of African countries throughout the 1960s and 1970s, seemed to gain a following in Ivory Coast from the 1980s onwards. The local press highlighted criminal cases involving foreigners, thereby encouraging the stereotype of immigrants as thieves, drug traffickers, or murderers. In the mid-1980s, the climate between Ivorians and immigrant communities deteriorated progressively. The social effects of the economic crisis and the plans for structural adjustment, which pushed up unemployment and poverty in 1981–82, provided the pretext for people to complain that immigrants had 'too great a role' in Ivory Coast. Such complaints took root in ever-larger social groups, especially in the middle classes. Numerous immigrants started to leave. Thus, from the mid-1980s onwards, there was increasingly large net emigration out of Ivory Coast. How is this change of attitude to be explained?

The explanation generally given lies in the long economic depression from 1980 to 1990 and the social crisis which accompanied it. Although this explanation is not wrong, it is insufficient. In reality, the effectiveness of the decisions taken in 1960 has not been evident for a long time, since the young state in fact had no control over migratory flows because it neither controlled its own borders nor had any overall policy for integrating immigrants where they settled. The response to these issues was merely pragmatic and, for the vast majority of immigrants, was based on economic considerations alone. The authorities paid no attention to the fact that the question of immigrants depends partly on how foreigners are perceived when they do not melt away into their new environment. There is therefore a constant tension between communitarianism and assimilation. To allow everyone to settle was not enough: social inclusion, and the desire to belong to the new community, had to be deliberately promoted.

Because immigrants to Ivory Coast come above all from neighbouring countries, it is always believed that they will return home once they have made their fortunes. The 'communitarian' choices of immigrants – their decision to live together in their own parts of town or in encampments, and their way of life – have reinforced their status as 'foreigners' in their country of residence. Some Ivorians were even concerned at the high employment rate among immigrants (73 per cent in 1993): indeed, even if there are plenty of immigrants in modern sectors of the economy, they tend to put in place networks of mutual assistance which have often allowed them to enjoy dominant positions for decades, especially in the informal economy and in agriculture.

It was in this context that rhetoric about national preference started to become popular in the mid-1980s. There was an insidious press campaign against the social marginalization allegedly caused by 'foreigners'.[18] It was also at this time that the notion of a 'threshold of tolerance' appeared; adopted from rhetoric about immigration in France at the time, this idea supported the positions of certain defenders of 'Ivorianness', a concept which made its début in public discourse in Ivory Coast in 1995 but which in reality dates from the 1970s.

The concept of 'a threshold of tolerance' was at the heart of a report published by the Economic and Social Committee in April 1999 which explicitly attacked the Ivorian policy of open borders. According to this concept, there is a certain percentage of foreigners above which integration is impossible and national cohesion endangered. This notion is grounded neither scientifically nor in any empirical analysis of facts and, in any case, it has no practical effect: the borders are so porous that Ivory Coast – like other countries in the region – is not able to fix its so-called 'threshold of tolerance' (even if such a thing exists), since none of these countries is in the position to measure the number of migrants coming in or going out of the country. Having recourse to traditional social networks based on cultural affinity,[19] immigrants simply bypass the border-crossing posts and move around more or less as they please, quite independently of official state policy or of relations with other states.

In the 1990s, a new factor made the immigration question even more dramatic. Attacks on civilians in the civil wars in Liberia and Sierra Leone caused involuntary or forced migrations which made a regional solution of the whole question of migration even more urgent than before. The exploitation of immigrants for political purposes[20] nourished the hostility of the receiving countries as much as the ethnicization of domestic political relationships, to the detriment of the nation-building project itself. Since the general election of 1980, but especially since the election in 1990, this was a recurrent theme in the struggles for power among the urban middle class in Ivory Coast.

For the immigrants, the fact that they have changed country itself reflects imbalances in their countries of origin. These include famines and poor harvests, low incomes, wars, and so on. But the immigrant always remains fundamentally attached to their community of origin: it is a fixed point in the strategy for survival immigrants have adopted by moving abroad more or less for the long term. When circumstances allow, networks of migrations are set up, involving financial transfers and even involvement in the political life of their country of origin and their country of residence. In the case of Ivory Coast, which is above all a country of immigration, financial transfers abroad by immigrants greatly exceed similar transfers by Ivorians back to their home country (see Figure 6). When legislation in immigrants' countries of origins allowed this in the 1990s, immigrants took part in the political life of their home countries without giving up their right to vote in Ivory Coast, awarded in 1980. In spite of themselves and as a result of the inadequacy of a policy based an exclusive understanding of the nation-state in young states, immigrants find themselves caught in a pincer grip between their reason for leaving home and the various issues raised by their presence in their country of residence. They are left with no choice but to cultivate their 'national' difference, according to circumstances, precisely there where they might have been able to build a new life. This situation is the same in Ivory Coast as in other West African countries.[21]

In Ivory Coast, this situation put great strain on the open-doors policy which had been in place since 1960. From the 1990s, measures were taken to limit the number of immigrants, notably by means of the introduction of a residence permit in 1991,[22] and also by limiting their economic rights. Access to the civil service was restricted in 1991; the law on the land ownership was introduced in 1998. Initially a budgetary measure, the residence permit soon became a means for selecting immigrants and for keeping out the poorest ones by preventing them from coming to Ivory Coast. This measure shows that the state intends to monopolize and to regulate people's right to move and settle.

The land reform of 1998 was the result of a desire to reduce the pressure on land but it took no account of 'customary' rules governing the access of newcomers to land, including Ivorians from different regions, territories, or families. The social effects of this private appropriation of the land, which was pushed heavily by the World Bank in 1991, was advantageous for local owners of capital who were in a position to relaunch the plantation economy (especially hevea and palm oil), that is, the middle classes. But the consequences were negative for the rural poor, especially young people and women, and for immigrants who had obtained land to cultivate. This decision, just like the exclusively citizen-based approach to politics, represented a break with tradition. Few people anticipated what a destabilizing effect it was have the social cohesion of this young state.

The abolition of the right to vote for immigrants in 1995 showed the determination to affirm the exclusively national character of civil and political rights in Ivory Coast, where the presence of common populations on both sides of state boundaries prevents any effective border control. Habits acquired during the period of one-party rule have remained in place even now that parties must win elections. This new departure reflects a domestic political debate about how to win power, a debate conducted very openly among the middle classes as a result of the chaotic transition to democracy and pluralism in politics and the trades unions. On the international level too, the debate is open: it concerns relations with neighbouring peoples of whom it is not certain whether they are friends or enemies, especially with the collateral effects of the civil wars in Liberia and Sierra Leone. At the same time, since the death of F. Houphouët-Boigny, Ivory Coast's position within France's strategy for West Africa has crumbled, just as France's role in the development of Ivory Coast has too. The ease with which the military coup of 24 December 1999 occurred illustrates the geostrategic situation in West Africa.

Thus the policy of welcoming and integrating immigrants has failed. This policy exploited immigrants within a policy framework inherited from the colonial period, although the exploitation was less overt under the one-party system. The crisis of the social and economic model caused severe tensions in the 1980s. Contradictory concerns were expressed: there was both the rejection of the idea that Ivory Coast should be a high-immigration country, lagging behind other countries in West Africa, and complaints that the country had no policy for integrating immigrants within a nation-building project, but also a desire for Ivory Coast to play an active role, or even to be a leader, in implementing a project for regional integration. These relations of connivance and repulsion nourished one another and weighed heavily on the immigration question in the young Ivorian national community.

The Issue of Migrations in the Construction of a West African Regional Sub-Group

From the point of view of the free movement of goods and persons, the attempts to create a regional group in the 1970s (the *Communauté économique de l'Afrique de l'Ouest* [Economic Community of West Africa – ECOWA] and the *Communauté économique des Etats de l'Afrique de l'Ouest* [Economic Community of West African States – ECOWAS]) would tend to suggest that, for the Ivorian authorities, the border was not the dividing line between two sovereign states but instead a zone of shared sovereignty in West Africa. The economic, civil and political rights which were *de facto* granted to all West African nationals settled in Ivory Coast created juridical ambiguities which were analysed differently by experts before 1990. For some people, the single party, the PDCI-RDA, wanted to exploit foreigners for political purposes; for others, it was the desire to realize African unity. After 1990, the change of policy with regard to immigrants was understood by some as the expression of a profound 'xenophobia' of the Ivorians or of an 'ethno-nationalism' which treated immigrants like scapegoats.

These interpretations are excessive. Those which develop the theory of 'ethno-nationalism' because they are close to ethnographic analyses which classify Ivorian society in terms of 'ethnic groups' and 'tribes', as in the colonial period, are simplistic. They bear little relationship to the sociological complexity of this country and to the profound effects of fifty years of economic changes and socio-demographic mutations. It is much more important to study social phenomena, notably the evolution of the political views of the middle classes which vary from those who support the complete opening of frontiers to those who want to reduce the role of foreigners in the country. When the social effects of policies of structural adjustment started to make themselves felt in the 1980s and when power struggles grew more acute, the hostility of the middle classes made itself felt more strongly, both in order to 'liquidate' a part of the popular base of the former single party (for example, the demand for the right to vote to be withdrawn from non-Ivorians in 1990), and also to better resist the general economic crisis by reducing the number of people who benefited from ground rent generated by the exploitation of the country's raw agricultural materials.[23]

It was at this time, at the beginning of the 1990s, that the idea of regional integration was relaunched, in the Francophone countries after the devaluation of January 1994 as in the whole of West Africa (the CEDEAO [*Communauté économique des Etats de l'Afrique de l'Ouest*] treaty was revised in 1991). The position of Ivory Coast in the West African economic area is more favourable than that of other countries (Ivory Coast has 40 per cent of the GDP of the West African Monetary Union and 27 per cent of the GDP of

West Africa, while it is the biggest trader within ECOWAS. The decision to break with the economic policies pursued between 1960 and 1989, a decision imposed by the middle classes, is therefore contradictory if not suicidal for the geopolitical and economic balance of West Africa. It will aggravate the overall situation in the region as in Ivory Coast itself if the logic of 'the exclusive nation' is not abandoned in favour of a policy directed towards the West African region as a whole. Such a regional policy would take account of the development of West Africa as a whole, as of changes which have occurred in the composition of the Ivorian population since at least 1950.

The attitude towards immigrants therefore makes foreigners foreign. All the fears and fantasies which are caused by the system of exchange developed in the twentieth century are projected onto the immigrant, to the detriment of most of them. It is essential to break out of this vicious cycle by adopting a regional approach to the question of migration and to the problems it poses.

A necessary condition for any policy on migration, whether it is national or regional, is an appreciation of the role of the immigrant (and therefore of the foreigner) in those regions where there are significant movements of population. This appreciation of the role of foreigners, in Ivory Coast as in the rest of West Africa, is based on the following elements:

- A better understanding of common populations and cultures. We have a memory of common struggles of West Africa in the recent past against all forms of subjection and domination, within the framework of a globalization process which has never been favourable to us since the colonial era. This can contribute to the establishment of a common historical consciousness in the West Africa region and it can help to improve the image of the inhabitants of each country in the eyes of its neighbours.

- Development of the role of schools and the media in improving the image of the foreigner migrant. It is through education that the negative images must be reversed, in favour of a more positive attitude towards other peoples.

- Policies for the integration of migrants should be formulated and harmonized, with reciprocal rights and duties in all places where they are present.

Such a policy would be an element in the overall strategy of regional integration. Migration policy should consist of:

- A regional policy, based on acts of regional integration
- Measures in favour of helping populations who have been damaged by the fact that they do have not adapted to the territory where they live or to

their methods of production. The notion of 'return' which is sometimes invoked in order to call into question the efficacy of policies directed at integrating migrants is, in fact, an illusion. With the exception of seasonal movements of population, there is no 'definite return' but instead a 'territorial plurality' in which immigrants come and go.[24]

- A regional policy for the management of territory which would take into account the true value of neighbouring populations with respect to one another on the basis of the fact that they are engaged in permanent and multiple exchanges with one another. By this means, strategies for the integration of peoples would achieve more than policies for the harmonization of state policies. The common implementation of policies for the management of territory, and the direct involvement of populations in this process, could be essential elements for such an integrationist policy. They would enable new links of development to weave together populations in the respect of dignity and human rights.

- The harmonization of legislation with a view to redefining state citizenship, the way it is implemented and the way it could be extended to create a regional citizenship, with rights and duties for all.

These propositions flow from the duty to pacify the situation, above all in areas where there is high immigration. The notion of the 'free circulation of goods and persons', which is currently based too much on economics, can only be suspect in the eyes of those who do not profit directly from it or of those who, in countries of immigration, think they lose more than they gain from migratory flows.

Conclusion

The manipulations and these reinventions of personal history or of the history of the community oblige us to have a critical attitude towards those written or oral sources, above all when the issue of foreigners and immigrants opens up questions of who is autochthonous, in a part of the world where there has been a long history of movements of population.[25] This situation means that the distinction between the right to citizenship for all those born in the country, and the right to citizenship only for those whose parents are Ivorian, does not apply in young political communities such as the states of West Africa.

The porosity of frontiers, inherited from colonization; the numerous common populations on both sides of state borders; the uncontrolled causes of migration, such as the demographic explosion, natural calamities, the pauperization of the masses as a result of bad economic policies, civil wars, and so on – all these cause permanent changes to the population map, especially in zones of immigration.

Ivory Coast has been a major destination for immigrants since the colonial era. It now finds itself trapped by a nation-building process based on the paradigms inherited from French colonization. In the West African region, this is an important factor behind the geopolitical reconfiguration now underway, both in terms of the movements of population and also in terms of the strategic choice made by states to move towards regional integration, that is, towards the decline of the nation-state invented after independence.

To understand this reconfiguration and its relationship to the question of intra-regional migration, it is essential to pay much closer attention to social factors, particularly the evolution of the position of the middle classes which varies according to the economic situation and to the balance of power within it. In the Ivorian case under examination here, we have seen that those positions went from one of support for borders open to all, to a restrictive immigration policy and a reversal of the earlier promotion of immigration once the colonial system entered into crisis, a crisis which only worsened after colonialism, between 1960 and 1990. Trapped by their own development strategy, which was based on the advantages of that system, these middle classes now refuse to see the country continue to accept immigrants. It is in this framework that power struggles are defined and accentuated, not the other way around.

Exploited by those who play on cultural differences and on concepts of citizenship which have not been internalized by the majority of them, immigrants find themselves caught in a pincer between the strategies adopted to solve the general crisis in West Africa. The implementation of a regional policy for migration in West Africa could enable us to reduce the causes of political conflicts, especially acts of violence committed against immigrant communities, as well as private conflicts such as those over land. The simple need to survive, and the simple facts of economics in a society which is generally very dependent on the outside world make regional solidarity more difficult, especially in a period of economic crisis, at least if we do not emphasize our common challenges of development, our cultural links and, above all, our shared human dignity.

Short Bibliography

Banegas, R. (2006), 'Côte d'Ivoire: Patriotism, Ethnonationalism and Other African Modes of Self-Writing', *African Affairs*, 105, 421, pp. 535–52.

Bendraogo, Z. (1999) 'Les migrants Burkinabé en Côte d'Ivoire: réflexions sur la participation au développement et l'implication dan le banditisme 1947–1997', International Conference at Université Paris 7, 'Etrangers et migrants en Afrique', Paris, 5–9 December.

Bredeloup, S. (1995), 'Tableau synoptique – Expulsions des ressortissants ouest-africains au sein du continent africain (1954–1995)', *Mondes en développement*, 23, 91, pp. 117–21.

CSAO/OCDE (2006), *Atlas de l'intégration régionale en Afrique de l'Ouest – série population* (Paris: CSAO/OCDE).

Dozon, J.-P. (2000), 'La Côte d'Ivoire entre démocratie, nationalisme et ethnonationalisme', *Politique Africaine*, 78, pp. 45–62.

Kipré, P. (1986), *Villes de Côte d'Ivoire 1890–1940*, tome 2: *Economie et société* (Dakar-Abidjan-Lomé: NEA).

Kipré, P. (2005), 'De l'immigration à l'intégration: le cas des villages burkinabé de la région de la Marahoué', in Ch. Chanson-Jabeur and O. Goerg (éds), *Mama Africa; Hommage à C. Coquery-Vidrovitch* (Paris: L'Harmattan).

Kipré, P. (2005), *Côte d'Ivoire. La formation d'un peuple* (Paris: SIDES-IMA).

Le travail en Afrique noire; edn spécial de *Présence Africaine*, Paris, 1951.

Mamdani, M. (1996), *Citizen and Subject: contemporary Africa and the Legacy of Late Colonialism* (Princeton, NJ: Princeton University Press).

Plan décennal de développement économique et social 1965–75 (1966) (Abidjan: Imprimerie nationale).

Sandlar, Chr. (2005), 'Les titrologues de l'ivoirité', *Outre-terre*, 11.

Statistical Manual of the United Nations Conference on Trade and Development, 2005.

Statistiques choisies des pays africains (2006) (Tunis: Banque Africaine de Développement).

Touré, Saliou (éd.) (1996), *L'ivoirité ou l'esprit du nouveau contrat social du président Henri Konan Bédié* (Abidjan: Presses Universitaires de Côte d'Ivoire).

Pierre-Aimée Kipré is a former professor at the Ecole Normale Supérieure in Abidjan, and former Minister of Education of Ivory Coast.

NOTES

1 In a debate between President Gbagbo and radio listeners, one listener, Mr Bernard Zongo, said that Ivory Coast owed everything to Burkinabes and condemned violence perpetrated against them in these words: 'They have been despoiled even though they have contributed to the

Figures

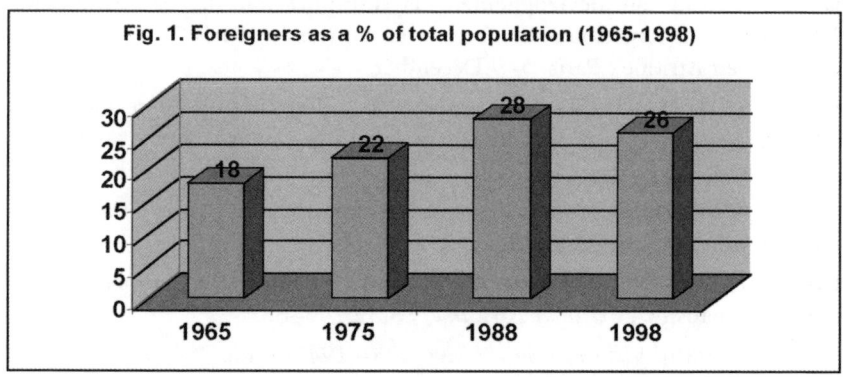

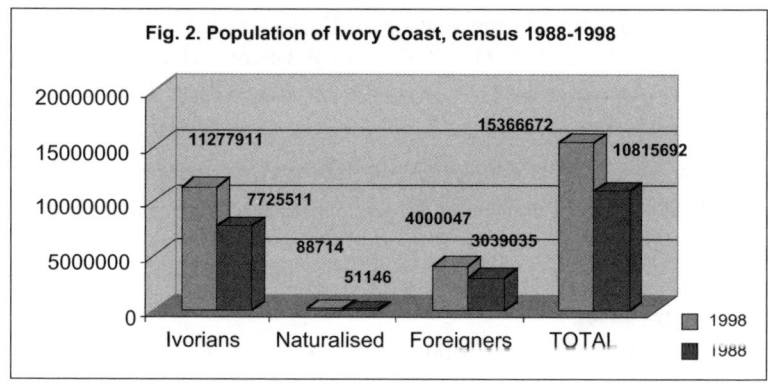

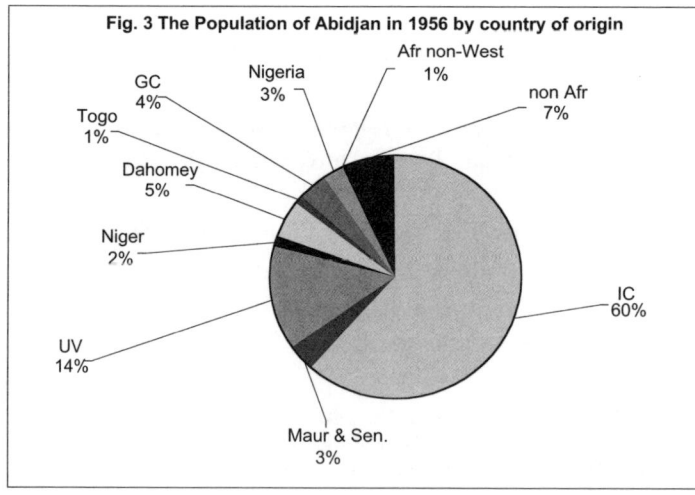

IC = Ivory Coast
UV = Upper Volta
Maur & Sen = Mauritania & Senegal
GC = Gold Coast (Ghana)
Afr non-West = from outside West Africa
Non Afr = Non-African

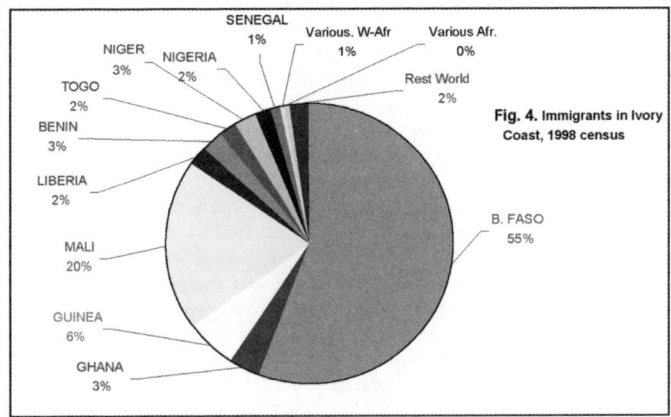

Fig. 4. Immigrants in Ivory Coast, 1998 census

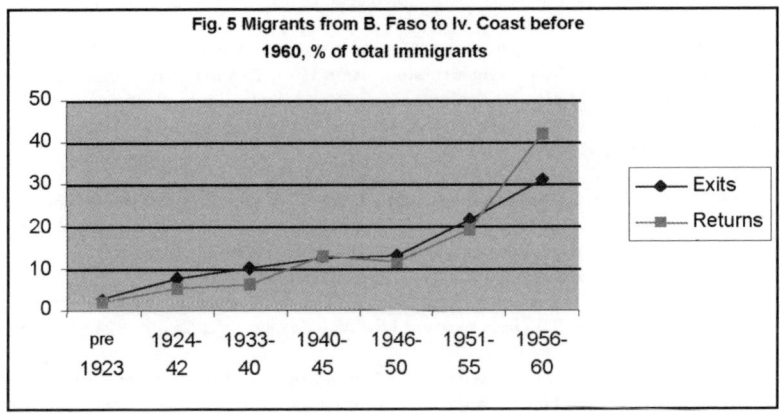

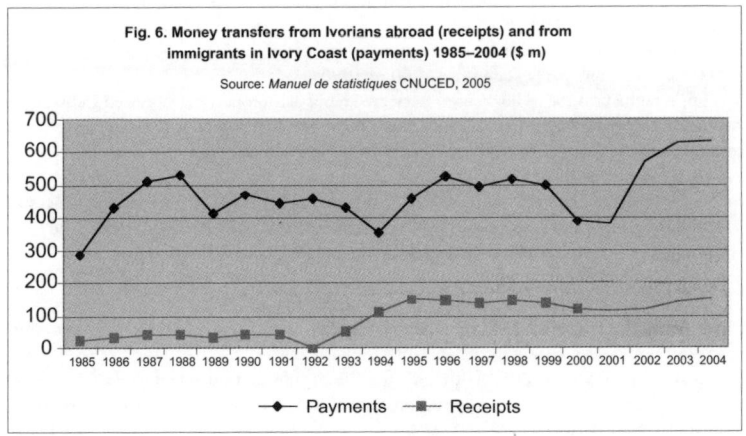

economy of Ivory Coast.' This listener concluded that Burkinabes should be given Ivorian nationality in reward for having worked in Ivory Coast, and that they should enjoy the same rights as Ivorian citizens. Among other things, President Gbagbo replied, 'If to be a worker in a contry was enough to claim its nationality, then France would be our homeland.' (interview with President L. Gbagbo, *Africa*, 1 (22 June 2006)).
2. The percentage of immigration was 3.7 per cent in 2005 and 4 per cent in 1990 (*Statistiques choisies sur les pays africains*, 2006, p. 7) and more than double between 1975 and 1980.
3. On these early associations, see Kipré (1986, vol. 2).
4. See the special issue of the journal *Présence Africaine* devoted to this theme at the time: *Le travail en Afrique noire*; edn spécial de *Présence Africaine*, Paris, 1951.
5. Immediately after decolonization, in April 1961, agreements were signed between Ivory Coast and Upper Volta (now Burkina Faso) to facilitate and improve the recruitment of seasonal agricultural labourers. Although these agreements were denounced by Upper Volta in 1962, this immigration became greater every year, with arrivals far outstripping those who returned home (less than 25 per cent of these labourers returned home after 1964). On average, between 1961 and 1985, 85 per cent of those who emigrate from Upper Volta chose Ivory Coast rather than Ghana (8 per cent) or any other country (7 per cent).
6. In 1988, foreigners born in Ivory Coast represented 49 per cent of immigrants; this figure had risen to 51 per cent by 1998.
7. Forty thousand Ivorians emigrated in 1990 to the rest of Africa and Europe, as against less than ten thousand (of which 80 per cent were students) in 1975. This is according to an unpublished report of the Ivorian Ministry of the Interior, drawn up before the general election in 1990.
8. Cf. Kipré (2005a)
9. Cf. Sylvie Bredeloup (1995)
10. On this aspect of colonial policies, see M. Mamdani (1996).
11. Trades unionists expressed their fear that it would adversely affect access to salaried employment if all nationals of the *Conseil de l'Entente* had the same rights everywhere. In the other countries of the *Entente* too, the project was opposed 'in the name of nation-building'.
12. 'The Ten-Year Plan for Social and Economic Development, 1965–1975'.
13. As a concequence of the present political crisis, the law of 17 December 2004 and the decree of application of 31 May 2006 have prorogued for twelve months the legal provisions of 1961 on obtaining citizenship.
14. Of the 306 ministers in the twenty-three governments between 1959 and 1993, 7 per cent were from non-Ivorian communities without there being an open criticism of this. However, the minutes of the Committee for the Reform of the Educational System in 1973 do reveal 'a desire to limit the presence of foreigners in national life'. Some of the participants at the Seventh Congress of the PDCI-RPA (*Parti démocratique de Côte d'Ivoire- Rassemblement démocratique africain*) in 1980 made similar suggestions but even more explicitly.
15. Ivory Coast was a one-party state from 1960 to 1990.
16. In 1996, Kragbé Gnagbé, who wanted to create an opposition party, denounced indiscriminately 'the theft of Bété peasants' land by foreigners with the 'complicity of the central government', 'French control of Ivory Coast,' 'the dictatorship of Houphouët Boigny', and so on. He even tried to organize an armed insurrection against the regime.
17. During a football match between Ivory Coast and Ghana, fights broke out in the Kumasi stadium. These were followed on 2 September 1985 by violent attacks in Abidjan against the Ghanaian community: ten thousand of the three hundred thousand Ghanaians living in Ivory Coast were then repatriated at their own request.
18. On the situation of Burkinabes in Ivory Coast, see Bendraogo (1999).
19. All the border regions of Ivory Coast share populations with their respective neighbouring state.
20. These immigrants can be a 'captive' electorate for the powers that be. This has been denounced by opposition parties all over West Africa since the beginning of democratization in the 1990s.
21. For various testimonies by immigrants on the theme of participation in elections in various countries in West Africa, see *PANOS infos*, 11, 2 septembre 2002.

22 This initiative was taken by the Prime Minister of the day, Alassane Dramane Ouattara.
23 The debate on land rights became sharper in the 1980s, above all in forest regions, at a time when numerous town-dwellers, particularly young people, tried to return to work the land in their villages. This 'return to the villages' increased in the 1990s.
24 The agreements signed in April 1960 between Ivory Coast and Burkina Faso, or between Burkina Faso and Mali in 1963, or again between Togo and Mauritania in 1965, were quick to fail because immigrants continued to move around anyway, according to their own assessment of the risks and advantages of moving abroad.
25 Cf. Kipré (2005b).

Is the Partition of the Ivory Coast a Consequence of Migrations in the Colonial Era?

CHRISTIAN BOUQUET

Several factors have been put forward to account for the crisis that the Ivory Coast has experienced since 1999 and the present article will not dwell on these (Bouquet, 2005). There is none the less one factor which has never been questioned and which is still a cause for concern: the impact of foreigners (Bouquet, 2003). For over half a century, the Ivory Coast has been a country where one inhabitant out of every four is not Ivorian. This situation eventually led to conflicts in two major areas – land appropriation and voting rights – as soon as exogenous models activated or reactivated processes that were founded on the right of blood and nationality.

In this article, I would first like to specify or correct some of the hypotheses that are sometimes put forward to account for the crisis. I will then consider this crisis in the context of the colonial era: intra-African migrations then influenced the population settlement of the current Ivory Coast and submitted the land to pressures that became unsustainable over time.

Two Misconceptions and an Omission

Two comments have often been heard: that the Ivory Coast was divided between a Muslim North and a Christian or Animist South; and that foreigners, in other words mostly migrants who came from the countries to the north (Burkina Faso and Mali), were the cause of the tensions. These two statements deserve closer scrutiny.

Foreigners did not colonize the north of the country; on the contrary, the majority of them settled in the south

The map taken from the latest census (RGPH 1998, published in 2001) does underline the importance of foreigners (26 per cent of the total population, which is to say 4 million out of 15.3 million in 1998). But foreigners have settled more densely in the south of the country, particularly in the south-east and the south-west which are both rural areas, rather than in cities – except for Abidjan, the economic capital, where one out of three inhabitants is a foreigner. In some coastal districts, foreigners represent over 50 per cent of the population.

One can actually see on this map the foundations of foreign presence in the Ivory Coast. Burkinabes (and to a lesser extent, Malians) have settled in the agricultural pioneering fronts and represent most of the workforce on the coffee and cocoa plantations. This situation is far from new and dates from the time when, before the Second World War, the colonizers had noticed that forest and littoral zones were under-populated compared to the Sudanese and Sahel zones. They had then been tempted to restore the demographic balance that was strongly encouraged by the agricultural potential of the south of the Ivory Coast.

Contrary to some of its neighbours, the Ivory Coast is not divided between a Muslim North and a Christian/Animist South.

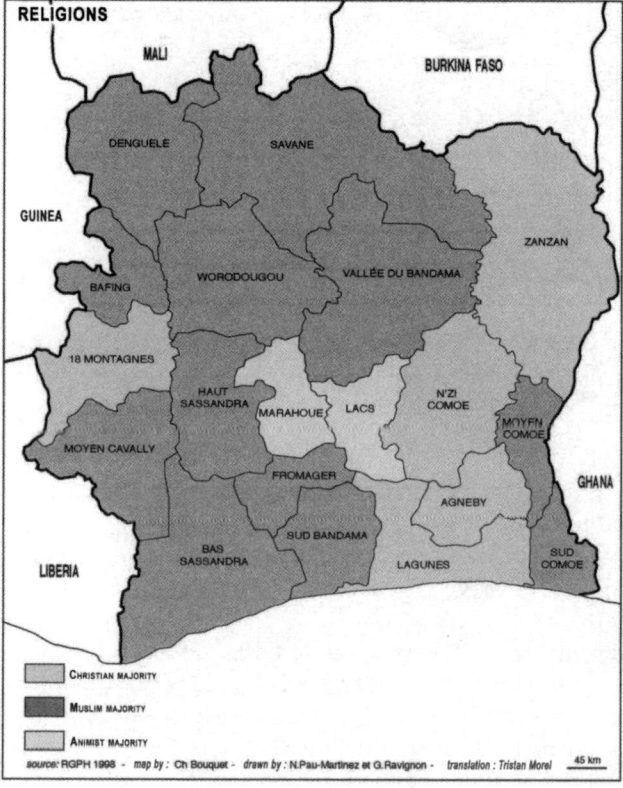

A Muslim North and a Christian/Animist South? This map, which is also from the 1998 census, qualifies this division to a large extent. Islam is the main religion almost everywhere and represents the first religion in the country. True, this situation is linked to the migration data mentioned previously. But there again, one should be cautious because if Malians are almost exclusively Muslims, Christian Burkinabes represent a very strong minority.

This division, which the Ivorian political class perpetuates more or less artificially, is not geographically true. Just as foreign populations integrated perfectly well in the host regions before independence, Islam has become present in every city and in some areas even in every village, without creating any serious conflict.

Born Ivorian in the Upper Volta?

Finally, one historical element seems doomed to remain hidden in the genesis of the Ivorian crisis: border delimitation. Everyone admits that colonial borders – in the Ivory Coast as elsewhere in Africa – are largely artificial and do not correspond to any cultural or physical reality. But since the times of independence, Africans have striven to create national feelings within these borders, which the Organisation of African Unity (OAU) proclaimed to be permanent in 1964.

And yet, today's Ivory Coast has not always been confined to the shape that we know. Thus in 1932, the colonial authorities decided to expand Ivorian territory by annexing two-thirds of what was then known as the Upper Volta, which became known as the Upper Coast.

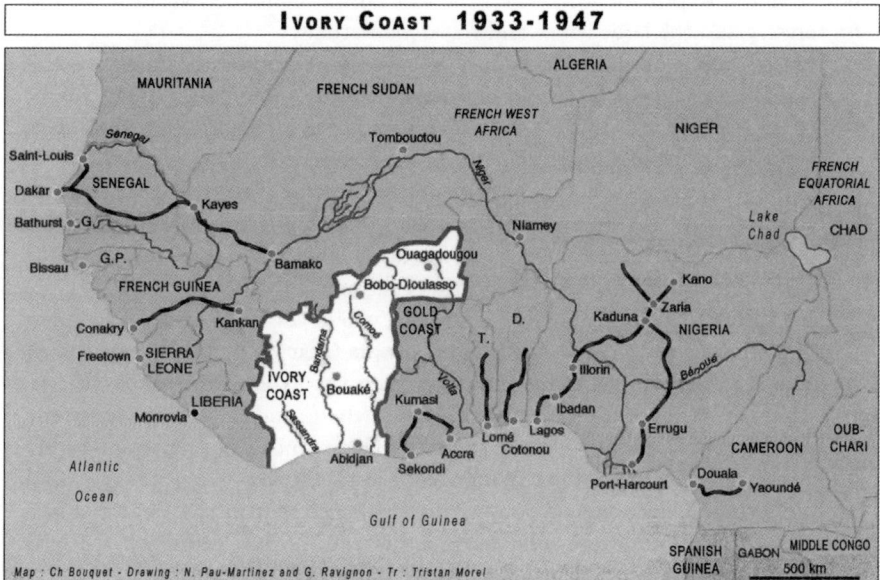

These remained the Ivory Coast's official borders for around 15 years, until 1947. It should be remembered for all those who were born in the Upper Coast between 1932 and 1947: are they not Ivorians?

Organized Migrations and Land Pressures

The reasons for this cartographic change highlight the two previous issues: the impact of foreigners from the north and the influence of Islam. Why did France feel the need (temporarily) to abolish the border between the Ivory Coast and the Upper Volta? Simply put, to make it easier for the migration of workers.[1]

Indeed, whereas Mossi plateau looked fairly densely populated but little favoured by the rain, the forest areas of the Atlantic coast were underpopulated and particularly suited for a colonial rent economy, especially for coffee and cocoa plantations.

The colonial authorities thus set up an organization for the importation of workers, initiating a large migratory flux between the north and the south. Some data was found in the archives of this organization: 420,000 Voltaic workers moved to the South before 1946 (when fewer than two million inhabitants lived in the Lower Coast); another 265,000 were registered between 1947 and 1959 by SIAMO (*Syndicat interprofessionel pour l'acheminement de la main d'oeuvre* – the Inter-professional union for the transportation of workers). The objective was to provide workers for the vast railway construction site as well as for forestry and the large coffee and cocoa plantations.

This historical reminder accounts for another important element in the Ivorian context: currently, almost one foreigner out of two was born on Ivorian soil (47.3 per cent). This early settlement has led to a paradoxical situation as far as the occupation of agricultural lands is concerned.

On the pioneering fronts indeed, the foreign workers who had initially been recruited to make up the workforce on the coffee and cocoa plantations have progressively cleared their own plots of land, in what was mostly virgin forest: they now represent almost the majority of farmers.

This means that in all of the six regions where the main coffee and cocoa cultivation areas are located, indigenous people (that is, those who were born in these regions) are in the minority, compared to Ivorian migrants (that is, those who are Ivorian, but who were born elsewhere) and those who have come from abroad. In the south-west of the Ivory Coast, indigenes represent barely 30 per cent of the population (28.2 per cent in the Moyen-Cavally region) and even less than 15 per cent in some areas (11.8 per cent in the Bas-Sassandra region). In the east, those who have come from abroad represent the vast majority of inhabitants (45 per cent in the Sud-Comoé region).

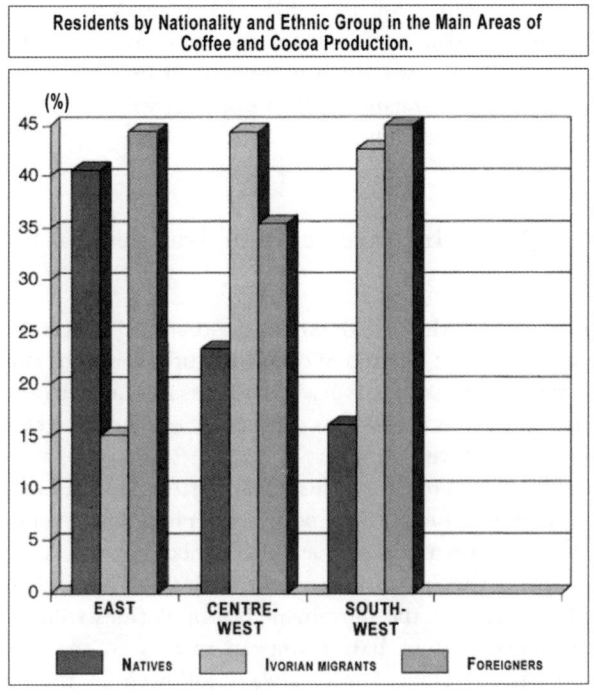

Residents by Nationality and Ethnic Group in the Main Areas of Coffee and Cocoa Production.

It is important to remember that in the coffee and cocoa plantations, foreign domination is not new. Two to three generations of plantation heads have occupied these lands. They have always considered them as 'their' lands, first because age-old African common law – reactivated by Félix Houphouët-Boigny, first president of the Ivory Coast (1960–93) – has long perpetuated the great utopian myth that the land belongs to whoever cultivates it, and second because many took the precaution of signing informal contracts with indigenous Ivorians, thereby claiming to establish a link with the initial owner.

But the Ivorian Utopia has had its day. In the 1990s, the World Bank obliged the countries that were under its thumb to face up to reality when it imposed a policy of 'land security' – in other words, a policy of appropriation based on the Western model, that is, on the right of the first occupier. This was a real revolution for all the Sub-Saharan African countries, because virtually no land had been officially registered and above all, because the concept of the 'non-working owner' had not really been integrated into local customs.

This is why the enforcement of the land law that corresponds to this international injunction had to be preceded by an operation called 'Rural Land Plan', partly financed by French cooperation funds. The objective was to choose sample zones and test the appropriation process that was about to be implemented (Chauveau, 2000 and 2001). The teams who participated in this exercise obviously learnt a lot about the best means to privatize land. Unfortunately, the land law, which was voted in in 1998, was enforced too hurriedly and conflicts immediately broke out between communities that had hitherto lived together peacefully.

Conclusion: On the Inconvenience of Transfers (of Populations and Models)

Without entering into the debate on the good and ill effects of colonization, and on the basis that the Bretton Woods institutions are the inheritors of the colonial era, the example of the Ivory Coast gives us – unfortunately for France – food for thought about the serious consequences that the reckless transfer of Western models can have.

True enough, when French colonists transferred populations between poor and better-endowed regions, they only accelerated movements which their own history had shown would have happened anyway, following Adam Smith's principle of the 'invisible hand of the market'. But in France, the great migrations (for example, the movement of populations from rural to urban areas, or the movement of East European miners to the mining areas of northern and eastern France) were always followed in the end by public policies and the law was adapted to the new situations.

In the Ivory Coast, two reasons may explain the fact that the problem of the nationality of the migrants was not dealt with more quickly: first, there was no sense of urgency, and second, cohabitation between communities, encouraged and celebrated by the founder of the Ivorian nation, appeared to be an African model in its own right.

On the contrary, the introduction of the principle of private land ownership (an expression which the World Bank is very careful to avoid) was a particularly destabilizing factor for societies that had functioned well under other arrangements for centuries. Because they wanted to apply the letter rather than the spirit of the law, the Ivorian authorities ran the risk of exacerbating a conflict which history will have difficulty proving was culturally unavoidable.

The international community, organized into world governance, demands that new countries convert to the rule of law. But whose law?

(Translated by M. Torrent)

Christian Bouquet is Professor of Political Geography and Vice-President, Michel de Montaigne University, Bordeaux.

Bibliography

Bouquet, Christian, 'Le poids des étrangers en Côte d'Ivoire', *Annals de Geographie*, 3 (2003), pp. 115–45.

Bouquet, Christian, *Géopolitique de la Côte d'Ivoire. Le désespoir de Kourouma* (Paris: Armand Colin, 2005).

Chaveau, Jean-Pierre, 'Question foncière et construction nationale en Côte d'Ivoire', *Politique Africaine*, 78 (2000), pp. 94–125.

Chaveau, Jean-Pierre, 'Mise en perspective historique et sociologique de la loi ivoirienne de 1998 sur la Domaine foncier rural au regard de la durabilité de l'agriculture de plantation villageoise', International conference on 'L'avenir des cultures pérennes,' Yamoussoukro, 5–9 November 2001.

NOTE

[1] In particular, to stop Voltaic migrants from going over to the British Gold Coast, where colonization was believed to be less harsh.

Emigration from Burkina Faso and the Ivorian Crisis

AUGUSTIN LOADA

Burkina Faso, a landlocked country situated in West Africa, is certainly well known for its history of migration. Millions of Burkinabe seeking a better life have no other choice but to emigrate to neighbouring countries. This phenomenon goes back even to pre-colonial times (before the 1890s). However, migration intensified during the colonial and post-colonial eras (full independence was declared in 1960). These massive numbers in migration went through three main phases: the colonial period which was characterized by forced migration, the period of economic expansion in the main countries of destination, and the phase of crises which resulted in the inversion of migration and more or less forced people to repatriate.

Burkinabe migrants usually originate from rural areas, particularly the Mossi central plateau and secondly, the West and South-Western regions of the country. The general population census in 1985,[1] the demographic survey of 1991,[2] the survey on 'migration and urbanization' of 1993,[3] and the general population and habitat census of 1996[4] all confirm these trends. The Burkinabe migrant to Ivory Coast is usually young, single and poorly educated. Thus it is not surprising that Burkinabe migrants occupy very low-level, poorly paid jobs. Nevertheless, over the last years we have noticed that there have been some changes regarding the Burkinabe migrant's profile. Some progress has been made in educational provision for children of migrants, as well as for children of those Burkinabe who have already settled in Ivory Coast. In addition to this, we also see a feminization of the phenomenon of emigration, which implies the development of migration of entire families.

Colonial Upper Volta was considered to be an inexhaustible reserve of labour to send to neighbouring countries, especially Ivory Coast.[5] After independence, what was now Burkina Faso continued to maintain its reputation as a source of migrant labour. This mass migration continued and increased despite severe climate conditions (continual drought from 1968 until 1974) with their consequences (increase of the population on the land).

Ivory Coast was the primary host country to such an extent that the archetype of the so-called "diaspo"[6] is the Burkinabe who speaks and behaves like an Ivorian. Moreover, when we discuss Burkinabian emigration, it means mainly the Burkinabe community which settled in Ivory Coast.

Having to deal with this massive migration which reduces the workforce but which is the same time is an economic opportunity, various governments in Burkina Faso have tried to prevent emigration and protect its workers. However, politically speaking, the authorities have adopted a wait and see policy and generally abandoned the Burkiabe emigrants to their own fate.[7] Various Burkiabe regimes are accused of having done nothing to help migrants or to set up a framework which would allow them to transfer money home or to settle back in their home country thereby allowing the state to benefit economically from their financial input or their experience. This absence of any national policy on emigration was also encouraged by the welcoming attitude of the former Ivorian president, Felix Houphouët-Boigny, who gave Burkinabe migrants in Ivory Coast more or less the same rights of those of the local population.

Nevertheless, at the beginning of the 1990s, when the era of Houphouët-Boigny came to an end, and when the crisis of "Ivority" erupted, a wave of xenophobia spread through Ivory Coast, encouraged by the Ivorian authorities who accused Burkina Faso of fomenting the rebellion of 19 September 2002. We then witnessed crimes committed against the Burkinabe community and the repatriation of millions of Burkinabe who became the scapegoats for a crisis which, from Burkina Faso's point of view, was a purely internal matter for Ivory Coast.

Faced with rhetoric directed against the large Burkinabe presence in Ivory Coast, the Burkinabe authorities never ceased to repeat that this presence had historical origins. (See Section I below.) Ougadougou indeed regards Burkinabe emigration as something which first the French colonial authorities and later the post-colonial government of Ivory Coast encouraged and supported. It is concluded from this that the only solution to this Ivorian crisis is to abandon the policy of 'Ivority' and to return to a more open conception of citizenship in order to integrate and protect the Burkinabe community. (II).

I. The Burkinabe Emigration in Ivory Coast: A Historical Fact

In Burkina Faso, there are many important studies which illustrate the historical and structural character of migration.[8].It has historical, social, demographic, and economic causes. Poverty is also a push factor: in Burkina Faso, there are fewer jobs than in neighbouring countries and in particular in

Ivory Coast. Emigration also has environmental causes: people migrate because of lack of rain or soil erosion. There are also socio-cultural factors: people can be banished from their community or village and this blocks innovation as well as emancipation among youngsters. The reason for choosing Ivory Coast as a destination is the fact that there is political stability in this country, economic growth a very good immigration policy. Ivory Coast has signed agreements on migration with its neighbours, and in particular with Burkina Faso.

Nevertheless, the size of the Burkinabe community in Ivory Coast also has a historical explanation. Burkinabe migration was encouraged by the colonial authorities as a source of labour. Migration was also immediately promoted by the Ivorian post-colonial state, as well as by policies of regional and sub-regional integration.

A historical fact which was provoked by the French state

The large-scale migration of Burkinabe to Ivory Coast can be explained by the colonial system which linked Upper Volta to Ivory Coast. The colonial administration tried to extent the areas of cotton farming into Upper Volta. This policy failed because of drought, popular resistance and the economic crisis of 1929. The decision was taken to concentrate on Ivory Coastand to get rid of Upper Volta altogether. In the words of two commentators, 'Upper Volta with its limited natural resources and large population was sacrificed during the economic crisis of the 1930s'.[9] The colony was dismantled in 1932 for the benefit of its neighbours, French Sudan, Nigeria, and Ivory Coast. Burkina Faso turned into a labour reservoir, something which particularly benefited Ivory Coast. The majority of the territory of Upper Volta was annexed to Ivory Coast, approximately 153 650 km2. Upper Ivory Coast was composed of the Bobo-Dioulasso regions as well as Gaoua, Koudougou, Ouagadougou et Tenkodogo: a total of 2,400,000 inhabitants or 77.4 per cent of the population. A series of measures were taken in order to direct the migration flows towards Ivory Coast and to discourage those who wanted to leave from going to Gold Coast (Ghana), a British colony which was booming. Bosses were given instructions to attract thousands of workers to Ivory Coast and to increase their salaries. The population of Upper Volta was then 3,000,000 while the population of Ivory Coast was 1,800,000.[10] Between 1920 and 1929, the emigrant Burkinabe workforce rose to 66,693. From 1934 to 1949, this number doubled to 128,683. From 1951 to 1959, 254,782 workers were recruited, without counting unofficial emigrants to Ivory Coast, estimated at 1,250,158.[11] Others estimate that between 1919 and 1959 there were 769,800 Burkinabe emigrant workers.[12]

The French government wanted to diminish the power of the (RDA) *Rassemblement Démocratique Africain*, the anti-colonist party, which was very influential in Ivory Coast,[13] and decided in 1947 to reconstitute Upper Volta. This caused a labour shortage in Ivory Coast which was greatly resented and which demonstrated how dependent the country was on its Burkinabe workers. Within this context, the *Syndicat Interprofessionel pour l'Acheminement de la Main d'oeuvre* (Inter-professional Trades Union for Directing the Work Force) was created in 1950 in order to remedy this situation.

A *historical fact, provoked by the post-colonial state*

Upper Volta's independence in 1960 did not stop the country from exporting workers to its neighbours. The new government of Upper Volta attempted to control the migration flows and guaranteed protection for people who wanted to leave. It tried to get economic benefit out of persuading them to stay. The Ivorian and Burkinabe governments signed an agreement on migrant labour in 1961 but it was never implemented. Moreover, a commission set up to regulate compensation, and which reported from 18 to 20 August 1975 could not agree on displacement compensation for agricultural workers. Migratory flows continued, in spite of differences within the government about migrants' contracts. A number of regional organizations of cooperation and integration were set up: the Entente Council in 1959, Economic Community of West Africa in 1972, the Economic Community of the States of West Africa in 1974. These all favoured the movement of people and goods. The export of labour Upper Volta therefore continued via professional recruitment agencies and especially thanks to the welcoming policy of the first Ivorian president, Felix Houphouët-Boigny. His policy could be summed up with the slogan: 'The land belongs to those who farm it'.

As a consequence, the number of Burkinabe migrants to Ivory Coast rose. Ivorian surveys of migration and the urbanization show that the Burkinabe represented 52.5 per cent of the 1,474,000 foreigners in 1975. In 1988, they were 51.5 per cent of the 3,039,000 foreigners in the country. Five years later they made up 52.9 per cent of 3,310,000 foreigner residents. The 1998 census estimated the numbers of Burkinabe in Ivory Coast to be 2,238,548.

Between 1950 and 1965, Ivory Coast had an annual growth rate of 9 per cent. For many Burkinabe this economic miracle is thanks to the Burkinabe workforce. However, since the 1980s, the Ivorian economy has been in crisis. Immigration continued, however, as economic opportunities were even worse in Burkina Faso. Migrants were encouraged by all the factors mentioned above and also by a desire to return to their villages with an improved social status as evidenced by their clothes, their food, their ownership of goods like radios, bicycles or mopeds. This in turn encouraged others to imitate them and to emigrate as well.

For Burkina Faso, emigration is an ambivalent variable. It hindered the country's development because the labour force haemorrhaged out of the country and there was a spread of illnesses[14] but it also had some economic benefits. Expatriates who return to the country with new skills and a sense of innovation have been an asset to the country, for instance in the transport sector and in the cultural field. Emigration is also an important source of money transfers into the country. Ivory Coast is the main source of such transfers, providing between 85 per cent and 93.7 per cent of the total inward transfers[15]. Businesspeople in Burkina Faso tried with limited success to get the Burkinabe to place their considerable savings into banks in Burkina Faso. Until 1998, these transfers were regular and stable but there was a big drop between 1999 and 2001, as a result of the crisis for Burkinabe in Ivory Coast since the 'Tabou events'.

The revision of the Ivorian Property code in 1999 resulted in the expropriation of many Burkinabe. This led to conflicts between foreigners and locals and ended up with massacres and an exodus of 19,647 people, according to the Council for Burkinabe Abroad (*Conseil Supérieur des Burkinabè de l'Étranger*).[16]. These were dramatic events which showed that there are many difficulties in integrating the Burkinabe community in Ivory Coast, and how urgent it was to have a migration policy in Burkina Faso.

II. Protecting the Immigrant Population: A Double Political Challenge

For many years, the Burkinabe community of Ivory Coast lived more or less in harmony with the Ivorian population, despite the occasional conflict or isolated act of xenophobia which were caused by the economic crisis. But when the crisis regarding citizenship was followed by the emergence of the concept of 'Ivority', and especially when a rebellion broke out in September 2002 in the Northern part of the country, the life of the emigrant community was completely overturned.

The Burkinabe community used to be appreciated and warmly welcomed because of its workforce. Nowadays, it is no longer desired by the rest of the population and it is exposed to actions of hostility leading to the repatriation of thousands of Burkinabe, even though Burkina Faso is unable to receive all these expatriate immigrants. In an attempt to deal with the situation, the Burkinabe government has tried to obtain guarantees and protection for its nationals abroad and also to develop a national policy on emigration.

The Ivorian authorities claim for a protection of the Burkinabe community

Many Burkinabe maintained close family ties with their country of origin. At the beginning, it was quite natural for them to return to their villages and many parents sent their children back to the village, thereby reinforcing ties. Over time, however, returns became less common, especially when migrants left with their wives. The Burkinabe community settled in Ivory Coast and became organised there. The new working conditions and the fact that their home country was far away reinforced their feelings of patriotism and of closeness between immigrants from the same region or town. These immigrants would meet, discuss and often create a framework of gathering and exchange. Such organisations were created based on place of origin, professional activity, and so on.

The official welcoming policy of the Ivorian State did not, however, prevent acts of xenophobia from being committed on a daily basis.[17] These attacks showed the lack of integration. In the mid-1980s, the end of the Ivorian economic miracle prompted the Ivorian government to establish a ' policy of Ivoirisation' which led to the eviction of several foreigners from several jobs, in particular Burkinabe. However, a large number also kept their jobs, typically low paid work in farming.

When the economic crisis worsened, the perception of foreigners changed. Now perceived as "invaders" who were stealing jobs from Ivorian nationals, the Burkinabe were accused of taking away jobs and resources, especially farmland. While being rejected by their country of residence, Burkinabe emigrants also felt that they were not being protected by their country of origin either with which in any case they had broken off ties. The Burkinabe government , aware of the size of the emigrant population and of its economic potential, set up the *Conseil Supérieur des Burkinabè de l'Étranger* on 7 May 1993, whose role was to unite Burkinabe communities in different countries abroad and to involve these communities in national construction at home. There are many doubt about this organisation's effectiveness.

The *coup d'etat* of 24 December 1999, was the first in the post-colonial history of Ivory Coast. It opened Pandora's box. Whereas the country was once admired for his political stability and for its policy as a host country for immigrants, this tradition was rapidly abandoned. The country even exported instability in the West African region with the army rebellion 19 September 2002. That crisis stirred up hatred towards the Burkinabe community, for Burkina Faso was accused of having supported the rebellion.

However, for the Burkinabe , the causes of the Ivorian crisis are internal. They lie in the end of the country's economic miracle but also in dangerous manipulation of the concept of 'Ivority'[18] by President Houphouët-Boigny's successor. This concept, established in 1995–96 and popularized by President Henri Konan Bédié, this concept was originally supposed to 'forge a common culture for those who live on Ivorian land, foreigners and locals alike.' In reality, it was produced to order because of the economic crisis and because people felt there were too many immigrants and that the country was getting too crowded. Ivorians feared they would be outnumbered in their own country.

This concept was also used for political shenanigans, such as the attempt to exclude from the presidential election a very important rival of president Bédié, the ex Prime-minister Alassane Ouattara, who was portrayed as a Burkinabe. The Ivorian Constitution of 2000 was also manipulated in order to prevent "foreigners" from taking power through universal suffrage[19].

The concept of 'Ivority' was also used as an instrument of social and economical exclusion against Ivoirians from the North of the country who were mainly Muslims. They were perceived to have "doubtful" Ivorian origins or to have adopted Ivorian nationality for convenience, and they were contrasted with "native Ivorians". Like the immigrants from elsewhere in the region, these people were accused of stealing the country's jobs, land and resources. Xenophobia was encouraged by politicians and led to robbery and atrocities committed against immigrants, especially Burkinabe.

Ever since the Ivorian rebellion of 2002, the authorities of Burkina Faso have been regularly accused by the government of Ivory Coast of complicity with the rebels. The Burkinabe government started a diplomatic offensive to counter the Ivorian accusations and to demand the protection of immigrants from acts of revenge taken by government forces or Ivorian militias. In order to deal with the consequences of such a conflict, the Burkinabe government established a group of repatriate volunteers and a reception centre for Burkinabe who fled atrocities.

For Burkina Faso, one of the most important things at stake in the Ivorian conflict is the freedom of movement of people and the protection of goods stipulated in the texts of the Economic Community West African States. Those texts were obviously flouted by their Ivorian counterparts. Fearing for its nationals abroad, the Burkinabe government initially adopted a very low profile, whenever it had to deal with atrocities, before raising its voice and calling on Ivory Coast to protect its immigrants. President Blaise Compaoré started to threaten his Ivorian counterpart Laurent Gbagbo with prosecution before an international tribunal along the lines of the one which prosecuted the Serbian president, Slobodan Milosevic. The Burkinabe president frequent public outbursts against his Ivorian counterpart went down very well with the voters who were exasperated by government's inability to prevent atrocities against the Burkinabe community.

The Burkinabe government was determined to defend the rights of its emigrants and therefore it did its best to gather all evidence referring to atrocities committed against immigrants. It turned to international and Burkinabe associations for the protection of the human rights. The Burkinabe government also played a role behind the scenes in negotiating the Marcoussis agreements between the two sides in the Ivorian conflict. There had been three fundamental questions raised among the Burkinabe authorities: the principle of the freedom of movement of people and goods; the question of land ownership; and citizenship issue. The effective implementation of the principle of the freedom of movement faced some difficulties and was subject to police harassment. The round table at Marcoussis condemned red tape and its victims, foreigners in particular, while recognising that foreigners in Ivory Coast have contributed to the national wealth and have given this country a special role and responsibility in the sub-region. The round table invited the Ivorian government of national reconciliation to suppress immediately all the residence permits provided for in Article 8 Paragraph 2 of the law of 3 January 2002 for all foreigners coming from ECOWAS, and to base immigration controls, recognised as necessary, on systems of identification which could not be tampered with. It also invited the Ivorian government to study every legislative and regulatory provision in an attempt to improve living conditions for foreigners and to protect people and goods.

Another important issue for the Burkinabe authorities was the question of land ownership. The Marcoussis agreement proclaimed that the law 98-750 of 23 December 1998 on rural land property, which was passed unanimously by the Ivorian National Assembly, constitutes a reference point for this juridically delicate and economically crucial issue. The government of national reconciliation was called on to accompany the implementation this text with a proper information campaign directed at rural populations in order to create a stable environment for land ownership. The Ivorian government is especially invited to agree to an amendment aiming at a better protection of the acquired rights of heirs and of owners, the holders of rights prior to promulgation of the law. This recommendation could not but please the Burkinabe government which wanted to remind Ivory Coast of international law on expropriation, which requires that people be given fair notice and fair compensation. On citizenship, the Marcoussis agreement recognized Ivory Coast's inalienable right to define precisely who its citizens are, and Burkina Faso adopted a very low profile on this subject. However, in view of the opportunity for naturalization which the agreement extended to certain long-term residents, Burkina Faso has, through its diplomatic and consular representations, the delegates of CSBE and the different associations in Burkinabe, invited its emigrants who wish to to take the necessary formal steps to acquire Ivorian citizenship.

For public opinion in Burkina Faso, Ivory Coast has partially developed its prosperity thanks to the sweat and to the blood of thousands of Burkinabe; the descendants of people who came to work in the country should have, they feel, the right to obtain Ivorian citizenship and to integrate into the country in the way Ernst Renan described. Renan wrote in 1882 that a nation does not need to be based on race, language, economic interest or geography but that instead a nation would be defined as *'a soul, a spirit ...It has a past, but it consists in the present in a tangible fact: the consent and the clearly expressed desire to live together. The existence of a nation is a daily plebiscite.'*[20]. But, it is precisely this conception of the nation which is being jeopardised by the policy of 'Ivority'. While hoping that this policy of encouraging xenophobia will be abandoned, Burkina Faso plans to develop a new policy of migration in order to protect its expatriate citizens.

Considering a new migration policy on the agenda of Burkina

There are only a few families in Burkina Faso that have not been affected by the Ivorian crisis. From a financial point of view, we see that this crisis had considerably reduced money transfers, especially since the Burkinabe Post Office decided not to honour payments from Ivory Coast. The crisis also led to a decrease in transfers towards households; this had dramatic consequences for nutrition, health, education and the economy as a whole. The state budget had a deficit of 20 million FCFA in 2003. From the very first hours of the crisis, there were interruptions in the supply of primary materials and finished products. Burkinabe towns bordering Ivory Coast (Banfora, Bobo Dioulasso) suffered. After the shock was over, the whole economic and social fabric of Burkina Faso had to readapt in order to break free of its dependence on Ivory Coast.

The crisis forced hundreds of thousands of Burkinabe to return to their country of origin. Some 18,000 Burkinabe were expelled after the Tabou events in 1999 and approximately 500,000 were repatriated during 2002 03. The Burkinabe government managed to direct those returnees to their home villages. When things calmed down, an large number returned to Ivory Coast, showing their persistent links to that country.

The Ivorian crisis also gave rise to various attempt to encourage people to emigrate to countries other than Ivory Coast. Many Burkinabe have emigrated to Europe (Italy, Spain, France, Germany, Switzerland), the United States and even the Maghreb (in particular Libya). Some of these destination countries were discovered only when Burkinabe without papers were expelled from them. Young Bukinabe people, indeed, are not discouraged by new anti-immigration policies or even by the risks and abuses from traffickers and smuggler networks.

The Ivorian crisis made the Burkinabe authorities realize that it was necessary to define a real migration policy. Burkina Faso, took some measures such as creating a web site (www.burkinadiaspora.bf), with the support of the International Migration Office (OIM), to maintain the contact with Burkinabe from abroad, as well as the broadcast of Burkinabe Radio and television programs (RTB) by the Intelsat satellite.

But the new desire to have a migration policy gave rise above all to a symposium on migrations in July 2006 in Ouagadougou, organized by the Burkinabe government. The results of studies on migration from Burkina Faso allowed the broad outlines of a migration strategy to be defined:

- Implement an observation post regarding migration;
- The adoption of a national policy which guarantees land ownership;
- Strengthening urban infrastructure and rural development;
- A debate about emigration and the rights of the emigrant workers.

Each one of these issues is being examined in detail at present through studies conducted under the aegis of the National Council of population. The symposium concluded that emigration was a priority for Burkina Faso. The goal is to maximize the positive effects of the phenomenon by defining a basic set of rights on which emigrants can rely. Therefore the symposium recommended integrating the international and regional provisions which apply to the free movement of people as well as to the right of residence and settlement. These provisions must however be included in the emigration policy of the state concerned. The *Conseil supérieur* was also at the centre of discussions, since its powers will have to be strengthened to achieve these goals. The symposium recommended the implementation of a structure which would represent Burkinabe civil society, which is called upon to play an important role in managing migration. On a diplomatic and consular level, Burkina Faso will strengthen its structures among the main immigration countries, so that protection can be assured to migrants in their host country. Finally, one of the most important suggestions of the symposium was undoubtedly the right to vote of the Burkinabe community living abroad. These people are deprived of their right to vote in the host country as well as in their country of origin. In Ivory Coast for example, President Houphouët-Boigny granted the right to vote to the immigrants in his country until 1990. Since foreigners were very grateful to this liberal policy, they naturally voted for the party in power. Notwithstanding hostility from the opposition parties, the political party in power, in order to maintain in power, counted on their votes. Under these circumstances, it was not surprising that the general elections of 1999 and of 1995 became a situation of conflict, and that foreigners were deprived from their right to vote with end of the First Ivorian Republic in 1999.

In some countries, such as Mali, citizens living abroad are allowed to vote in embassies. This is not the case in Burkina Faso, where nationals living abroad may vote only if they are happen to be in the country, and even then they have to be physically present in the home constituency for seven days before the vote to get onto the electoral roll.[21] The political party in power and the Opposition seemed to agree so far on the decision not to give the right to vote to the Burkinabe from abroad. It was felt that the right to vote for emigrants abroad would create huge organisational difficulties. Moreover it was felt inopportune, given the huge size of the Burkinabe diaspora and the xenophobia it faced in Ivory Coast. But with the Ivorian crisis, several parties of the Opposition, as well as associations from the civil society, are now campaigning to give the right to vote to Burkinabe from abroad. The government, which used to be reluctant, now seems more open to this demand. This needs to be implemented in the light of considerations as to whether this would make it easier or more difficult for emigrants to integrate in their host country, since it would emphasize their link to their country of origin.

Augustin Loada is Professor of Public Law and Political Science at the University of Ouagadougou, Burkina Faso.

1   National Institute of Statistics and Demography (INSD), Ram Christophe Sawadogo, Communication on Integration of Emigration in the Procedure of the Development in Burkina Faso, at the Second Assembly of the Superior Council of Burkinabe from Abroad, 8–10 December 1998.
2   'Analysis of the Results of the Demographical Survey of 1991', Ouagadougou, 1995, pp 207–208.
3   Cf. Désiré Konaté, 'The Burkinabe Migration: Its Origins, its Importance and its Contribution to the Development of the Country', the contribution of Burkina Faso at the Inter-Regional Meeting on the Participation of Migrants in the Development of their Countries of Origin, Dakar, 9–13 October 2000.
4   'Analysis of the Results from the General Population and Housing Census' RGPH, vol. II (1996), p. 156.
5   Cf.Issiaka Mande, 'Les migrations de travail des Voltaiques' (The Migration of Work of the Inhabitants of Volta: A Panacea for the Ivorian Economy in 1919–1960), in Gabriel Massa and Georges Madiega (eds), *La Haute Volta coloniale* (The Colonial Upper Volta: Testimonies , Researches, Views) (Paris: Karthala, 1995), p. 313.
6   'Diaspo,' from 'diaspora', is the nickname in Burkina Faso for Burkinabes who have lived abroad, especially in Ivory Coast.
7   For example, in 1997, an association called 'Le Tocsin' was formed, whose members were essentially Burkinabe who had lived abroad; the group campaigned for the rights of the Burkinabe abroad. The actual acknowledgement of the right to vote has recently become a major part of its campaign.
8   Demographic survey in 1960–61; national survey in 1974–75 on migratory movements; general census of the population in 1975 and in 1985; demographic survey in 1991; surveys on migration and urbanization from 1993 and in 199.
9   Cf. Massa and Madiega (eds), *La Haute Volta coloniale*, p. 17.
10 Cf. Robert Delavignette, *Afrique occidentale française* (Paris, SEGMC, 1931), p. 177. Aujourd'hui, la population de la Côte d'Ivoire dépasse de loin celle du Burkina Faso.
11 Cf. J.L.Domba, cité par Traoré Karamogho, *Problématique de la gestion des migrants burkinabè victimes de la crise ivoirienne du 19 septembre 2002*, Mémoire de l'ENAM, mars 2004, p. 21
12 Cf. Ram Christophe Sawadogo, Communication, *op. cit.*

13 'Do not forget that the country of Mossi is a big labour reservoir in Ivory Coast and when one controls the Mossi country, it also controls the planters of Basse-Côte and we could make Houphouët admit his errors', quoted by Mandé, 'Migrations', p. 328.
14 Among the French-speaking countries of West Afrika, Burkina Faso has the second highest number of people affected with AIDS, after Ivory Coast.
15 Cf. Konaté, 'The Burkinabe migration', citing Ram Christophe Sawadogo, *Following the population movements through the money circulation and transfers* (BCEAO, January 1995).
16 A report on the events in Tabou, CSBE, 2001.
17 For many Ivorian, all Burkinabe are *mossi*. Despite that, when somebody is called 'mossi' in Ivory Coast, it implies that this is an insult, it is an insult meaning you are a ' peasant ' with all the negative connotations this might entail.
18 About the discussion on Ivority, cf. Jean-Pierre Dozon, 'Ivory Coast, between democracy, nationalism and ethnonationalism', *African Policy*, 78 (June 2000); Richard Banégas and Bruno Losch, 'La Côte d'Ivoire au bord de l'implosion', *African Policy*, 87 (October 2002); Marc le Pope and Claudine Vidal (eds), *Côte d'Ivoire. L'année terrible 1999-2000* (Paris: Karthala, 2002).
19 'The president of the Republic is elected every five years by direct universal suffrage. He can only be re-elected once ... *He must have Ivorian origins; both parents must be Ivorians as well. He must have always retained his Ivorian nationality. He must never have applied for another nationality* ...' (Article 35 from the Ivorian Constitution of 2000). For an analysis of the Ivorian Constitution, cf. Alban Alexandre Coulibaly, 'Articles 41 and 50 of the Ivorian Constitution: a potential obstacle for the implementation of a democratic alternation', *Revue juridique et politique-indépendance et coopération*, 1 (January–March 2003), pp.39 ff.
20 Renan, stated by Guy Hermet et al., *Dictionnaire de la science politique et des institutions politiques* (Paris: Armand Colin, 2nd edn, 1996), p. 179.
21 Cf. Article 59 of the electoral roll.

Experiences at a Consulate in Africa

ALEXANDRE LEITÃO

I used to be a diplomat in Dakar, where I was chargé d'affaires *ad interim* at the Portuguese embassy for two years. I was involved in economic cooperation, the commercial environment and above all in consular activities. These gave me some useful tools with which to analyse the political and social situation of the country, since I was called upon to take decisions on more than ten thousand visa applications and to undertake about a thousand interviews. This gave me an X-ray insight into the insides of Senegalese society and, in some cases, into the nine other countries to which we were also accredited.

This X-ray view concerned especially the reasons why people made applications for visas, asylum, residence and citizenship. The fact that we had to assess how determined the various candidates were to emigrate forced us to confront difficult realities and to understand that the rest of the world is probably growing distant from Europe, its way of life and its values, more than it is approaching them.

In the first place, it is important to point out the very high number of visa applications for states which belong to the Schengen system. From Senegal alone, there have been between 60,000 and 100,000 such applications per year over many years. There are some variations from one year to the next but the overall trend is upwards.

One of the reasons, but not the only one, for this high figure is that Dakar is a transit point for Europe and the world. There are daily flights from Dakar to Paris, Lisbon, Brussels, Madrid and Milan. However, other capital cities in the region also have similar levels of visa applications.

The majority of applications are for short-term so-called tourist visas, which are valid for a maximum of ninety days. In reality, close analysis by the consular services of all EU member states has confirmed that half of these applications for tourist visas in are in fact hidden attempts at emigration. The idea is to get into the Schengen area in order to stay there beyond the date when the visa expires in order to find work, often found by a relative or a friend, or in the hope that the receiving country will decide on extraordinary measures of naturalization.

It is an established reality that previous such measures exercise an enormous power of attraction. Information about this kind of initiative circulates quickly by word of mouth and in the press. We were able to verify that the decision announced in 2005 by the Spanish government to naturalize hundreds of thousands of illegal immigrants caused a rise in visa applications and an explosion in the phenomenon of 'boat people' crossing from West Africa to the Canaries.

Anyone who has lived recently in Senegal or who has surfed the net to look at the local newspapers will know that there is a widespread belief that it is sufficient to get to Europe, by whatever means, in order to be regularized sooner or later, by one means or another. This image of a Europe in which everyone lives well is largely responsible for the influx of illegal immigrants in Europe.

Thus, applications for visas lodged with the Portuguese embassy in Dakar declined by 12.2% in 2003, with respect to the previous year, and by 18.9% in 2004. From 2005 onwards, applications rose again and broke all records in 2006, reaching levels which are approximately 25% above those of 2002. Extraordinary regularizations have not been the only factor but they have been a contributory one.

These data, combined with the figures for arrivals in the Canaries, Mauritania and Morocco demonstrate that attempts to obtain a Schengen visa remain the preferred means for those who wish to emigrate.

This enormous pressure forces missions to organize and cooperate with one another. The consular services of the Schengen states meet frequently to share experiences and to try to harmonize their procedures. This is necessary, because if one embassy has a reputation for being more lax than the others then it will immediately be deluged with new applications.

How do we know that a large part of these applications for tourist visas are in fact attempts to emigrate? First, an enormous percentage of the documents and declarations presented to consulates are false. Some of the forgeries are obvious but there are also agencies which specialize in making very good forgeries. Second, the discrepancy between the written declarations of applicants and their behaviour in the interview is often embarrassing or sad. One sees very quickly that the person in front of you in fact wants to emigrate.

Conversely, the number of applications for full emigration has fallen while the applications for short-stay visas has risen. Why would this be the case if there was such a desire to emigrate?

There are numerous motives, including the inevitable fact that it is burdensome to acquire all the necessary documents and the fact that the controls in consulates and in the destination countries are more and more strict. But the numbers of applications which are granted poses raises political questions, especially about the role of immigrants in the development of the European economy and what sort of partnership we want with developing countries, especially as far as education is concerned.

The problem of long-stay visas is not only a consequence of European policies on immigration and cooperation; it concerns as well the authorities of the countries of origin since there are very serious problems with forged documents.

For instance, for family regroupment, we asked for documents confirming applicants' civil state but we soon established that many of these documents are false but authentic. In other words, the document had indeed been issued by the competent authority but the information it attested was false. In 2004, 40% of the documents which were submitted to Schengen state consulates did not correspond to the persons in whose name they had been submitted. In these conditions, there is a lack of credibility and this make it difficult for us to open up our policies, especially at a time when European public opinion is tending more towards closing them down.

Visas for students also come up against the fact that many of the students never return to their country of origin at the end of their studies or when their scholarships run out. I cannot comment on the students' motives but it is obvious that this phenomenon causes the confidence of European authorities to diminish and thereby for everyone who makes applications subsequently to be penalized.

As far as working visas are concerned, we come up against the lack of a common European policy on the matter, each EU state having its own approach. Personally speaking, I am convinced that the absence of a single market and of a single clear European policy on immigration is what pushes cunning people to seek alternative means of getting into Europe, by means of short-term visas or illegal crossings. It is the reason why the current government of my country has changed its legislation in this area, making immigration more flexible in order to control it better.

Let us very briefly look at issues related to illegal crossing and trafficking.

First, visas: in order to get onto an aeroplane, you have to have a valid visa. Most people who wish to emigrate thus try to get one, any one and at any price. Less documentation is required for short-term visas and it is cheaper and easier to falsify hotel bookings or airline reservations. Often, would-be emigrants forge bank statements or employ deceitful tactics like taking out short term loans from friends or family so that large sums appear on the balance.

But let us be clear: the majority of those who try to deceive the consular services try their luck but are not truly dishonest. Sometimes these attempts are naïve, while one has to understand that the notion of deceiving the administration does not represent the same taboo every where, especially when a state is not consolidated or where it has earned little respect.

The recourse had to trafficking is more serious. We have to combat such activities. As a citizen, I defend the rule of law and therefore I do not like violation of the law. Besides, I know that the principal beneficiaries of such trafficking, which often involves people, drugs and weapons altogether, are neither the simplest nor the poorest people. The people itself does not profit from this trafficking and you only have to see how the countryside is being depopulated and the sudden disappearance of a large number of Senegalese fishermen in two years to understand that economic systems which allowed hundreds of thousands of families to live can completely collapse.

Many fisherman abandoned their traditional activities to take part in people trafficking to the Canaries. Each boat measures between 12 and 35 metres long and can transport between 22 and 100 people. Given that the fare charges is between €700 and €1,000 and that the investment necessary to buy such a boat is between €15,000 and €20,000, motor and GPS included, it is easy to see that trafficking represents good money for fishermen, even though a large percentage of their income is creamed off by the organizers of the networks.

Such trafficking obviously has collateral effects. The industries connected to fishing – transforming and exporting fish products – which are among the last remaining industries in Senegal, are experiencing serious difficulties in obtaining primary materials. They are therefore struggling to survive. This development therefore helps practically no one and certainly not efforts to keep the Senegalese economy growing. The government has begun to understand this and is starting to take action to break this vicious circle.

With respect to false documents, I should also mention false passports and real passports onto which are pasted visa stickers granted to other people. Neither their quantity nor their price is negligible. According to very recent information, a passport with a Schengen visa is worth €2,500 on the black market and €5,000 if there are two Schengen visas. These figures are not even confidential because the free advertising newspaper *Tam-Tam* recently published on its front page a sale of "guaranteed" visas at a special summer price of €8,000.

Illegal crossings are concerned, which usually take place by sea rather than over land across the Sahara – a route which has become more difficult since the Mauritanian and Moroccan authorities have started to become more vigilant – are cheaper, as we have seen. This could explain why there has been such a sharp increase in a method in which success is not guaranteed and which are, more and more frequently, huge organized con-tricks in which the victims have no chance at all of arriving at the right port.

The figures involved are enormous with respect to an average salary of below 150 euros. The consequences for those who do not arrive in Europe are sometimes very serious, since it becomes very difficult indeed to reimburse the family or the village friends who have put up the money.

There is also a paradox. These sums would enable people to set up businesses which, although small to start with, would enable people to have a higher standard of living than that which an illegal immigrant in Europe can expect. This is why organizations like the Movement of Businesses of Senegal try to help people obtain small amounts of credit.

I should also mention the risks which illegal means of emigration and the decline in living standards in Africa pose for stability, peace and the war on terror. Terrorist networks are well versed in the means of penetrating Europe secretly. However, I would like to conclude with a few general remarks about what we should do to combat this problem. Here are a few personal and semi-random thoughts.

1. One of the things that has worried me most is the complete absence of a debate on birth control. It is a taboo. I can understand this but I cannot help but regret it because, for as long as the growth rate in the population remains as high as it is, it will absorb the fruits of economic growth even when that growth is relatively high, as it is in various members of the Economic Community of West African States. As a result, this real economic growth, about which the governments in the region make a good deal of propaganda, is not felt by the populations and their frustration contributes to their desire to leave.
2. Europe cannot receive hundreds of millions of immigrants since it does not have the jobs or the population to deal with such an inflow, the means by which to give immigrants the same rights as we enjoy, or the possibility of offering a serious prospect of effective integration. Such mass immigration would result in a degradation in everyone's standard of living and a change in social-cultural relationships which would be unacceptable in the eyes of Europeans. If nothing changes, Europeans will vote, if necessary, for extremist political parties.
3. Europe is in any case living through a crisis in its identity. It has no project and its economic growth is low. You only have to read the French press to see that the quality of life of the middle classes is declining. The middle classes are afraid of globalization and open markets. On the other hand, post-colonial complexes and the abuses of political correctness have created a strange dialogue between Europe and Africa in which it is very rare that each side speaks frankly to the other. Our African friends, incidentally, would prefer such frankness more than we sometimes think.
4. None the less, Europe genuinely needs fifty million immigrants over the coming decades in order to have young human capital, a spirit of enterprise and social security systems which work. It is essential for our competitiveness to define who should come and how, but the fact is that we need immigrants.

5. Within the framework of the Lisbon strategy, which aims at making the EU more competitive, Europe has the right and the duty to make choices about the people it needs. This does not necessarily mean a brain drain because we have plenty of people in Europe with degrees. What we need instead are people with competitive degrees. We need to educate young Africans and help them to return to their countries of origin to create competitive jobs. In Portugal, my government calls this 'training but not draining'.
6. In my opinion, illegal immigration into Europe will cease either when Europe ceases to be attractive, or when Africa becomes attractive. Since we do not desire the first outcome, we must work to achieve the second. This indeed is the position of the Portuguese government: as the Portuguese foreign minister has said, "It is essential to understand the question of migration globally and multilaterally. We must base this understanding on the promotion of an increased development and cooperation with the countries of origin, in order to combat the causes and the negative effects of illegal immigration both within the countries of origin and also in those of transit and destination." (Speech by the Minister of Foreign Affairs and Cooperation, 28 October 2005, Lisbon: 'Portugal, Morocco and the future of the Barcelona process.')
7. In this regard, it is incomprehensible that Africa should be the only continent with which there is no high-level political dialogue. This is why, during the Portuguese presidency of the European Union in 2000, we organized the first EU–Africa summit, in Cairo, and we will do out best to ensure that a second one takes place. It is desirable that this second EU–Africa summit should adopt, at the highest level, a joint strategy which would represent an important step towards a Euro-African partnership based on shared responsibility and cooperation.
8. Our Presidency will therefore accord considerable importance to the dialogue with Africa and the whole Mediterranean basin.
9. It seems to me that Europe will have no geo-strategic future if she does not establish a partnership between the first and the last developed continents. Others are not standing by: Russia has her own objectives and a clear strategy, the Middle East is unstable and the Pacific is becoming more and more an axis of progress. China has understood this and has just organized its first summit with Africa which has culminated in an aggressive strategy for commercial penetration and the export of resources at an unprecedented level. We have advantages in such a dialogue, we have only to understand that, faced with China and the United States, even in Africa individual EU member states, whatever their history, are not big enough to act alone.

Alexandre Leitão is a Portuguese diplomat.

PART VI
War and Migration

PART VI
War and Missionary

Conflicts in Chad and Darfur

DOUAL MBAINAISSEM

Introduction

Since 2003, there has been constant conflict on the eastern border of Chad. This situation has had major consequences for the life of Chadian people: there have been massive waves of refugees from both sides of the border and armed groups regularly invade eastern cities like Goz Beida, Am Timan and even the capital, N'Djamena.

According to the Chadian government, the country is the victim of Sudanese aggression, as Khartoum's intention is to expel the black population from Darfur and to spread Islam among the Chadian populations.

Chad: Human Geography and Political Development

After Sudan, the Democratic Republic of the Congo and Libya, Chad is the largest territory in Africa, covering 1,284,000 square km, with nine million inhabitants – seven inhabitants per square km.

The Chadian population is a real ethnic mosaic, divided into several groups. Most people live in the south of the country where the main activities centre on agriculture and rent cultures like cotton. The main linguistic groups can be divided as follows: Sara (30%), Arabs (15%), Mayo-Kebbi (12%), Kanembou (9%), Ouaddaï (15%), Hadjaraï (8%), Toubou/Gorane (6%) and Zaghawa (1.5%). These figures give the following spatial repartition of the population: 7 % in the north, 33% in the central area and 38% in the south.

In April 1900, France conquered Chad, which led to many rebellions in both the north and the south of the country.

Political life in Chad followed from the country's engagement on the side of Free France and the Brazzaville Conference in January and February 1944. Three political parties dominated on the political stage at the time:

- The Chadian Democratic Union (*Union Démocratique Tchadienne*), established in 1947, was the local branch of the French People's Assembly (*Rassemblement du Peuple Français*, RPF) and was supported by the

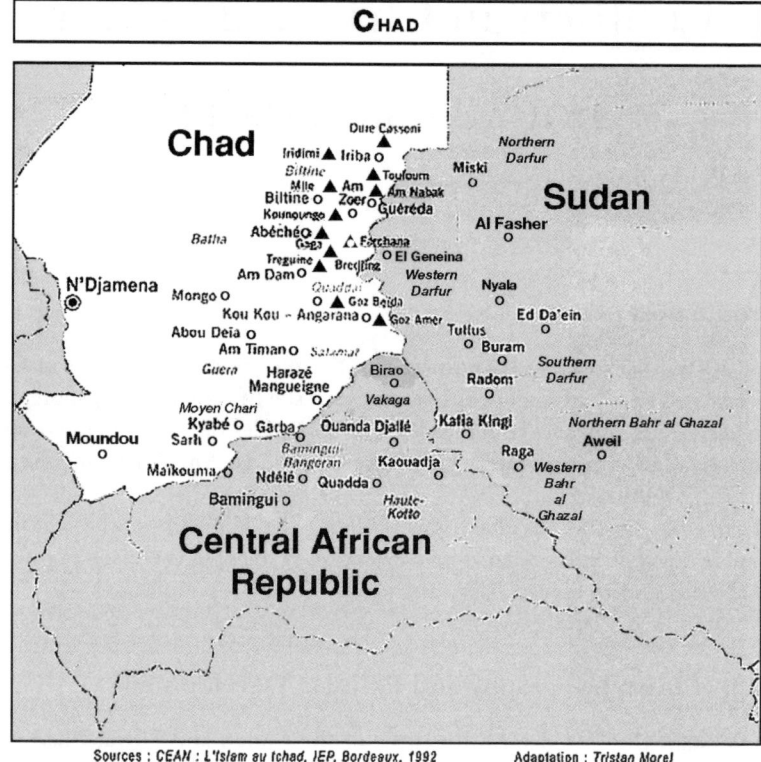

Sources : CEAN : L'Islam au Tchad, IEP, Bordeaux, 1992   Adaptation : Tristan Morel

administration. It was the party of the notables (sultans and district chiefs) and remained the major political movement in the north of the country until 1952.

- The Chadian Progressive Party (*Parti Progressiste Tchadien*, PPT) was the local branch of the African Democratic Assembly (*Rassemblement Démocratique Africain*, RDA) and was settled in the cotton-producing areas of the south. Most of its recruits were administrative officials. It was linked to the French Communist Party. The PPT became the major political player in Chad after 1952, with François N'Garta Tombalbaye as one of its leaders.

- The African Socialist Movement (*Mouvement Socialiste Africain*, MSA), founded by Ahmed Koulamallah, had links with the French Section of the Workers' International (SFIO). It was hostile to independence at first but changed positions under the influence of the SFIO.[1]

CONFLICTS IN CHAD AND DARFUR

THE MAIN ETHNIC GROUPS IN CHAD

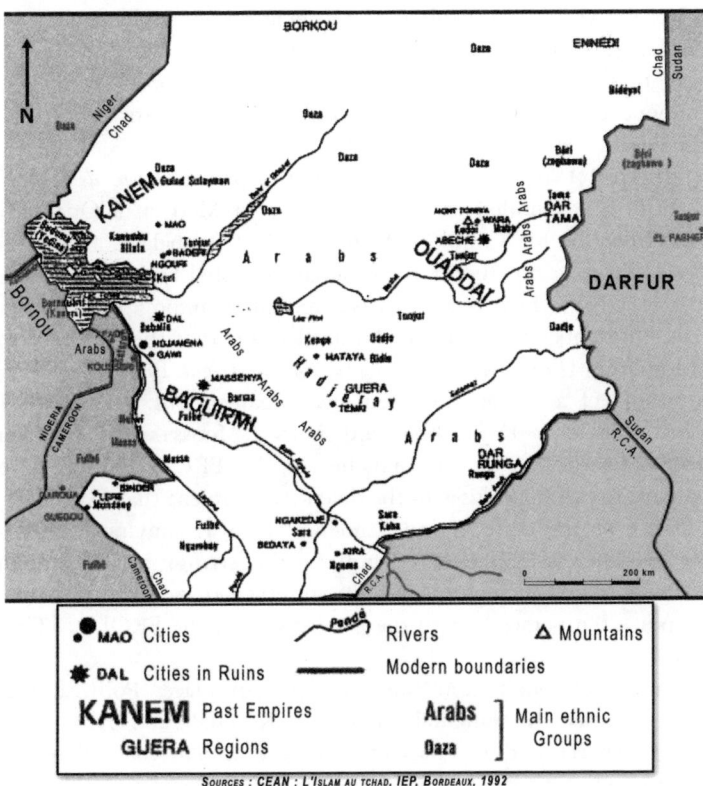

Sources: CEAN: L'ISLAM AU TCHAD, IEP, BORDEAUX, 1992

On 29 March 1959, François Tombalbaye became prime minister and his party, the PPT-RDA, won the general elections in May 1959. After independence was declared on 11 August 1960, Tombalbaye became Chad's first president. In the Muslim regions, the victory of the PPT was seen as the victory of the South and of the Sara, not as the victory of the entire country. Indeed, the ethnic-religious factor has played a major role ever since the eve of independence.

In keeping with the popular doctrine on the African continent at the time, the single party was considered to be the cement of national unity and all parties other than the president's were abolished in 1962. The opposition-repression cycle then began in Chad's public sphere. Several political officials were arrested, sentenced to death and later pardoned; some chose exile and ended up in Sudan.

Darfur, the Sanctuary of the Opponents to the Chadian Regimes

Sudan has always been an ideological pole of attraction and a sanctuary for the Chadian elites of the North in their struggle against central power. They believed they had been unjustly denied leading posts because of the Francophone orientations of the Tombalbaye government. Diplomas from Arab countries were of no use in Chad.

On 20 April 1965, the Chadian Liberation Front (*Front de Libération du Tchad*, FLT) was founded in Nyala by Ahmed Moussa, a Ouaddaï and a renegade from the MSA, and by Mohammed El Baghalani, a Djellaba from Sudan and a former member of the Chadian National Union (*Union Nationale Tchadienne*, UNT). FLT recruits were recruited among the thousands of Chadian emigrants. It was also in Nyala (Darfur) that the Chadian National Liberation Front (*Front National de Libération du Tchad*, FROLINAT) emerged as the result of a merger between two organizations: the UNT, led by Ibrahima Abatcha (a Kanembou) and Ahmed Moussa's FLT. The guerrilla conflicts between the various branches of the FROLINAT led President Tombalbaye to use the clauses in the defence agreement that Chad had signed with France at the time of independence: he was convinced that foreign powers (Sudan and Libya) were pulling the strings and organizing the aggression in Chad. French troops thus intervened in April 1969 and were able to put a temporary halt to the progression of the FROLINAT's various factions.[2]

A policy of reconciliation was then put in place. Political prisoners, including Ahmed Koulamallah, Djibrine Kerralah and Abo Nassour (a Zaghawa), were released. In a flash of lucidity, President Tombalbaye declared: 'The populations who have rebelled are merely people who want to be looked after, who want wells and schools to be built and tracks to be opened for their cattle to pass. They are citizens who want to pay their civic taxes only once.'[3] But such a policy could not be carried out and the country was once again caught up in the opposition-repression cycle, leading to a military coup on 13 April 1975.

With the intervention of Sudan and France, the new authorities signed a treaty with Hissène Habré's Command Council of the Armed Forces of the North (*Conseil de Commandement des Forces Armées du Nord*, CCFAN), one of the branches of FROLINAT, on 22 January 1978 in Khartoum. Habré became prime minister and the French Army began to recruit, train and equip a new army for him. Why so much solicitude from the French authorities for Hissène Habré? The French senior officers resented President Felix Malloum for having criticized the defence agreements in October 1975 and having expelled French troops from Chad.[4] From then on, FROLINAT would appear as the alternative solution to N'Djamena.

The country thus found itself with two armies, with two leaders as the head of state, and the conflict turned into a civil war. In 1979, Goukouni Oueddeï formed a National Union Transition Government (*Gouvernement d'Union Nationale de Transition*, GUNT) where all the branches of the FROLINAT and the Chadian armed forces were represented.[5]

This institution did not last and fighting broke out again in the capital. Libya's support helped two FROLINAT branches, the Popular Armed Forces (*Forces Armées Populaires*, FAP) and the Revolutionary Democratic Council (*Conseil Démocratique Révolutionnaire*, CDR) to defeat Hissène Habré's CCFAN on 15 December 1980. Habré fled to Darfur, which he used as a base to regain power in Chad. The authorities in Khartoum, Egypt, the United States and France, which opposed Gaddafi's Libya, helped him in this. After Goukouni Oueddei requested a partial and planned withdrawal of Libyan troops from Chad, Libya decided to pull out immediately, on 3 November 1981. What followed was the capture of N'Djamena by the FAN on 7 June 1982 and the creation of a FAN state, which lasted until 1 December 1990.

Based in Libya and Bornou, GUNT troops then renewed hostilities against the FAN state and put Hissène Habré's forces under severe strain on several occasions. Only the intervention of Zairian troops and the Manta Operation of the French troops in 1983 managed to secure the survival of the FAN state.

The forces that would destroy the FAN state emerged from within. The rivalry between the major ethnic groups of the FAN's leaders – Hadjaraï, Zaghawa and Toubou/Gorane – was demonstrated by the abortive coups, as instigators fled to Darfur again. There was first Captain Maldoum Bada Abbas' attempt to overthrow Habré in late 1988 as he and his group, the Hadjaraïs, protested against the murder of their relative, Idris Miskine. They then fled to Darfur. Then, on 1 April 1989, a group of Zaghawa officers – Hassan Djamous, Mahamat Itno and Idriss Déby, the chief of staff – failed as well; they fled to Darfur, though only Idriss Déby managed to reach safety.

The Emergence of Zaghawa Power in Chad, and its Consequences

The arrival of Idriss Déby, the chief of staff on the run, and his efforts to put together a military force, led to fundamental changes in the political situation in Chad, Darfur and later in the whole of Central Africa. The current conflict in Chad, the Darfur region and the Central African Republic cannot be understood fully without bearing this factor in mind.[6]

Déby has Bideyat origins and derived support from his ethnic group, the Zaghawa, from the Arab members of the CDR, from Maldoum Bada Abbas' Hadjaraïs, and from Libya and the authorities in Khartoum. This new rebellion targeted eastern Chad. In retaliation, Hissène Habré's troops stepped up incursions into Darfur and destroyed villages and camps indiscriminately. They also provided arms to the Fur, a tribe who were in conflict with their own Zaghawa and Arab enemies.

This policy led the rest of the Sudanese Zaghawa to join Idriss Déby's ranks. The presence of the great Zaghawa-Kobé families from Chad – Abbas Koty and the descendants of Sultan Haggar, Mahamat Ali Abdallah – strengthened mobilization within the group.

In June 1990, Hissène Habré fell out with President François Mitterrand who had said in his speech at La Baule that French help would be conditional on democratic progress. The Chadian head of state considered that a former esclavagist country like France was ill-placed to teach Africa lessons in democracy: if anything, France ought to pay African countries damages to compensate for the slave trade and colonization.[7] As a result, Mitterrand loaned the French Special Forces, under the authority of Colonel Fonbomme, to Idriss Déby. FAN power fell like a ripe fruit on 1 December 1990.

With the accession of Idriss Déby to power, Chadian political life entered a new cycle: the famous Bureau of Documentation and Security (*Direction de la Documentation et de la Sécurité*, DDS) was dissolved, and an investigation commission on Hissène Habré's crimes was set up while associations and political parties could be created freely. Under French pressure, Déby later agreed to hold a Sovereign National Conference (*Conférence Nationale Souveraine*, CNS) from 15 January to 6 April 1993. This conference enabled the Chadians to initiate dialogue and to reach national consensus on new democratic institutions and on a transition period of three years. A new constitution was elaborated and voted in in March 1996; the presidential mandate was limited to two five-year terms. Presidential elections were organized in June–July 1996, in 2001 and 2006.

As soon as Idriss Déby came to power, he made sure he thanked the Darfur Zaghawa by giving them strategic positions (customs, Cotontchad – which controlled all of Chad's cotton production and export). His group supervised customs and thus imports and exports of vast quantities of goods (hydrocarbons, manufactured goods and foodstuffs) with Nigeria, Sudan and Libya. The country's economy was thus endangered and several companies were shut down. Several tens of thousands of men from this ethnic group were integrated within the Chadian armed forces and the Republican Guard; they roamed the country in groups of ten to twenty in their Toyota pick-ups, looking for victims to kidnap for ransom. Because state revenues were too limited to be widely distributed, they were reserved for the close friends and allies of the Déby family. Power gave the tribe the authorization to feed on the country. The Zaghawa thus raided the Ouaddaï region; they seized cattle and the territories controlled by the Chadian Arab groups, who fled to Darfur. For the Chadians from other ethnic groups, the arrival to power of a Zaghawa led *de facto* to the multiplication of special favours and many Zaghawas obtained positions of accumulation, whether in the economy strictly speaking or through a more direct system of predation (customs or extortion). These practices, and the impunity displayed by the Zaghawas in Chad, crystallized the resentment of Chadian populations against them and the power of Idriss Déby.

Chadian Power in Crisis

The current war in Chad is first and foremost an internal affair, despite the fact that the authorities choose to emphasize its external origins (that is, Sudan and Saudi Arabia's project to spread Islam in the country). The President and his party broke the national consensus and the Republican pact that were inscribed in the Chadian Constitution. After gaining power by force on 1 December 1990, Idriss Déby was legitimized by the 1996 and 2001 elections. The results of the presidential elections were at first contested by the opposition parties but later accepted to preserve the peace. The President's opponents ended the contestations over the elections because he promised to respect the Constitution, in particular the clause that limits the President's mandate to two five-year terms. Idriss Déby made this promise in his inaugural address at the beginning of his second term: 'I will not stand for the presidential elections in 2006. I will not change the Constitution even if I thought I had a one hundred per cent majority! I say it loud and clear: what is left for me to do during my final term is to prepare Chad for a democratic, peaceful and smooth changeover.'

Opponents to the regime were hoping that the authorities would reward their republican attitude, that the President would respect his public promises, that the 2006 presidential elections would be organized in all transparency and that Idriss Déby would not stand. Such a scenario would enable Chad, for the first time since independence in 1960, to stop the vicious cycle of repeatedly violent political transitions. But unfortunately, Idriss Déby and his French protectors thought differently.

In November 2003, the President's Patriotic Salvation Movement (*Mouvement Patriotique du Salut*, MPS) met and decided to alter the Constitution to enable its leader to seek another term in office. In mid-May 2004, the National Assembly, where the MPS had a large majority (115 representatives out of 155), adopted the constitutional revision that abolished the clause limiting the President to two terms. During an official visit to Chad, the French Minister for Cooperation, Xavier Darcos, renewed France's infallible support to this perjury: 'France supports the position of President

Déby, who was elected democratically twice and praises the fact that the National Assembly approved the revision of the Constitution, with a larger majority than usual.'[8] *Dixit*! This constitutional reform was adopted by referendum in June 2005 – many boycotted the event, as political and civil society opposition had urged them to do so.

The violation of the Constitution is the origin of the violence that Chad is going through. Indeed, the MPS barons hoped that the derelict management of state affairs would end and that they would one day become the supreme authority. But they had to either give up their ambitions or opt for armed conflict. Thus members of the first circle initiated a coup on 16 May 2004: the action of these members of the Republican Guard failed because the French Special Services intervened. No sanction was taken against the perpetrators because they belonged to the President's circle.

However, several Zaghawa officers and personalities failed to lend support and formed various armed groups in Darfur. One can mention Yaya Dillo Djerrou's Platform for Change and Democracy (*Socle pour le Changement et la Démocratie*, SCUD) and Abakar Tollimi's Rally for Peace and Justice (*Rassemblement pour la Paix et la Justice*, RPJ). These two movements merged to form the Rally for Democratic Forces (*Rassemblement des Forces Démocratiques*, RFD), led by one of the Edimi twins, Timan, the President's nephew and the ex-director of his civilian cabinet. In 2005, the Rasputin Timan and Tom Edimi broke away from the authorities in place.

On 13 April 2006, the United Front for Democratic Change (*Front Uni pour le Changement Démocratique*, FUCD), an armed opposition group led by the Tama Captain, Mahamat Nour, entered the capital, N'Djamena. This offensive was immediately halted, thanks to the logistic support and the involvement in the fighting of the French troops stationed in Chad. Once again, France intervened to save brave soldier Déby, outside all legal framework. Indeed, there has not been any defence agreement between Chad and France since 1976, when General Malloum's military junta revised the documents signed at independence. France and Chad's military cooperation agreement states (in Article 4) that the French military personnel who help to train the Chadian Army '... cannot, under any circumstances, be directly involved in the execution of war operations, in keeping or restoring order and legality...'.

Most Chadians and Africans wonder why Ivory Coast, Chad and the Central African Republic are treated differently. There is a defence agreement between Ivory Coast and France. But the democratically elected Ivorian authorities, who believe they are being attacked with the support of neighbouring countries, have been asking France in vain, for four years, to use the text to guarantee the integrity of their national territory. And for four years, France has been happy to see a Maginot line across Ivory Coast. Simultaneously,

France is prompt to save regimes that have imposed themselves by force and that maintain their authority by force, both in Chad and in the Central African Republic.

The Coordination of Political Parties for the Defence of the Constitution (*Coordination des Partis politiques pour la Défense de la Constitution*, CPDC) appeared in the political arena and played a major role in the boycott of the constitutional referendum and presidential elections in 2006. No candidate from the major parties took part in the presidential elections, despite pressure from France, the European Union and the United Nations. The leaders refused to act as foils for Idriss Déby. Faced with this degenerating political and military situation, the President tried – with some success – to convince the French authorities and part of his public that Chad was being attacked from Sudan in order to impose a radical form of Islam and to ill-treat the black population! With the help of the Françafrique (that is, the unofficial collusion of France and its African interests), the entire Chadian diplomacy instrumentalized the 'Arab threat'. But informed observers well know that many actors in the Darfur conflict held high posts in the Chadian Army and special forces, and that President Idriss Déby himself has promoted the spread of radical Islam in the South of the country, ever since he came to power.

The Crisis in Darfur and the War in Eastern Chad

In February 2003, Sudanese Zaghawa soldiers and officers from the Chadian Republican Guard took over the village of Golo (in northern Darfur), led by a former comrade of Idriss Déby's, Colonel Abdallah Abakkar. They called themselves the 'Sudan Liberation Army' (*Armée de Libération du Soudan*, ALS). The ALS's military equipment was smuggled across the Chadian border with the complicity of some highly placed Chadian state officials, though not necessarily with the President's approval. The ALS took over Tiné, the headquarters of the main Zaghawa sultan, and defeated the Sudanese forces on several occasions. On 24 April 2003, the ALS entered El Fasher, the capital of northern Darfur, took control of the airport and seized General Ibraim Bushra, the commander of the air base in this town.

President Omar al-Bashir then decided to take the bull by the horns: he appointed new civilian and military authorities to the region and transferred the southern Sudan units to Darfur; the new military authorities enrolled Arab militias, where many belong to the Chadian tribes who fled as the Zaghawa devastated the Salamat and Ouaddaï regions. At the diplomatic level, the Sudanese President asked his Chadian counterpart to act as a mediator. The latter managed to get the ALS to sign a cease-fire agreement on 3 September in Abéché. The cease fire was soon violated and war broke out again but the Sudanese Army managed to retake the main centres in Zaghawa country (Tiné and Kulbus) and to execute Colonel Abdallah Abakkar, the ALS's military leader.

A second group was then formed in Darfur, under the name of the 'Movement for Equality and Justice' (*Mouvement pour l'Egalité et la Justice*, MJE). Its leader is an official from the People's National Congress of Hassan al-Tourabi, Dr Khalil Ibrahim Mohamed, of Zaghawa/Kobé origins. President Idriss Déby tried several times to arrest the MJE's leader and deport him to Sudan but every time, he faced opposition from the Republican Guard officers who share family ties with him – Khalil Ibrahim Mohamed's mother is the sister of Déby's stepbrother, Timan, the sultan of Iriba.

The military defeat of the ALS led part of the Chadian armed forces to organize a coup against Idriss Déby in May 2004. One of the main demands of the coup leaders was that Chad needed to send troops to help the ALS forces. According to the Sudanese authorities, the rebels had the support of the Chadian high authorities, including the President. The belief is based on material evidence found in the theatres of operation: Chadian military papers, weapons and transport equipment from the stocks of the Chadian Army. Sudan thus decided to support Chadian exiles in Sudan; Khartoum once again welcomed all those who wanted the man from Fada to leave.

Sudan provided the various groups (FUCD, RAFD, the Union of Forces for Democracy and Development – UFDD) with arms and logistic support. War and serial attacks began once again, destroying towns and lives. Once again, Chadians in Adré, Am Timan, Abéché, Biltine and N'Djamena suffered physical and psychological damage, while the resources that should have been used to improve the living conditions of the Chadians served to buy weapons and benefited arm traders and intermediaries.

Chadian Perceptions of the Crises in Darfur and Chad

The President used all his relations to spread the idea that Chad was being attacked by the Sudan via eupatrids and mercenaries in order to spread Islam and Arab cultures among the black population. As for the Chadians, they cannot understand why their President has initiated an international showdown in Darfur with the sole aim of saving the Zaghawa. Nor can they understand how Chad can accuse Sudan and Saudi Arabia of Islamization.

Indeed, it is under the current president that radical muslim sects (Wahhabism, the Muslim Brotherhood), different from Chadian Islam, based on confraternities (*tidjaniya*, *semousiya* and *ahmadiya*) proliferated with a new intensity and that mosques opened in all the hamlets in the south of the country.

It was also Idriss Déby's government that signed several agreements with Muslim international NGOs so that they could settle in the country and promote the expansion of Islam, for example, the Libyan NGO *Dawa Alamiya Islamiya* (World Islamic Call Society), whose headquarters are in Tripoli, whose president is Libyan and whose financial sources are also Libyan. Their objective is to build Koranic schools, to train teachers, to print textbooks and to be involved in the mosques. The same government signed a treaty with Sudan to create the *Dawa Alamiya* (World Call Society): its president is Sudanese; its vice-president is Chadian; its headquarters are in N'Djamena; Saudi Arabia, Kuwait and the Sudan finance it; its objective is to promote Islam by building mosques and training imams. It is thanks to this NGO that mosques were built in southern Chad; it is also involved in the recruitment of young people from the south of the country, who are sent to Omdurman in Sudan or to Karachi in Pakistan for a year of training, before they return to their country on a 'mission'. It is this same government that today tries to get the West to react and defend the black populations from a forced Islamization of the Arab part of Sudan!

Politicians and members of Chadian civil society have chosen not to take sides between the government and the armed opposition factions. On 24 October 2006, General Wadal Abdelkader Kamougué declared on RFI: 'The rebels intend to take over power. But we, who belong to political parties, wish to get power through the ballot box. So of course, another coup would change the regime, but it would once again throw us into the unknown.' Similarly, Saleh Kebzabo, the leader of the National Union for Development and Renewal (*Union Nationale pour le Développement et le Renouveau*, UNDR)[9] said: 'Our position regarding the rebellion is extremely clear. The UNDR has always condemned taking power by force. It is not because we say we understand those who have taken up arms that we support them. It is because people feel oppressed in this country. There isn't a single little window that could bring in some air.' Or again, the CPDC's spokesman, Ibni Omar Mahamat Saleh, said in the press:

> [We] lament and condemn the huge loss of lives on both sides and the material damage for which President Déby and all those who encourage and support him in rejecting democracy and good governance are primarily responsible; Deplore the position of France who supports by all available means – military, diplomatic and financial – one camp against another, encourages the Chadians to kill one another, to destroy national cohesion and to jeopardize economic and social development; Consider that while the interference of neighbouring countries should be condemned, all the participants should condemn the subversive transborder activities of Déby's regime, which has encouraged armed rebellions in the Central African

Republic, over into Cameroon and Darfur, openly supporting the MJE and the MLS which are prosperous and highly respected in N'Djamena and in many places in eastern Chad.

Or again Massalbaye Ténébaye, a member of the Steering Committee for the Call for Peace and National Reconciliation in Chad (*Comité de Suivi de l'Appel à la Paix et à la Réconciliation nationale au Tchad*, CSAPR):

> The Chadian government is involved in the situation in Darfur since it was asked to act as a mediator in the conflict. One should also note that the United Nations report in December stated that Chad was one of the arm traders in Darfur. Moreover, the Chadian government, by trying to internationalize the national crisis, chose to oppose the Sudanese government rather than deal with its internal problems. We believe one should oppose these attempts to internationalize the Chadian crisis because it would only play in the hands of the various powers involved in the crisis in Darfur. Chad thus deserves criticism but it is also very vulnerable.

Mahamat Saleh Haroun, a Chadian filmmaker, focused on the option of violence in his latest film, *Darrat*:

> Every time it is the vizier who wants to become caliph ... by using violence. So it becomes impossible to find a way out and no proposal can put an end to this violence. This is why I thought it was important to deal with this issue, which handicaps Chad the most. The majority of Chad's nine million people are the hostages of this violence. A violence that is generated by a few groups who fight over power.

Prospects for the Future

The conflicts in Darfur and Chad can be explained by both exogenous and endogenous reasons. Internal political factors (the absence of liberties and of a peaceful political alternative, as well as predations and revenues derived from the control of the state apparatus in both countries) are the main causes of the current war.

It would be useful if Idriss Déby could meditate on President Goukouni Oueddeï's declaration when he left the CNS in 1993:

> When we went to the national conference, everybody was afraid of everybody else. And after a month, serenity prevailed. Through

dialogue, people realized that they were all Chadians, they all belonged to the same country and defended the same cause. I for one felt this way. *If we manage to establish democracy and build an army to serve the nation and not one clan only*, the North–South rivalry will die out. In other terms, I believe that once a president is elected under democratic conditions, he will gain the support of the entire people. And he will be able to save Chad.

If we're back to square one, sixteen years after Idriss Déby came to power (wars in the East, political opposition, tribal army), the only man responsible for this can only be the President himself.

What prospects are there to end this situation of conflict?

- The vicious cycle of a necessarily violent transition must be broken (Idriss Déby must go, according to a legitimate constitution).

- The demands of all the populations in Darfur must be taken into account.

- The French Army must understand that it cannot blindly support authoritarian regimes.

Political forces and armed groups should come together and enter into an all-inclusive dialogue. It would be important to discuss, among other issues:

- Initiating a period of transition

- Restoring the 1996 Constitution

- Setting up a new consensual electoral commission

- Abolishing voting rights for nomads

- Defining constituencies which would represent populations

- Building a truly national army, using the demographic data on Chad.

Doual Mbainaissem is an agricultural engineer and a consultant in strategic and territorial projects.

NOTES

1 Some members left the MSA and founded the Chadian National Union (*Union Nationale Tchadienne*, UNT). The future leaders of FROLINAT (for instance, Ibrahima Abatcha) are among its founding members.
2 Cf. Christian Bouquet, *Tchad: genèse d'un conflit* (Paris, L'Harmattan, 1982), pp. 136–42.
3 Cf. Ibid., p. 141.
4 Cf. A. Sanguinetti, 'Les interventions françaises', *Tricontinentale*, 1, *La France contre l'Afrique*,

François Maspero (ed.), 1981, p. 97: 'This desire to emancipate is unforgivable in the sense that it can generate other initiatives, and France will make sure he pays for this.'

5   The outcome of this war could have been different if Habré's troops had not had the support of General Forest's troops (the Moussoro-Ati dam was open to Goukouni's columns, the French battallion in Abéché gave its support to the FAN troops), cf. Bouquet, *Tchad*, pp. 156–9.

6   Cf. R. Marchal, 'Soudan: d'un conflit à l'autre', *Les Études du CERI*, 107–108, September 2004; ibid., 'Tchad/Darfour: vers un système de conflits', *Politique africaine*, 102 (2006), pp. 134–53; International Crisis Group, 'Tchad: Vers le retour de la Guerre', *Rapport Afrique*, 111 (June 2006).

7   Cf. L. Feckoua, 'Tchad, la solution fédérale', *Présence Africaine* (1996), pp. 73–4.

8   Agence Frances-Presses (AFP) news bulletin, 28 May 2004.

9   *N'Djamena bi-hebdomadaire*, 991 (9–12 November2006).

The Migration and Displacement of Assyro-Chaldeans in Iraq

FRANÇOISE BRIÉ

Assyro-Chaldeans converted to Christianity at the very beginning of the Christian era. They are divided into two groups: a minority of the so-called Nestorian tradition, and a Chaldean majority attached to the Roman Catholic Church.[1] They form the main Christian community in Iraq and they also claim recognition as a nation and an indigenous people.

Around 1921, when the Iraqi state was founded, nearly all the Christian population was located in northern Iraq, mainly in rural areas. Their numbers had increased due to the influx of several tens of thousands of refugees following the genocide in the Ottoman Empire.[2] There were more than one million of them in 1987; 800,000 in the 1990s; 500,000 since 2003. This represents a genuine haemorrhaging of the population over twenty years, if one compares these figures with those for other communities. Most studies agree that Christians represented between 3 and 5 per cent of the total population of Iraq until the 1980s. Yonadam Kanna,[3] elected to the Iraqi Parliament on the National Rafidain list, reckons that one million Christians have gone into exile since the beginning of the 1970s. They are estimated to represent 80 per cent of all Iraqi refugees in the United States, in particular in Detroit and Chicago – their foreign 'capitals', as it were.[4] The diaspora also settled in Canada, Australia, New Zealand and Europe (especially Sweden).

The post-2003 period has been especially harsh. A UN study carried out on a sample of Iraqi refugee families in Syria demonstrated that 19.3 per cent are Assyro-Chaldean in origin.[5] They are also said to be the most numerous amongst the families that arrived in the Jordanian capital of Amman in the first quarter of 2006. Internal displacements in Iraq do, however, remain the greatest migratory problem, with close to 17,000 Christian internally displaced persons recorded by the International Office of International Migrations at the end of October 2006.[6]

Christian churches, especially the Catholic Church, and their representatives in politics have called for the implementation of plans to resolve this crisis on both national and international levels. According to Iraq's

Human Rights Minister, Wijdan Mikha'il, if such steps are not taken, it will lead in the short term to the complete disappearance of Christians from Baghdad and Mosul, of whom half are said to have left already. More generally, it could lead to the disappearance of the whole Christian population of Iraq in twenty years.[7] American bishops have gone as far as requesting Condoleezza Rice for special protection for the religious minorities who are the victims of ever-increasing deliberate attacks.[8]

The Main Waves of Migrations and Displacements of Assyro-Chaldeans since the Founding of Iraq

Internal and international conflicts, as well as a policy of forced displacements institutionalized by the Baath party when it came to power in 1968, lie at the root of the main migratory movements in Iraq that affected Assyro-Chaldeans, Turkmens and other minorities, as well as the Kurds.

From 1920 to 2006, emigration and internal displacements of Assyro-Chaldeans continued with peaks in 1933, 1975, in the 1980s, in 1991 and since 2003. Between 1920 and 2006, there were two periods of relative calm: from 1935 to 1960 (under the monarchy and at the beginning of Brigadier Qasim's regime), and then from 1988 to 1990, between the end of the Iran–Iraq War, and the beginning of the First Gulf War.

In 1933, the rise of Arab nationalism put an end to Assyrian territorial claims

At the end of the First World War, resolutions in favour of the protection of minorities in the documents governing international relations, as well as the recruitment by the British of a thousand or so Assyro-Chaldeans as an auxiliary force during the British mandate, sharpened radical nationalism, especially in the military.[9] Assyrians were stigmatized by the press and considered to be traitors.

In the summer of 1933, several thousand Assyro-Chaldeans, mainly civilians settled in the region for generations, were massacred by the military and their Kurdish and Arab reinforcement troops in Semel, next to Dahuk. This massacre led to the emigration of 9,000 people to Syria. This event marked the decline of Assyrian nationalism, which nevertheless lasted until the beginning of the 1970s.

From 1960 to 1991, conflicts between Kurds and Iraqi governments led to the evacuation of Assyro-Chaldeans from the mountainous regions

In 1960, Assyro-Chaldeans lived for the most part between Zakho (at the time 45 per cent of that city's population were Christians) and Amadiyah, all the way to Diana; some also lived down in the plains and in the city of Mosul. As they lived in territories claimed by both Kurdish and Arab nationalists, they could not escape from the permanent conflicts involving both groups. Starting at the end of the 1960s, Saddam Hussein made the Assyrians' plight even worse by imposing a policy of forced Arabization and by arbitrarily dividing the different territories. A majority Kurdish governorate was established in Dahuk, while the Nineveh governorate was designated a majority Arab governorate. Saddam Hussein then proceeded to distribute Assyro-Chaldeans administratively into both communities. This promoted their assimilation whilst exacerbating the political divisions based on their place of residence. In 1977, Assyro-Chaldean was eliminated as an identity from the census and they were forced to choose between a Kurdish or Arab ethnicity. If they wanted to be protected and recognized as individuals, they had to commit to the party in power, or to that of the majority in their place of residence. This refers mainly to the Kurdistan Democratic Party (KDP),[10] to which Assyro-Chaldeans supplied many fighters and military or political leaders, such as Malek Hormuz Tchiko, who ended up as an emblematic figure of the Kurdish struggle.

On the other hand, the Baath party sought to gain the support of Assyro-Chaldeans against the Kurds by attempting to form Assyrian militias and through cultural and political openings.[11] Those who chose to stay in the mountains were subjected to reprisals, such as raids conducted on a regular basis by the 'Djash' (Kurdish mercenaries working for the government), or they were excluded for amnesty in favour of individuals having surrendered to the military.

War and exodus against the backdrop of the disintegration of the PDK in the mid-1970s, the establishment of a security area along the borders following the Algiers Accord between Iraq and Iran in March 1975, and finally the al-Anfal operation of 1988–89, led to the destruction of almost all northern Iraqi villages and to the flight or deportation of their inhabitants. Ali Hassan al-Majid, for example, Saddam's cousin known as 'Chemical Ali', ordered two chemical weapons attacks in a region where more than 15 per cent of the villages are Assyro-Chaldean.[12] Several thousand villagers fled to Turkey. Elsewhere, expropriations organized by decree by the Revolutionary Command Council (the Iraqi government) led to the settlement on these lands of Arab populations. After the repression against the uprisings in the provinces in 1991, following the First Gulf War, several thousands fled and the exodus included Christians from Dahuk, Zakho, Erbil and Kirkuk.[13] The migratory wave, 200,000–300,000 people in the 1990s, was the largest of its kind since the First World War.[14]

Their migration towards the centre of the country weakened the alliance with the Kurds and accelerated the Arabization of Assyro-Chaldeans. At the end of the 1980s, 80 per cent of them had abandoned their historical homelands and taken refuge in large cities, mainly Baghdad. This explains why there were about fifty churches in the Iraqi capital in the 1990s, as compared to only six in 1909; the building of the Shorja and de Bataween neighbourhoods, as well as Baghdad Al-Jadida, marked the beginning of the settling of 300,000–400,000 Christians in Baghdad. The Nineveh governorate, in particular, on the plain of Nineveh, formed the last zone where the Assyro-Chaldeans had lived constantly since ancient times.

The Causes of the Exodus since 2003

The political parties that were engaged in the fight against Saddam Hussein were in favour of the return of refugees and of the right to vote for exiled Iraqis and for those deprived of their nationality. These demands were major ones for Assyro-Chaldeans, although returning to Iraq was rather problematic due to the general security situation in the country.

The terrorist attacks which, according to the Iraqi delegation on its visit to Paris in November 2006, were the work of former pro-Saddam groups united with Islamic Sunni groups close to Al Qaeda, targeted Christians as well. Mowaffak al-Rubaie, the Iraqi National Security Adviser, stated that the latter groups did take part in attacks against Christian churches: thirty attacks from 2003 to October 2006.[15]

Minorities[16] were also the direct target of several fundamentalist Islamic groups seeking to eradicate all non-Muslim groups from Iraq, and more generally all those that do not support them. Post-September 11 religious themes, the case of the Danish cartoons, Pope Benedict's speech in Regensburg and the permanent suspicion of collusion with the West fanned the flames of hatred against Christians.[17] In Mosul, Christian families are forced to support Sunni guerrillas financially or else convert to Islam. In the spring of 2003, Moqtada al-Sadr, the young radical Shiite leader, demanded application of the Sharia law, including mandatory use of the *hijab* headscarf for women;[18] in Basra, sentences as severe as capital punishment were pronounced against Christian alcohol retailers and barbers. Sheikh Muhammad al-Fartusi, close to Sadr, also attacked non-religious individuals, and the wealthy middle class, as well as moderate Shias, or Shias belonging to other parties.

Christians are vulnerable because of their political and social weakness and because of their minority status which means they have very little tribal or militia support. Most of them believe that they have always been considered as second-class citizens in spite of their courteous relations with their Muslim neighbours.[19] The Catholic Archbishop of Baghdad denounced a 'context of hidden dhimmitude' (the old system under which religious minorities were governed according to Islamic law) in both legal and practical terms,[20] the corollary of which is a brain-drain. The preamble of the 2005 constitution does not mention Assyro-Chaldeans as an ethnic group. Politically divided – there were eight competing lists in the January 2005 elections – and therefore poorly represented within the new parliamentary assembly, they have two ministers entrusted with Human Rights and Industry, but belonging respectively to Iyad Allawi's Iraqi National List and to the Kurdish Alliance.

Organized kidnapping leading to asset seizure in the case of departure generates significant income for several Mafia-like Islamic groups or militias affiliated with political parties; such practice targets mainly the middle class of which Christians represent an important component.

The exacerbation of community and religious divisions since the attack against the Samarra mosque in February 2006 and the battle for control of some neighbourhoods affect Baghdad Al-Jadida or Al-Dawra, a part of the capital where the Christian population had already decreased from 50 to 20 per cent.[21]

Several attacks and murders perpetrated in Baghdad and Mosul, including the beheading of the Syriac Orthodox priest, Paulos Iskandar, in October 2006, are said to be reprisals against individuals having sided with the Kurds or having accepted financial assistance from them.

Finally, the lack of economic development support for the plain of Nineveh and initiatives taken by the Assyro-Chaldean diaspora, such as the US Chaldean Federation aimed at integrating Chaldeans into American society,[22] do not motivate Iraqi Christians to stay in their homeland.

Possible Solutions to Stem the Exodus of Iraqi Minorities

As early as 1991, tens of thousands of Assyro-Chaldeans started to rebuild the villages that had been destroyed since the 1960s in the Kurdish region above the 36th parallel, villages which were no longer controlled by Saddam Hussein.[23] Those efforts were hampered by the battles between the Turkish Army and the PKK (the Kurdish Workers Party) that took place in the 1990s. Furthermore, the unfair redistribution of the proceeds from the UN's 'oil for food' programme, as well as the confiscation of land and villages, added to the difficulties of reconstruction. The representatives of the KDP – there are few Christians in the area controlled by the Patriotic Union of Kurdistan (PUK) – affirm that there are now no outstanding restitution claims.In spite of openings on the freedom of the press, freedom of association and the use of the Syriac language in education, Conciliation Committees are failing and the relations between the two communities are deteriorating. Five members of the

ADM (Assyrian Democratic Movement) were assassinated in the 1990s, including Francis Yusuf Shabo, a former well-known opponent of Saddam Hussein's regime and a member of the Kurdistan Parliament, where he had spoken in favour of a law to prevent confiscation of assets. The exact circumstances of his murder were never established.[24]

By the end of 2005, the regional government of Kurdistan had made some progress on the reconstruction of Christian villages. There was still a will to weaken Assyrian parties, as opposed to sponsored organizations, but the Kurds now do want to secure the support of the Christian population during this very important period when the borders of the federated region might be withdrawn to include Kirkuk and Nineveh. Some Kurdish officials are trying to thwart the strategies of the Islamic Union of Kurdistan, which withdrew from the Kurdish Alliance at the last elections, by overriding their opposition to the opening of two evangelical churches.[25] The support for Christians can also be explained by the will to react to international criticism: the Kurds are criticized for having abandoned the northern Assyro-Chaldean villages and yet that is precisely where all the displaced persons flee to from Baghdad, Mosul and Basra. The Regional Federal Government is said be holding back US funds destined for these areas.

Should there be a Governorate for Minorities?

The Assyrian Democratic Movement and other parties such as the Assyrian Democratic Organization (ADO) speak in favour of the establishment of a governorate associating other minorities (Shabak[26] and Yezidis[27]). It would include the districts of Qaraqosh, Tell Qayf, Sheikhan, possibly the south-western part of the Dahuk governorate (the district of Semel including the Selevani plain) and the Sinjar district.[28] The idea is that it would help stem the exodus of Iraqi Christians by guaranteeing to them both security and economic development.[29] The ruins of the former Assyrian capital can be found on the Nineveh plain, close to Mosul and Erbil, and they are of course of utmost importance to the Assyro-Chaldeans. There are still several entirely Christian villages there, and towns with no Kurds at all, with the exception of the Sheikhan district which is mainly populated by Yezidis.[30] However, like the city of Mosul, Sheikhan is contested by Kurds and Arabs alike.

A new draft constitution debated in the Kurdish Parliament at the end of 2006[31] takes the Assyro-Chaldeans' proposals on board, but unites the requested autonomous governorate to the Kurdish region. Two Christian ministers in the regional government, Sarkis Aghajan, in charge of Finance and the reconstruction of Christian villages, and Nimrud Baito, Secretary-General of the Assyrian Patriotic Party (APP), stated that the support of the

Kurds remains essential for the achievement of the overall Kurdish project and that the funding for a 1,000-man unit to protect Christian villages has already been pledged.[32] Sunni Arabs, however, consider that this territory is part of the Mosul governorate.[33] According to the Assyrian Democratic Movement, the achievement of the Kurdish project would immediately unleash violence and the ADM foresees increased attacks against Christians.

Conclusions

Given the particularly fragile situation the Christians find themselves in, it will be difficult to solve the minorities' issue without a stabilization of Iraq. The assassination of Isoh Majeed Hedaya, President of the Syriac Independent Unified Movement, who had presented to the Central Government a request for the autonomy of four districts on 31 October 2006, is a perfect illustration of the situation. There is also some uncertainty about what attitude the Kurds will adopt, since their present regional constitution does not provide for the possibility to form autonomous entities within Kurdistan.

Christians, supported by secular parties, will also have to establish relations with moderate Muslims and convince the Shiites in particular of the importance of their presence in Iraq to help rebuild and develop the country.

Furthermore, the United States and Europe have yet to make either a political or an economic commitment on the issue of minorities in Iraq.

Françoise Brié is a Doctoral student at the Sorbonne, Paris.

NOTES

1 Divided on the nature of Christ, Syriacs, the Christians of the Orient, separated in the fifth century into two branches: the Nestorians, also called Assyrians, and the Monophysites, or Jacobites. A second schism took place in 1552 between Chaldeans and Catholic Jacobites. The Syriac language remained in use in the form of various Aramaic dialects including Sureth, spoken mainly in Iraq.
2 A quarter of a million Assyro-Chaldeans are said to have been executed; amongst others the survivors of the Hakkari Nestorian tribes are said to have been part of the refugees.
3 Kanna is the Secretary General of the Assyrian Democratic Movement (ADM), represented within the 2003 Interim Government and then part of Iyad Allawi's government.
4 Other indicators underline the scale of the emigration of this minority: in the 2005 elections, the Assyro-Chaldean list National Rafidain was one of the lists which won the greatest number of votes in Syria, the United States (respectively 29 per cent and 26 per cent of expatriates' votes), Australia and in Canada. See The Independent Electoral Commission of Iraq, *Iraq Out-of-Country Voting 2005 Transitional Assembly Voting Provisional Results*.
5 UNHCR, UNICEF, WFP, *Assessment of the Situation of Iraqi Refugees in Syria*, March 2006.
6 International Office of Migration, *Emergency Assessment – Displacement due to Recent Violence (post 22 Feb 2006) Central and southern 15 Governorates*, 30 October 2006.

7   Wijdan Mikha'il is one of the representatives of Iyad Allawi's Iraqi National List within present-day Iraq. See Mark Lattimer, 'In 20 Years, There will be no more Christians in Iraq', *The Guardian*, 6 October 2006.
8   Cf. AsiaNews.it, 'US Bishops call for "Specific Measures" to Protect Iraqi Christians', *Asia News*, 31 October 2006.
9   Perceived as foreigners, and on top of it Christians, they were the first target group for radical nationalist Arab groups and for Rachid Ali al-Gaylani's government propaganda. After that came the repression of Yezidis in 1935, the pogroms against Jews in 1948–50, the deportation of Fayli Kurds from 1969 to 1980, the genocide of Kurds from 1975 to 1990 and the Shiite massacres of 1991.
10 Headed by Mustafa and later by Massoud Barzani. They have also been very active in the Communist Party since its foundation.
11 For example, the restoration of the Nestorian Patriarch's citizenship and the restitution of the assets of the Nestorian Church which had been confiscated following the events of 1933.
12 Middle East Watch, *Genocide in Iraq, the Anfal Campaign against the Kurds* (Paris: Karthala, 2003). Out of the 782 missing persons registered during that last phase, 150, that is, about 20 per cent, were Assyro-Chaldeans.
13 In the Silopi refugee camp, on the Turkish border, half of the refugees were Assyro-Chaldeans.
14 The religious leaders in place under Saddam Hussein's dictatorship saw the embargo as the main cause of emigration in the 1990s; The Assyrian Democratic Movement however, emphasized economic disadvantages inspired by Islam and introduced to gain the support of Muslim states, such as the closure of shops selling alcohol, the monopoly over which was held by the Christians. Religious propaganda, launched during the 1980–88 war against Iran, was reinforced in the 1990s and led to the inscription 'Allahou Akbar' on the national flag.
15 Mounia Daoudi, 'Appel à l'unité après les attentats anti-chrétiens', *RFI Actualités*, 2 August 2004. Ayatollah Sistani condemned the attacks against churches, cf. AFP, *Irak: Ali Sistani qualifie de 'crimes terribles' les attentats anti-chrétiens*, 2 August 2004; Nimrod Raphaeli, 'The Plight of Iraqi Christians', *MEMRI, Inquiry and Analysis Series*, 213 (March 2005).
16 That is, Mandaen Sabeans, Yezidis, Shabaks and Kakais.
17 Cf. Jonathan Steele, 'We're Staying and We Will Resist', *The Guardian*, 30 November 2006. So-called 'Brigades for the elimination of Christian spies and agents' target individuals working for the coalition.
18 Al-Sistani does not wish to impose such obligation. See Juan Cole, 'The United States and Shia Religious Factions in Post-Baathist Iraq', *Middle East Journal*, 8 October 2003.
19 They have to pay what is called a *fasl*, a practice often imposed by tribal elders in the settlement of dispute.
20 Mgr Jean Benjamin Sleiman, *Dans le piège irakien: le cri du cœur de l'archevêque de Baghdad* (Paris: Presses de la Renaissance, 2006).
21 See Martine Gozlan, 'Comment un quartier de Bagdad a sombré dans la guerre civile', *Marianne*, 465 (18–24 March), and 'Interview with Yonadam Kanna, ADM Secretary General', *Outre-Terren*, 14, *Arabies malheureuses II*, pp. 182–6 (Toulouse: Editions Erès, March 2006).
22 <http://www.chaldeanfederation.org>: In August 2006, the US Chaldean Federation succeeded in getting all restrictions lifted on the immigration of Christian Iraqis. The federation believes that the targeted violence against Christians does not allow them to return to their country. The Chaldean Federation launched the R4 Operation (Research, Rescue, Relief and Resettlement) and fundraising in the US Chaldean Community has been organized for the resettlement of Iraqi Christian immigrants. Some anonymous sources among Iraqi refugees in Jordan talk of an organized deliberate exodus of the Assyro-Chaldeans; others say that the insecurity in Iraq is also deliberate and organized by the Mossad, the CIA and Iran.
23 30,000–50,000 depending on sources; two to three families per day, often from wealthy backgrounds, went to church in Dahuk in 1992–93.
24 Amnesty International, *Iraq: Human Rights Abuses in Iraqi Kurdistan since 1991* (London: Amnesty International, 1995).

THE MIGRATION AND DISPLACEMENT OF ASSYRO-CHALDEANS IN IRAQ 351

25 See Chris Kutschera, 'L'exode des chrétiens d'Irak', *Le Monde 2*, 28 October 2006.
26 A Shiite community of Iranian origin in the district of Qaraqosh.
27 Heather Maler, 'Iraq: Christian Minority Seeks Haven From Violence', *RFE/RL*, 18 October 2006.
28 Al-Sheikhan is a mixed area of Yezidis, Assyro-Chaldeans (especially in the city), Kurds and some Arab families; Sinjar remains mainly Yezidi with a Kurdish population.
29 See Michael Youash, 'An Assyrian Administrative Unit Ending the Exodus of Iraq's Most Vulnerable' <www.nineveh.com>.
30 Some of the Yezidis do not see themselves as being part of the Kurdish community.
31 In contrast to the situation before 2003 when only the Zimar subdistrict, in the Nineveh governorate, was included in the Kurdish region, see Chris Kutschera, 'The Kurds' Secret Scenarios,' in *In the Shadow of War: Iraq, Israel, Palestine, Middle East Report*, no. 225, Winter 2002.
32 This project led to a memorandum supported by the APP, the Bet-Nahrain Democratic Party (BNDP), the Bet-Nahrain Patriotic Union, the Chaldean Democratic Forum, the Chaldean Cultural Association and the Chaldo-Ashur Organization, cf. <http://www.zindamagazine.com> vol. XII, 23, 20 November 2006.
33 The ethnic cleansing of Kurdish inhabitants from the left bank of the Tigris, in Mosul (Sumer, Tahrir, Intisar and Sinaa neighbourhoods) displaced thousands of people.

Algiers agreement and Al-Anfal campaign

Sources: The AMAR International Charitable Foundation, 2001; Middle East Watch *Genocide in Iraq*, 1993.

Assyro-Chaldeans Autonomous Governorate and Place of Origin

THE MIGRATION AND DISPLACEMENT OF ASSYRO-CHALDEANS IN IRAQ 353

Return of displaced Assyro-Chaldean population

■ Assyro-Chaldeans: return to regions of origin, 1990-present

⌒ Cease-fire line between Kurdish and Iraqi forces after the Second Gulf War (1991)

Sources: Humanitarian Information Center for Iraq (HIC), 2003

Displacements and Deportations in Iraq

FRANÇOISE BRIÉ

In 2005, Iraq had the fifth largest number of internally displaced persons in the world (after Sudan, Colombia, Congo and Uganda). The United Nations High Commissioner for Refugees (UNHCR) estimates that in spite of the advice given by the government and international agencies, more than 253,000 Iraqis returned to their country between 2003 and 2005, which constitutes one of the largest movements of repatriation on the planet.[1] Economic difficulties and the security situation did not prevent returns, in particular from Iraq's immediate neighbours where the living conditions for refugees are especially precarious. However, 1,800,000 people are said to remain in exile, above all in Jordan and Syria but also in Egypt. However, to the contrary of what was predicted, the war in 2003 did not cause a massive flight of people out of Iraq or large-scale displacements within the country. From 2003 to 2006, only some 200,000 became internally displaced. This does not include short-term displacements. These population movements were caused by fighting between international and Iraqi forces, on the one hand, and militia on the other; they were also populations who had been displaced under the old regime returning to their homes. This phenomenon worsened following the bombing of the Shiite mausoleum in Samarra on 22 February 2006.[2] This event marked the aggravation of sectarian violence and reprisals between Sunnis and Shiites. Within a few months, at least 300,000 people were forcibly displaced.[3] The United States recognized the magnitude of the problem and a year later, Condoleezza Rice formed a working group within the State Department to study the situation of Iraqi refugees and the problem of inter-communitarian violence across the country.[4]

Deportation, the Baath Party's Supreme Weapon

It was the Baath Party which institutionalized the forced displacement of populations as soon as it was definitively installed in power in 1968. This is the date which the Iraq Property Claims Commission (IPCC) of January 2004 (itself replaced in March 2006 by the Commission for Resolution of Real Property Disputes [CRRPD]) has taken as the starting-point for the historical phenomenon of deportation.

Each time, the aim of these forced migrations was to compensate for the demographic weakness of the Sunni Arab minority against the Shiite Arab majority which represents some 60–65 per cent of the population. The idea was to assimilate populations to an Arab identity, to forbid all forms of regional autonomy and to disperse groups suspected of not supporting the regime, and to mobilize them in the case of international conflicts. Expulsion was methodical and rapid, sometimes conducted with extremely violent military campaigns using embargoes, conventional or non-conventional weapons to eliminate populations, mass executions, and the transfer of populations to concentration camps. At the same time, there was a policy of encouraging immigration by Sunni Arab workers who were given financial and housing advantages. For example, a million immigrants, mainly Egyptians, were employed in Iraq.

Ethnocide I: The Fayli Kurds

The Fayli Kurds are a Shiite population living along the Iranian border but which also make up a significant percentage of the towns of Al-Kut, Amara and Baghdad. They were expelled into Iran in two waves, by the thousands in 1969–71 (together with Shiites who were considered Iranians, such as theology students from the *hawza*, the teaching centres of Najaf, and, more generally, Shiites from other holy places) and again in the spring of 1980, less than one year after Saddam Hussein assumed absolute power and a few months before he launched his war against Iran. The Fayli were expelled from Baghdad, Basra and Amara, at the same time as tens of thousands of Shiite Arabs and other Iraqis 'of Persian origin', whom the regime regarded as a fifth column. The total number of people expelled at this time was 200,000.[5]

The Arabization of North and Central Iraq

Saddam never abandoned his policy of Arabization, including when he signed a deal with the Kurds to create the Kurdish Autonomous Region on 11 March 1970. Further expulsions followed in the 1970s, mainly by means of a process of territorial division, as follows:

- The old province of Kirkuk was split into two. On the one hand, the governorate of Al-Ta'mim, including the city of Kirkuk proper and zones with a significant Arab population; on the other, the new governorate of Salah Al-Din (Saladin) with Tikrit as its capital and including the zone of Tooz Khurmatoo, populated by Kurds and Turkmen. The districts of Chemchemal and Kalar were annexed to the governorate of Suleimaniya.

- The creation of the governorate of Dahuk, which meant that the governorate of Mosul became majority Arab with a strong presence of Assyro-Chaldeans (Christians) and Turkmen (who speak a variant of Turkish).

- The districts of Khanaqin and Kifri were annexed to the governorate of Diyala.

- The governorate of Kerbala was reduced in size and the territory transferred instead to the governorate of Al-Anbar.

Arabization was also practised on the zone along the Iranian, Turkish and Syrian borders, from Khanaqin to Sinjar.[6]

In the 1970s and above all in the 1980s, hundreds of thousands of nomads from the semi-desert plane of Al-Djezireh to the south-west of Mosul, or Sunni tribes loyal to the regime from the governorate of Salah Al-Din were sent into parts of the country marked for Arabization where they were put in charge of guarding the oil fields, just as the Al-Hadidi were sent to the district of Sheikhan in 1974–75, or the Shammar to the governorate of Kirkuk. Research in 2003 revealed that these transfers of population were planned and systematic.[7] Between 1957 and 1977, in the city of Kirkuk proper, the Kurdish and Turkmen populations declined respectively from 48.3 to 21.4 per cent and from 37.43 to 16.31 per cent of the population. The Arab population, meanwhile, increased from 28.2 to 44.41 per cent.[8]

This strategy of Arabization continued after the Second Gulf War in 1991 with the encirclement of the Kurdish regions which were no longer under the control of Baghdad, all along the cease-fire line. Hundreds of villages were cleared out. Hundreds of thousands of people were expelled by means of a series of laws and decrees from the Revolutionary Command Council. Titles to property which had been in force under the Ottoman Empire were invalidated or confiscated without compensation and the land, which became state property, was rented to Arab populations who were given financial incentives to build on it. Kurds were also displaced from Khanaqin, Kifri and Mandali in the governorate of Diyala, both within the governorate itself or towards Suleimaniya. They were replaced by Arab families.

In 2001, Decree No. 199 was introduced the concept of 'corrected nationality' in favour of the Arab populations. Whoever did not want to submit to this new ruling was expelled to the Kurdish Autonomous Region. The Turkmen abandoned Kirkuk and Tall Afar for Erbil and Turkey. The

Assyro-Chaldeans who had been displaced from the mountains to Baghdad, Mosul and Basra in the 1970s returned to the governorates of Dahuk and Erbil. In total more than 100,000 persons were expelled or fled over the 'green line' after the Gulf War in 1991.

Military Conflicts and the Ethnic Recomposition of Iraq

After the agreement signed in Algiers between Iran and Iraq on 6 March 1975, which provided for a security zone of between five and thirty kilometres along the Turkey–Iraq and Iran–Iraq borders, 250,000 Kurds fled to Iran. Two hundred thousand more were displaced to the centre and south of Iraq, into the governorates of Al-Anbar, Qadisiya, Thi-Qar, Missan and Al-Muthanna. The destruction of several hundred Kurdish villages and of about sixty Assyro-Chaldean villages depopulated the whole of the border area. These massive population movements were reported in the Iraqi newspapers and presented as legitimate measures taken to protect the revolution and the national interest.[9] The operation was completed in 1983–84.

During the First Gulf War of 1980–88, border areas essentially emptied out. In the south, the marshes around Thi-Qar, Missan and Basra became strategically important and their conquest by Iran caused the population to flee. In the governorates of Wasit and Thi-Qar, but above all in Missan and Basra, tens of thousands of people had to leave their houses because of the fighting, but also because the regime suspected them of disloyalty.

By 2005, about 80,000 people from the Shatt Al-Arab, Al-Khaseeb and Fao districts were living in camps or public buildings.[10] Twelve thousand families expected to return to the interior of the governorate of Missan. People displaced by the war with Iran were registered in Kerbala, Najaf and Babylon.

Ethnocide II: Al-Anfal

Four months after the census of October 1987, the Al-Anfal campaign started, with eight offensives in Iraqi Kurdistan, which was also being punished for its collaboration with Iran. This culminated in the destruction of two large towns, Halabja which was bombed with chemical weapons in 1988 and Qaladiza the following year.[11] At least 100,000 people disappeared. Two thousand villages and a dozen towns and administrative centres were destroyed. The mountain-region governorates of Dahuk, Erbil and Kirkuk were emptied of their population.

It is estimated that the number of people displaced in the three governorates of northern Iraq in 2000 is 805,500, more than half of whom were put into camps set up by the regime in the 1970s and 1980s.

After the cease-fire of the Second Gulf War, signed on 28 February 1991, there was insurrection in fifteen of Iraq's eighteen provinces. Although the Allies had called on Iraqis to rebel, their forces allowed the Republican Guard and Saddam's helicopters perpetrate terrible repression in Shiite parts of the country and an implacable attack in Kurdistan. This latter operation caused a huge exodus of two million Kurds into Turkey and Iran. These people returned to the northern Iraq after UN Security Council Resolution 688 was voted on 5 April 1991, which provided for Kurds in the north and Shiites in the south to be protected by Operation Provide Comfort. This massive return of people required Kurdish militias paid by the regime to hunt down fighters from the Kurdistan Democratic Party (KDP) and the Patriotic Union of Kurdistan (PUK) and thereby to participate in the destruction of villages. Eight thousand families, mainly from Akre and called 'Al-Qilaa' after the military camps where they had been held, went towards Mosul. The Arab tribes installed in Kurdistan as part of the Arabization project started to withdraw from the autonomous region, just as the Al-Fahd had done, who, originally from Al-Kut in the south, moved to Khanaqin in 1997 after spending twenty years in the governorate of Suleimaniya.

Ethnocide III: The Marshlands

The draining of the marshes, a project which had originally been conceived in the 1940s and then again in the 1970s, was eventually decided on in 1984 during the Iran–Iraq War. It was implemented by means of the building of dams in Turkey and Syria. A letter dated 30 January 1989 and discovered during the uprising in 1991 confirms the existence of this plan. The inhabitants of the marshes were organized in several dozen different tribes: Albu Mohammed, Bani Lam, Abu Salih, Bani'Isad, Bani Hasham, al-Juwaibiri, al-Shumaish, al-Musa, al-Rahma, and others. The regime thought of them as politically, ethnically and religiously pro-Iranian, just as it did of the Shiites in the south and they were suspected of protecting deserters and hostile elements.[12] An attack by 5,000–10,000 men from the Badr Brigades, trained by Mohammed Baqr Al-Hakim's Supreme Council for the Islamic Revolution in Iraq based in Iran, led to the expulsion of the 'Madans' whom the newspaper *Al-Thawra* called 'Simians' and foreigners.[13] Seventy settlements were destroyed between December 1991 and October 1992 and 50,000 persons from the surroundings of Chabaish were evacuated to Nasiriya: first the towns of Maimuna, Salam and Adil, later the governorate of Missan including Qala't Salih, and finally the marshes of Hammar and Amara. The operation was continued in 1993–94. The marsh Arabs were displaced more than a dozen times both within and outside their territory and today they are among the most impoverished members of the Iraqi population. This Shiite minority was estimated at between 400,000–500,000 people, mainly farmers, in 1950. At the beginning of the 1990s, there are only 250,000 left, and in 2003, their numbers had fallen to 83,000.[14]

Institutional Measures in Favour of Displaced Persons and Refugees

At the end of 2003, a Ministry of Displacement and Migration was created after tensions erupted between Kurds and Arabs in Kirkuk and Baghdad. (The ministry's responsibilities were defined by Regulation No. 50 passed by the Interim Coalition Authority which entered into force on 11 January 2004.) The new government appointed on 28 June 2004 includes within it an organization working with the United Nations, and Working Group No. 8 for Iraq. On 8 March 2004, Article 58 of the Transitional Administrative Law stipulated that the new government would 'act expeditiously to take measures to remedy the injustice caused by the previous regime's practices in altering the demographic character of certain regions ... by deporting and expelling individuals from their places of residence, forcing migration in and out of the region, settling individuals alien to the region, depriving the inhabitants of work, and correcting nationality'. The law also required that the government act in accordance with Article 10 of the Iraqi Property Claims Commission 'to ensure that such individuals may be resettled, may receive compensation from the state, may receive new land from the state near their residence in the governorate from which they came, or may receive compensation for the cost of moving to such areas'.

The Position of the Authorities and Political Parties before the End of the *Ancien Régime* and the question of Kirkuk

Having begun as soon as the Baathist regime was destroyed in 2003, the return of refugees and deportees has provoked turmoil, especially in Arabized areas of Iraq. Without underestimating the number of expellees who have returned to their homes of their own accord, it is obvious that the Kurdish political parties wanted to regain control of all the regions claimed by Arabs since 1961,[15] by encouraging Kurds to return to them, accompanied by the *peshmerga* (Kurdish militia), especially in 2003. Kurdish leaders demanded that return populations would restore the situation which had existed before Arabization.[16] Naturally, these conflicts are particularly sharp where the oil-rich city of Kirkuk is concerned, which the Kurds dream of controlling. Kirkuk is especially important for the Kurds who consider the territory to be historically theirs. Kirkuk has always been a permanent bone of contention between successive Iraqi governments, and agreements on autonomy were abrogated in 1970. As soon as thousands of families arrived in Kirkuk and the commissions tasked with examining ethnic cleansing and applying Article 58 of the Transitional Administrative Law were created, the situation rapidly became explosive. Thousands of Kurds who no longer have any faith in the negotiations between the Iraqi government and the Kurdish regional leaders are threatening to seize back their lost lands and homes themselves, if the law is not applied. By the same token, Arabs suspects the Kurds of sending people into Kirkuk who were not from there in the first place, precisely in order to 'Kurdify' the city. The Shiite Arab followers of the radical cleric, Moqtada Al-Sadr have accused the Kurds of apostasy; many of them have fled from Samarra to Kirkuk or Suleimaniya. The Turkmen side with the Arabs against the Kurds in this: after January 2005, Orhan Ketene, the representative in the United States of the Turkmen Front, which is supported by Turkey, accused 'Kurdish terrorist groups' of controlling the centre of Kirkuk.[17]

In 2006, the numbers of Kurdish families returning to the governorate of Kirkuk were lower than in previous years. The Kurdish authorities are above all trying to stabilise the status quo which is in their favour, as they await the referendum on the disputed territories which, according to Article 140 of the constitution, must be resolved by 31st December 2007 at the latest. A census planned for before the elections in January 2005 has been postponed several times because of differences within the government and because of the difficulty of re-settling displaced people, is also supposed to be held before this date. It was no surprise, therefore, if one of the two Kurdishish leaders, Massoud Barzani, denounced the Baker-Hamilton report of December 2006 which suggested postponing the application of this part of the constitution. The Kurds have also called for the districts of Kifri, Chemchemal, Kalar and Tooz Khurmatoo, separated from the governorate of Kirkuk in 1976, to be reannexed to it. It is expected that there will be a rise in ethnic violence if either the referendum or the census is not held. The Committee for the Normalization of Kirkuk signed an agreement to distribute financial incentives to Arabs leaving Kirkuk for their regions of origin. This sparked off a violent reaction by the Association of Muslim scholars in Iraq which condemned the process as one of ethnic cleansing. Turkmen are hostile to any change in the balance of power in Kirkuk and the plan has also attracted the ire of Sheikh Ra'ad Al-Najafi, the representative of the exiled radical cleric, Moqtada Al-Sadr.

Partition via Displacement?

As Arabs leave Kirkuk and Diyala, Kurds are leaving the left bank of the Tigris in Mosul, following the murder of several mayors in 2005.[18] To the south of Baghdad, Sunni insurgents are carrying out attacks and kidnappings in an attempt to force Shiite Arabs to leave. This is particularly the case in the so-called 'triangle of death' between Al-Latifiya, Al-Mahmudia and Al-Yussufiya where Sunni Arabs were settled for strategic reasons.[19] In Baghdad, people are moving from mixed districts into homogenous ones. Shiites are moving from the centre to the south; the Sunnis from the south to the centre and especially to the governorate of Al-Anbar; the Kurds are leaving the governorate of Baghdad for the three northern governorates, Kirkuk, Diyala, or moving from within Diyala. Especially vulnerable minorities like the Christians are moving towards the Nineveh plain, Erbil, Dahuk from Baghdad. The Turkmen are moving around within the governorates of Saladin and Nineveh or from Nineveh to Kerbala and Kirkuk. The governorates to which the largest number of people moved in 2006 are Baghdad and Diyala, as well as to the north of the country where the situation is relatively calm.

The new security plan which sets up shared chains of command between Sunnis and Shiites, the purge of criminal elements within the army and the police,[20] the international conference organized at Baghdad with representatives of Syria and Iran in support of national understanding, and the agreement of the government on a bill on the management and control of oil,[21] are all certainly going to lead to reconciliation. But these displacements may prove to be the precursor to territorial division.

Françoise Brié is a Doctoral student at the Sorbonne in Paris.

NOTES

1   UNHCR, '2004 Global Refugee Trends: Overview of refugee populations, new arrivals, durable solutions, asylum-seekers, stateless and other persons of concern to UNHCR' (Geneva: UNHCR, 17 June 2005).
2   IRIN, 'IRAQ: Another million people could flee homes this year', 6 March 2007.
3   International Office for Migration, *Iraq Displacement 2006, Year in Review* <www.iom.int/>.
4   David Shelby, USInfo, 7 February 2007.
5   *Iraq Press*, 30 January 2004.
6   See Ali Babakhan, 'Les Kurdes d'Irak, leur histoire, leur déportation par le régime de Saddam Hussein' (Lebanon: no publisher, 1994), p. 114.
7   Human Rights Watch, 'Claims in Conflict, Reversing Ethnic Cleansing in Northern Iraq', 16, 4 (August 2004).
8   Nouri Talabani, 'The displacement of Kurds and minorities in the Kirkuk region', International Conference on Refugees and Displaced Persons in Iraq (Manteqat Kirkuk, Baghdad: no publisher, 1999; 2nd edn, July 2002).
9   Ephrem Isa-Yousif, *Une chronique mésopotamienne (1830–1976)* (Paris: L'Harmattan, 2004), p. 223.
10 Global IDP Project, 'Iraq: Insecurity and Lack of Shelter Exacerbate Internal Displacement Crisis' (Geneva, 11 July 2005), p. 3.
11 Middle East Watch, *Génocide en Irak, la campagne d'Anfal contre les Kurdes* (Paris: Karthala, 2003).
12 See Christopher Mitchell, 'Assault on the Marshlands', in Peter Clark and Sean Magee (eds), *The*

Iraqi Marshlands: A Human and Environmental Study (London: AMAR International Charitable Foundation, 2001), pp. 42–69.
13 Ibid.
14 Emma Nicholson, 'Destruction et génocide dans les marais du Sud de l'Irak', in Chris Kutschera (ed.), *Le livre noir de Saddam Hussein* (Oh Éditions, 2005), pp. 279–96.
15 Human Rights Watch, 'Claims in Conflict', p. 58.
16 Dexter Filkins, 'Iraqi Kurdish Leaders Resist as the U.S. Presses Them to Moderate Their Demands', *New York Times*, 21 February 2004.
17 'Turkmen Front: "We have lost Kirkuk"', *Turkish Weekly*, 15 July 2005.
18 'Kurds' Mass Immigration from Mosul to Kurdistan Region' <www.almendhar.com>, 26 July 2005.
19 Salam Farraj, 'Au sud de Bagdad, les chiites fuient le triangle de la mort', *La Croix*, 23 May 2005.
20 Khouloud Al-Amiry, 'Interview with the Iraqi PM Nouri al-Maliki', *Al-Hayat*, 1 February 2007.
21 Zalmay Khalilzad, 'A Shared Stake in Iraq's Future: How the Oil Agreement Points the Way Forward', *Washington Post*, 3 March 2007.

Borders and displacements

DISPLACEMENTS AND DEPORTATIONS IN IRAQ

Repopulation by Kurds

Return of displaced Kurdish population, 2003

Source: UNOPS-IDP, OIM, 2003-2005.

THE LONG MARCH TO THE WEST

Ethnocide II and III

- Al-Anfal campaign, 1988
- The Marshlands in 1985
- The Marshlands in 2000
- Border of the Kurdistan Autonomous Region, 1975
- Forced displacements of populations and collectivisation
- Flight of population

Sources: The AMAR International Charitable Foundation, 2001; Middle East Watch *Genocide in Iraq*, 1993.

DISPLACEMENTS AND DEPORTATIONS IN IRAQ

Populations in Iraq, 1960

- Sunni Arab majority, 1960
- Sunni Kurd majority, 1960
- Shiite Arab majority, 1960
- Cease-fire line between Kurdish and Iraqi forces after the Second Gulf War, 1991
- Border of the Kurdistan Autonomous Region, 1975

Sources: Hanna Batatu, *The Old Social Classes and the Revolutionary Movements of Iraq*, 1978; Middle East Watch *Genocide in Iraq*, 1993; Humanitarian Information Center for Iraq (HIC), 2003.

PART VII
East Meets West

PART VII

The Impact of East European Immigration on the UK

ANDREW GREEN

The political impact of immigration from eastern Europe to the UK has been more significant than its economic effect because:

- It came on top of unprecedented levels of immigration from the rest of the world.
- The government's forecasts were wildly wrong, affecting their credibility.
- East Europeans, though distributed all over the country, clustered in significant numbers in some regional towns so that their presence in these areas was widely noted.

Net immigration to the UK has been rising sharply – from 47,000 in 1997 to 185,000 in 2005, slightly down from 222,000 in 2004. The main increase has been from Asia and Africa. Until May 2004, net immigration from the EU was low and broadly stable.[1]

A Home Office sponsored study, published in June 2003,[2] predicted that net immigration from the first eight new members from eastern Europe (the A8) would be between 5,000 and 13,000 a year – a figure on which the government relied. This estimate was based on the historical record of immigration from some fifty-seven countries, none of them in eastern Europe. Furthermore, the estimate took no account of the possibility that other EU members would impose transition arrangements as provided in the accession treaties.

In response to public concern, mainly about 'benefit shopping', the government introduced the Workers Registration Scheme (WRS). This required employers to register workers from the A8 countries and closed off non-work-related benefits for a twelve-month period. However, it does not apply to the self-employed, nor, crucially, is there any record of departures.

The most recent quarterly report of the WRS shows that 486,000 workers registered between May 2004 and September 2006 inclusive. However, ministers have acknowledged that the true figure, taking account of the self-employed, could exceed 600,000. Meanwhile, the number of dependants registered remains low with 23,000 dependant children and 22,000 dependant adults.

The WRS provides interesting information on the geographical distribution and wage rates of the East European immigrants. Most African and Asian immigrants join existing communities in London and other city centres, at least initially. By contrast, only 15 per cent of East Europeans work in London while the remainder have spread all over the country. Some cities and towns, such as Edinburgh, Southampton, Boston and Crewe, have several thousand immigrants who are making a visible impact on public services, notably education. They have thus alerted a much wider section of the public to the practical consequences of present levels of immigration.

The other salient feature of this wave of immigration is their low wages. According to the WRS, 80 per cent earn less than £6 (€9 an hour) and 95 per cent earn less than £8 (€12 an hour). There is anecdotal, but not yet statistical, evidence of British workers being replaced. Both employment and unemployment have increased in the period. Unemployment has risen from 4.7 to 5.5 per cent in the year to October 2006; youth unemployment has also risen since 2004. The claimant count is approaching the politically sensitive level of one million.

Migration Watch UK calculations, based on these earnings levels, suggest that eastern Europeans contribution to GDP per head is broadly neutral.[3] That is to say that their addition to GDP is similar to their addition to the population. Their fiscal contribution has been slightly positive – at least while the number of dependants is low. More generally, their main effect on the economy has been to moderate wage inflation – especially in certain sectors such as the building trade.

Preliminary work by the Department for Work and Pensions (DWP)[4] has found a close correlation between GDP per head in the sending country and the numbers registering in the WRS. This may have contributed to the Government's decision to impose transitional arrangements on immigrants from Romania and Bulgaria – two countries with even lower GDP per head than, for example, Poland.

The outlook for net immigration from eastern Europe now depends on:

- decisions by EU partners,
- developments in East European economies, and
- decisions by individuals as to how long to stay.

Knowledge of the English language, now quite widespread in eastern Europe, will remain an important factor in choosing a destination, but the opening of other EU labour markets will spread the migration more widely.

In the longer term, the economic incentive to emigrate will decline as the economies of eastern Europe grow. When GDP per head in Portugal, Spain and Greece reached about 70 per cent of that of the UK, migration came into balance.

In the medium term, however, net immigration will be mainly affected by how long East Europeans stay in Britain. The only evidence so far is from the WRS survey itself. Fifty per cent said that they intended to leave within a year, 10 per cent planned to remain longer than a year, and 40 per cent left the question blank or answered 'do not know'. If most have come only for a year or two, then the outflow will balance the inflow fairly soon. But if they stay for four or five years there will be a significant net inflow for the medium term. This will have significant political as well as economic consequences. An opinion poll conducted by Harris Interactive[5] found that 76 per cent of Britons thought that there were too many immigrants in the country. The survey also found that the British were most likely to state that both immigrants in general and recent migrants from the new EU members have made a negative impact on the economy – 46 per cent and 50 per cent respectively.

Sir Andrew Green is the chairman of Migration Watch UK.

NOTES

1 <www.migrationwatchuk.org> PowerPoint presentation, p. 4.
2 Home Office Online Report 25/03 'The impact of EU enlargement on migration flows', <wwwhomeoffice.gov.uk/rds/pdfs2/rdsolr2503.pdf>.
3 <www.ind.homeoffice.gov.uk/aboutus/reports/accession_monitoring_report>.
4 <www.migrationwatchuk.org> Briefing Paper 1.12.
5 DWP Working Paper 29 of 2006.
6 *Financial Times*, 20 October 2006.

Migration and Migration Narratives in the Era of Globalization

ATTILA MELEGH

Introduction[1]

Globalization has been a major new phase in the history of world capitalism which, directly and indirectly, has had a major impact on international migration. Globalization is not an absolutely new phase of development, but a new cycle in the history of world capitalism. Previous systematic relationships between migration and the functioning of the world system have not been changed substantially (Chase-Dunn, 1999). In the early 1980s and late 1970s, globalization as a new discursive and social order based on the integrity of the global financial system and the conditions for transnational corporate capitalism ended a historic epoch, a period based on the competition of capitalist and socialist modernization projects organized in the framework of nation-states and their block alliances like the Comecon and EEC (Melegh, 2006a). This change was not just a phenomenon effecting the so-called West, but also the 'socialist world' too, due to its strong connections with world capitalism (McMichael, 2000). We can even say that the socialist experiment ended as a result of the penetration of capitalism into the economic system of centrally planned economies and the related social arrangements. In these dramatic changes migration played an important role, as the movement of people was a direct factor in the rearrangement of global power structures. (On globalization and migration, see among others: Staring in Kalb et al., 2000; Sassen, 1996, 1999, 2001; Orozco, 2002; Mittelman, 2000; Lutz, 2002; Phizacklea in Koser and Lutz, 1998; Okólski, 1999; Forsander, 2002; Böröcz, 2002; Melegh, 2003; Baumann, 1996; Beck, 2000; Appadurai, 1996.)

The key point is that world capitalism, as it emerged in the early modern period, has always been a hierarchical system in which different areas of the world have been configured into core areas, semi-peripheral areas and peripheries. The aim of this article is to show how migration operates within such a system, or rather within the hierarchical system which re-emerged in eastern Europe from the 1980s when East European state socialism first decomposed and was then replaced by a certain pattern of capitalist economy with very low saving rates and a high dependence on foreign investment. In other words, our concern is with migration and migration narratives as individual reflections on such life-changing events in East European countries which have been reintegrated into the hierarchical system of world capitalism which they partially left during the state socialist period. The concrete major issues for us are the following:

- As a new phase of world capitalism, globalization is a powerful macro-structure increasing global inequality and hierarchy. It is linked to migration. We must then consider the different links between the movement of capital and labour.

- Under the auspices of the state and capital, class, race/ethnicity and gender as a combined power structure contain the room for manoeuvre for migrants.

- How do macro and micro structures formulate the life course and the life-course perspective of migrants?

Reintegrating Eastern Europe into Global Hierarchies 1981–2001

The late history of state socialism can be told as a gradual reintegration of planned economies into the capitalist world system under the East–West civilizational ideological umbrella. Financial and international debt links were established between state-socialist systems and the 'world economy' in the context of the energy crisis of the 1970s, which then led to market-oriented reforms and 'dual dependency' of smaller East European countries (Böröcz, 1999). In this respect the change in the global political economy (from a modernizationist political economic scenario based on the nation-state to globalization and the dominance of financial aspects) provides the proper context for understanding the fall of socialism.

As Figure 1 clearly indicates, all East European states suffered a severe economic crisis between 1989 and at least 1996. In this period, the key point is that it led to a major collapse and restructuring of the state socialist economy, especially industry. This collapse 'freed' a lot of employees, especially in the 'countryside', and some of the local communities started building up migratory networks in order to secure casual labour either in the West (the most eastern border in this respect Romania, western Ukraine and Moldova) or in the East, namely Russia.

The East–West border can be clearly located around Ukraine, Moldova and

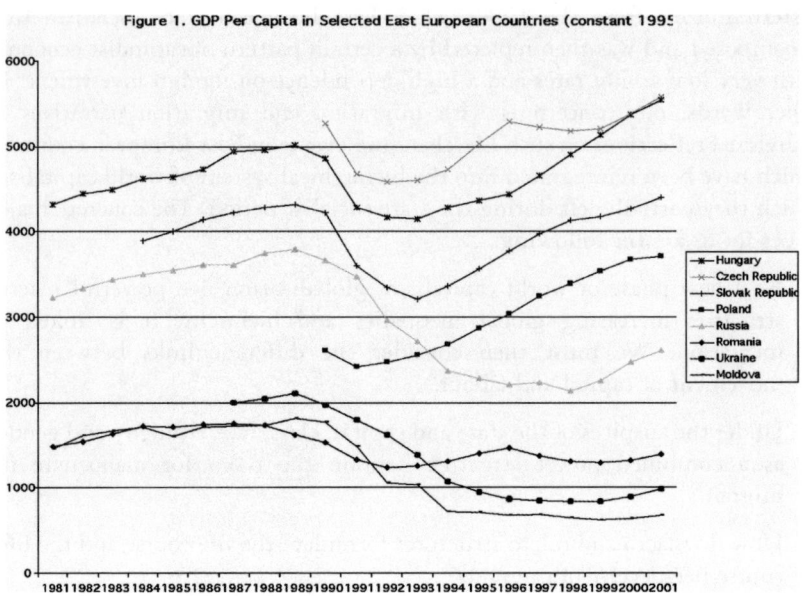

Figure 1. GDP Per Capita in Selected East European Countries (constant 1995

Source: *World Development Indicators*, World Bank, 2003.

Romania. They are the countries which are at the bottom of the league (at least during the 1990s, early 2000s) as inequalities in gross domestic product widen. In the 1980s, there were gaps between the state-socialist economies but these gaps have increased, *and it seems that these countries are on a gradual, differentiated 'slope' of economic well-being.* East European countries have been reintegrated into a hierarchical slope of world capitalism, which has also been clearly reflected on the ideological construct of an East–West slope (Melegh 2006a).

Global and Local Inequality Versus Migration from 1995

The above described differentiation is almost directly linked to migration. For the sake of the reintegration of the state-socialist economies' industries, huge groups of people have been 'abandoned', some of whom have become involved in transnational migration. Countries like Hungary, the Czech Republic, Slovakia and Poland (which are also countries of emigration) have become targets of immigrants coming from Ukraine, Russia, Romania and even from further east (China, Vietnam). Migration is an extremely complex phenomenon and historical links are crucial of course, but it seems that global inequality plays a major role in this, and thus the reintegration of eastern Europe into a more differentiated global capitalist hierarchy has been a major engine of migration.

We can illustrate this with the relationship between Romania and Hungary, two countries bound together historically. As Figure 2 illustrates, in Hungary with regard to Romanian citizens the immigrating population (flow) and labour permit follow changes of GDP differences.

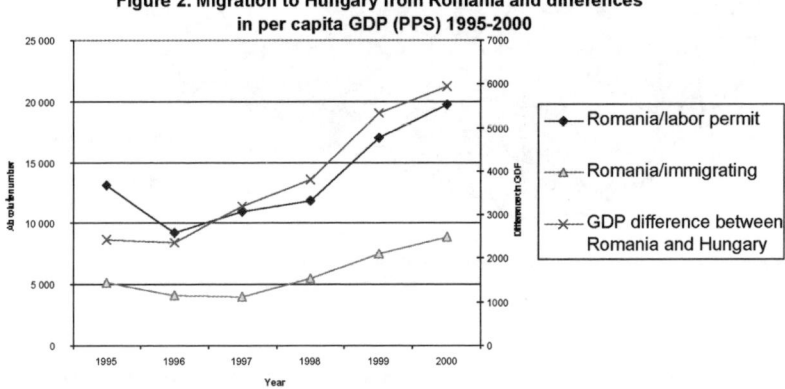

Figure 2. Migration to Hungary from Romania and differences in per capita GDP (PPS) 1995-2000

Source: Central Statistical Office, Budapest, Hungary.

Local inequality is not to be to be separated from global inequalities, and with the more intensive integration of local economies into global capitalism there are certain patterns which influence local regional patterns of migration. As it has been pointed out above, East European economies are very dependent on foreign capital, especially on the ruins of the state-socialist economies which have collapsed due to the reintegration into global hierarchies.

As can be shown in the case of Hungary (a rather small and homogenous country), foreign subscribed capital and the regional distribution of resident international migrants are strongly related, even if the link is not without complications. The central region with the highest per capita foreign subscribed capital is the region in which international migrants reside in the highest ratios compared to the resident non-migrant population. None the less, there are regions in which migrants are not so numerous but foreign capital is heavily invested and vice versa. Thus capital and migration are definitely related to each other, and we can even argue that the local

Figure 3. Regional distribution of foreign subscribed capital and foreign residents in Hungary in 2001 on a sub-regional level (per 1,000 people).

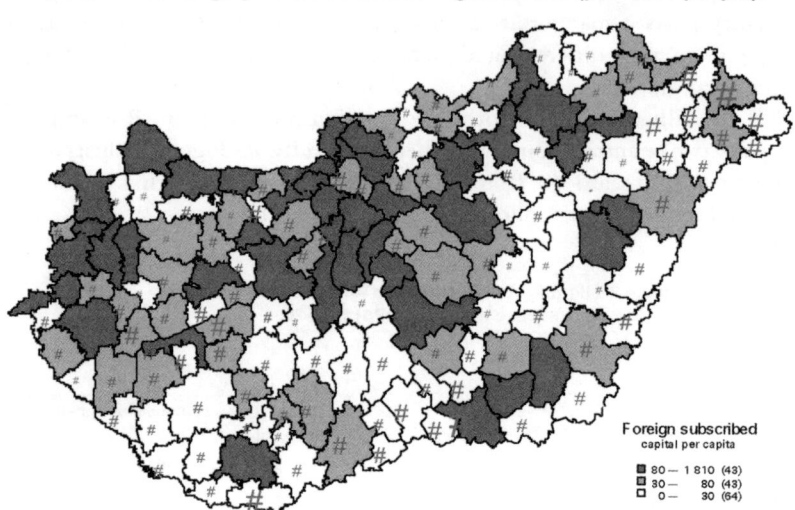

Source: Regional database of the Central Statistical Office, 2001.

consequences of global inequalities are also a structural factor in international migration in eastern Europe.

But global inequalities also appear in other ways in immigration patterns in eastern Europe (Melegh, 2003). Looking at the immigrating population in terms of occupational structure with regard to resident migrants and work-permit holders, we can also observe the consequences of global inequalities. A country like Hungary receives low-skilled and deskilled labour migrants from the East (Ukraine, Romania) or, in a special niche of the global/local economy, the Chinese diaspora. From the West, a kind of a secondary elite comes and resides in eastern Europe, often with the aim of buying property, looking for marriage partners, or seeking job opportunities. Central Europe is a meeting place for these different migrant groups, even if they rarely meet physically.

Cognitive Mechanisms and Migration

Public Discourses in Central and Eastern Europe: Hierarchical Imagination in a Hierarchical World

Migrants are put into a web of discourses and motivated by them. Discourses are major centres of power in the sense that they provide interpretative frameworks for social action and identities. Migrants are especially vulnerable, in the sense that they move from one area to another and they must adapt to different discourses. With regard to eastern Europe, the most important cognitive shift, related to changes in the political economy, is that from the early 1980s the geopolitical and geo-cultural imagination has been recaptured by the idea of a 'civilizational' or East/West slope, providing the main interpretative framework for reorganizing international and socio-political regimes in the eastern part of the European continent. In this radical 'normalization' and 'transition' process, almost all political and social actors in 'East' and 'West' identify themselves on a descending scale from 'civilization to barbarism', from 'developed to non-developed' status. This discursive structure appears in very different forms and areas of knowledge, and is utilized by very different speakers ranging from the European Union to restaurant owners and migrants (Melegh, 2006a). This idea of an East/West slope has a major impact on the perception and management of migration. Migrants are perceived and perceive themselves as who is coming from where, and in which direction, on the slope. The migrant, and in this respect the receiving social self, is constructed in the web of different perspectives on the slope. The identity and the life-course perspective of migrants is a really inventive work, as migrant individuals must find ways in which they can legitimate themselves in the web of different perspectives hierarchically related to each other (Melegh, 2006b; Hegyesi and Melegh, 2003).

This hierarchical imagination can be seen very clearly in the following caricature on labour migrants from the 'East' to Hungary.

'Before we go to work, both of us should sing our own national anthem'
(2001, *Magyar Nemzet*)

The context of this caricature was allowing Romanian citizens to work in Hungary irrespective of ethnicity. The hierarchy is very clear in contrasting a 'clean' and hard-working Hungarian peasant (an identity which does not exist any more in post-communist society) to a drunken, badly dressed Romanian rural figure. Migrants are supposed to counterbalance in their legitimacy strategies such hierarchical images.

Hierarchical imagination also appears in narratives of migrants when they present their life-story (Kovács and Melegh, 2001, 2004) The most important point is that there are only a limited amount of patterns which migrants find legitimate in presenting their life-story. Among these patterns, they rarely use assimilation or even integration perspectives, even though these are a major requirement of the 'civilized' receiving countries in their own discourses on migration. Instead they suffer in retelling stories of subordination, suppression and traumas.

One of the most important findings is that migrants (men and women alike) present their migration story in a passive manner, that is to say, as being 'taken' to the target country which is a major sign of social and cultural oppression. Also they often 'rediscover' their ethnic origin, and in the narrations they actively work on presenting an ethnic identity which they did not have before migration. Thus we can see that the 'fundamentalism' of migrants is a post-migration creation (interaction between migrants and host societies) and not something originally taken from the sending society. It is important to note that they often struggle to present a refugee story, since it is probably the only legitimate way of self-presentation in European societies. If they succeed in presenting themselves as refugees, then they can give a much more active impression of themselves. *None the less, subordination also appears in the reporting on the fight against discrimination and very importantly in 'paper narratives' in which they organize their life-story according to sagas of gaining legitimate status in the host society.*

Conclusion

European discourses on migration are extremely simplified, in the sense that migrants are seen as guests whom it is necessary to select. The above article shows that with regard to eastern Europe, such political discourses are misleading, as migration is a complex phenomenon which is driven by major changes in the global political economy (an interplay between global/local capital, the state and the migrant and their sending society). In our analysis, we therefore must see the whole picture, or else, out of sheer political will, we will not only miss important points but will actually push migrants into the a position of a semi-criminal, semi-civilized person who wants to disturb us.

Taking a more macroscopic perspective, and also looking at the related individual struggles of migrants, we are able to settle some of the emerging cultural and social conflicts which increasingly 'unsettle' European societies.

Bibliography

Appadurai, A. (1996), *Modernity at Large: Cultural Dimensions of Globalization* (Minneapolis, London: University of Minnesota Press).

Baumann, Z. (1996), 'From Pilgrim to Tourist – a Short History of Identity', in Stuart Hall and Paul Du Gay (eds), *Questions of Cultural Identity* (London: Sage).

Beck, U. (2000), *What is Globalization?* (Malden, MA: Polity Press).

Böröcz, József (2002), 'A határ: társadalmi tény', *Replika*, 47–8 (június), pp. 133–42.

Chase-Dunn, Christopher (1999), 'Globalization: A World-Systems Perspective', *Journal of World-Systems Research*, 5, 2 (Spring) <http://csf.colorado.edu/wsystems/jwsr.htm>.

Forsander, A. (ed.) (2002), *Immigration and Economy in the Globalization Process*, Sire Reports Series. No. 20. Vantaa, Finland.

Gödri, Irén (2005), 'The Nature and Causes of Immigration into Hungary and the Integration of Immigrants into Hungarian Society and Labour Market', *Demográfia*, special English edn.

Hegyesi, Adrienn and Melegh, Attila (2003), '"Immár nem mi vagyunk a szegény rokon a nemzetközi világban" A státustörvény és az Orbán-Nastase-egyezmény vitájának sajtóbeli reprezentációja és diksurzív rendje' ('We are not anymore the poor relatives in the world'. The press representation and the discursive order of the status law and the Orbán–Nastase pact), in Erika Sárközy and Nóra Schleicher (eds), *Kampánykommunikáció* (Campaign Communication) (Budapest: Akadémiai), pp. 135–71.

Kalb, D., et al. (eds) (2000), *The Needs of Globalization. Bringing Society Back In* (Lanham, MD, Boulder, CO, New York, Oxford: Rowman & Littlefield Publishers).

Koser, K. and Lutz, H. (eds) (1998), *The New Migration in Europe. Social Constructions and Social Realities* (London: Palgrave, Macmillan Press Ltd).

Kovács, Éva and Melegh, Attila (2001), '"It could have been worse, we could have gone to America" – Migration Narratives in the Transylvania-Hungary-Austria Triangle', in Nyíri Pál et al. (eds), *Diasporas and Politics* (MTA Politikai Tudományok Intézete Nemzetközi Migráció Kutatócsoport Évkönyve), pp. 108–38.

Kovács, Éva and Melegh, Attila (2004), 'A vándorlást elbeszélo narratívák neme avagy nök és férfiak elbeszélései – nöi és férfi elbeszélésmódok' (The Gender of Migration Narratives. The Narratives of Men and Women – Male and Female narratives), in: Andrea Petö (ed.), *A társadalmi nemek képe és emlékezete Magyarországon a 19.–20. században* (The Image and Memory of Gender in Hungary in the nineteenth and twentieth centuries) (Budapest: Nök a Valódi Esélyegyenlöségért Alapítvány), pp. 175–98.

Lutz, H. (2002), '"At Your Service Madam!" The Globalization of Domestic Service', *Feminist Review*, 70 (Spring), pp. 89–104.

McMichael, P. (2000), 'Globalization: Myths and Realities', in: T. Roberts and A. Hite (eds), *From Modernization to Globalization. Perspectives on Development and Social Change* (Malden, MA, Oxford: Blackwell), pp. 274–92.

Melegh, A. (2002), 'Globalization, Nationalism, and Petite Imperialism', *Romanian Journal of Society and Politics*, 1, 2 (May), pp. 115–29.

Melegh, A. (2003), *Perspectives on the East-West Slope in the Process of EU Accession* (Paris: OGRE).

Melegh, Attila (2006a), *On the East/West Slope. Globalization, Nationalism, Racism and Discourses on Eastern Europe* (Budapest, New York: CEU Press).

Melegh, Attila (2006b), 'Globalisation and Migration in Eastern and Central Europe', in Julianna Traser (ed.), *A Regional Approach to Free Movement of Workers: Labour Migration Between Hungary and its Neighbouring Countries* <http://www.ecas.org/1192.pdf>.

Mittelman, J.H. (2000), *Globalization Syndrome. Transformation and Resistance* (Princeton, NJ: Princeton University Press).

Okólski, M. (1999), 'Migration Pressures on Europe', in D. van de Kaa et al. (eds), *European Populations. Unity in Diversity* (Dordrecht, Boston, MA, London: Kluwer Academic Publishers).

Orozco, M. (2002), 'Globalization and Migration: The Impact of Family Remittances in Latin America', *Latin American Solitics and Society*, 44, 2 (Summer), pp. 41–66.

Said, Edward (1978), *Orientalism* (New York: Vintage).

Sassen, S. (1996), *Losing Control? Sovereignty in an Age of Globalization* (New York: Columbia University Press).

Sassen, S. (1998), *Globalization and its Discontents* (New York: The New Press).

Sassen, S. (1999), *Guests and Aliens* (New York: The New Press).

Sassen, S. (2001), *The Global City* (New York, London, Tokyo, Princeton, NJ, Oxford: Princeton University Press).

Sik, E. (szerk.) (2001), *A migráció szociológiája* (Budapest: Szociális és

Családügyi Minisztérium).

Sik, E and Tóth, J. (2002), 'Joining to EU Identity – Integration of Hungary or the Hungarians', unpublished manuscript.

Todorova, M. (1997), *Imagining the Balkans* (Oxford: Oxford University Press).

Wolff, L. (1994), *Inventing Eastern Europe. The Map of Civilization on the Mind of Enlightenment* (Stanford, CA: Stanford University Press).

Zolberg, A.R. (1999), 'The Politics of Immigration Policy: An Externalist Perspective', *American Behavioral Scientist*, 42, 9 (June–July), pp. 1276–9.

Attila Melegh is an associate of the Demographic Research Institute of the Central Statistical Office, in Hungary.

NOTE

1 Parts of this article have already been published in Melegh (2006b). They have been re-edited and substantially revised for this article.

East European Emigration to the West: A 'New South' or an 'Anti-South'?

BRUNO DRWESKI

Since the enlargement of the European Union (EU) on 1 May 2004, to include eight countries of central and eastern Europe, the issue of regular migration from these countries to the West – and equally perhaps from countries closer to the East with which these countries have traditionally had close cultural and personal relations – is in the news. A simple glance at the British or Irish press, or the press of the eastern and central European countries, should suffice to demonstrate this.

According to certain observers, the 'East' will hereinafter replace the 'South' in terms of a labour reservoir, one which will supposedly be better adapted to the specificities of western countries. For others, the East will join the South to make a more flexible labour market in a globalized Europe. Some think that the East is a 'new South' which brings to the West new classes of destabilized populations who are not completely different in their social, political and trade union behaviour from those of the South, being themselves characterized by differentiated trends and requirements. Some even compare the 'fundamentalist' religious behaviour of East European Catholics with the Muslims of the South – a view which, they say, is substantiated by the recent constitution of political coalitions in Poland and Slovakia which are described as conservative, even extremely right-wing.

On this issue, it must be recalled that in the mid-1930s, when immigrant workers were repatriated by force from France following the crisis of 1929, the French press and reports from *préfets* indicate that people thought that Polish, Greek, Catholic Ukrainian and Jewish immigrants from the East were prone to crime, as well as being nationalist, fundamentalist and incapable of assimilation.[1]

Two years are obviously not long enough to assess the lasting effects of the new migration process which is under way, though massive migration from East European countries had begun well before 2004 and, in the case of Poles and Yugoslavs, even before 1989.

It is also too early to estimate the magnitude of the new migratory flows under way today or to appreciate the consequences of the dismantling of the Iron Curtain, the enlargement of the EU, and the process by which eastern borders are now closed, thanks to the Schengen Agreement. However, we may already note that profound changes are under way, allowing us to project some long-term consequences and make some assumptions.

What is the current state of the discussion about migrations in the societies concerned? Who are the winners, and the losers, in today's migration and immigration processes, since the countries concerned are often both exporting and importing labour? What is the social and political impact of these processes, including on future relations between nations?

Does migration from eastern and central European countries mark the beginning of migration from further afield, from the former Soviet Union or even more distant countries like Vietnam or North Korea? What is the impact of these migrations on public opinion in the countries concerned? And what comparisons can be drawn between the immigrants from the South in western Europe, from the South in eastern Europe, and from the East in western Europe?

The Importance of Recent Migration between East and West

In this article, I concentrate mainly on the Polish case, given that among the ninety-nine countries of emigration studied by the World Bank, Poland belongs to the group of those ten countries which have sent the greatest number of emigrants,[2] and because Poles are particularly represented among migrants in the new EU.

Certainly, among the eight countries of central and eastern Europe which joined the EU in 2004, the proportion of Poles who have emigrated is less than the Lithuanians: the percentage of the population from Lithuania which has emigrated is double that of Poland. This is true for Latvians and Slovaks who also emigrate more than the Poles, while Czechs, Hungarians, Estonians and Slovenes emigrate less.[3] However, because of the size of its population, nearly 38 million, the impact of Polish emigration on western Europe is, in absolute terms, incomparable to that from all the other countries put together.

This leads to another question: what will happen after the accession of Romania and Bulgaria to the EU, since Romanians, and to a lesser extent Bulgarians, are already massively present on the (black) labour market in the EU? And what will happen to foreign residents in the states that have just joined, when these states also join the Schengen Agreement? We stress that this is mainly about populations coming from all the countries of the Commonwealth of Independent States (CIS) and the Asian socialist countries, in particular from Vietnam (but note too the employment of North Korean workers at the naval site in Gdansk). These immigration trends in the enlarged EU sometimes affect other Third World countries (China, South Asia, the Middle East, and so on etc).

As regards immigration from central and eastern Europe to western Europe, the British authorities officially counted 260,000 Polish immigrants among the 427,000 registered immigrants coming from the countries that have recently joined the EU.[4] These numbers seem incomplete and must be compared to the two million Poles who have entered Britain since 2004, a large number of whom have probably stayed in the country to work, whether temporarily or more or less permanently. For the whole of the EU, the figures given by the Polish Ministry of Foreign Affairs at the end of 2005 vary between more than 700,000 Polish working legally in various countries, either on a seasonal or permanent basis, in other words about 300,000 in Britain, 400,000 in Germany (out of which 300,000 are seasonal workers for three months of the year), 200,000 in Ireland and 80,000 in Sweden.[5]

***

In late 2006, Finland, Spain, Italy, Portugal and Greece were added to the list of the EU member states which have opened their labour markets completely, with a subsequent flow of immigrants from the eastern and central European countries. Some immigrants made little effort to formalize their residential status, which had previously been illegal, in these countries. This is why it is difficult to determine precise migration statistics. In the older EU member states (Germany France, Austria, the Netherlands, Belgium, Denmark), which continue to regulate the recruitments from the new member states, the influx of citizens from the new member states is less numerous but important; this is all the more true, as family and friendly contacts facilitate their often illegal absorption into the labour networks, which often existed well before the enlargement of the EU.

In the older EU member states, the labour conditions may be harsh but the salaries may in other cases be more attractive than the average, including on the black market. For example, construction workers who emigrate to France may prefer to work in the black market in France rather than work legally, because of the employers' charges. Therefore, there are no easy answers, and the social, financial, political and emotional consequences will be very different, as things stand. Our role here is to sketch the various possible consequences, without attempting to map the dominant trends, which are too early to establish.

Globally, the estimates on the number of Polish immigrants in the fifteen older member states of the EU vary: the European Citizen Action Service based in Brussels gives a number of 1.12 million Polish immigrants, the Polish Catholic Church (a source of information we would be wrong to neglect), estimates the number of recent Polish immigrants at around one million, whereas most Polish experts estimate the possible number of emigrants between one and two million.[6] According to a specialist on migrations of the University of Opole, two million Polish people work either permanently or temporarily abroad, particularly in EU or EU-related countries such as Iceland, Switzerland and Norway, that have also opened up their borders to new migrants.

Wojciech Lukowski, of the Centre of Migration Studies of the University of Warsaw, proposes that the number of two million could be an underestimate and that it should even be doubled.[7] The remaining problem, therefore, is to find out a way to register people who in most cases, do not state whether they have emigrated or not, and who continuously drift between indefinite and fixed-term contracts of employment and illegal employment. In many cases, the migrants themselves have not definitely decided to emigrate and remain for many years in a situation in which they keep waiting and looking for opportunities offered to them in both their country of origin and in western Europe. At the moment, Polish immigrants represent clearly more than half or even two-thirds of the immigrants coming from the new member states of the EU. However, this situation could change, according to the decisions to be taken in the future, concerning Romania and Bulgaria, not to mention Turkey, the former Yugoslavia, Albania, or Ukraine.

We also cannot dismiss the idea Russia may try to establish precise migration rules in the future, with all its neighbours in the West being more or less remote, within a European partnership framework that both Russia and Brussels have expressed a willingness to establish. Romanian immigrants, like the Russian and Ukrainian or Armenian citizens will hereinafter form a relatively numerous population in western Europe, even if today, this population concentrates in jobs in the black market. It should be reminded that like Poland, Romania is a young and populous country.

Differences between Migration and Movement

The public utterances of the leaders of western countries regarding the very distinct ideas of migration and movement are usually non-committal. The freedom of movement is a right recognized in principle, *inter alia* by the Treaty of Helsinki; however, the right to the freedom of movement is different from the right to immigrate to a country and to legally find a job there. Within the new borders of the EU, the various member states have substantially applied the right to the freedom of movement and made provision for the migration processes, provided for by the Schengen Agreement, even if this is for the moment unsatisfactory for some people, as far as the right to work is concerned. On the contrary, for the countries which are not members of the EU, to the east and the south, the Schengen Agreement is a violation of the principles of the right to free movement. Moreover, the migration policies decided by the member states of the EU very often contradict the principle of the mutual interests of the countries of emigration and the countries of immigration.

And this is where the increasing international tensions on this issue are derived from; terrorism or trade and health issues could even serve as a pretext to prevent the more powerful states from facing this issue frankly. This is what was highlighted by the recent international dispute on the principle of chosen immigration. The denial of the western European countries to draw up regulations allowing the generalization of the freedom of movement for all the populations on the planet, with a parallel adoption of stricter but mutually advantageous laws between the countries of emigration and immigration of the East or South, allows the preservation of a black labour market, exercising pressure on salaries in all countries and making the position of migrants more fragile. There cannot be free movement of capital on the one hand, and on the other, punitive controls, incarceration, or putting the migrant population in a state of illegality, on the other. Sooner or later, compromises must be negotiated between the free-market principles serving the interests of large multinational companies and the social principles which serve the marginalized populations of the South, the East and the West. The limits to the freedom of movement also prevent many migrants from taking part in the regular exchange flows with their countries of origin, confining them to a geographical immobilization and precarious employment conditions without possibilities of development in their countries of reception.

The opening of the EU to ten new member states also exacerbated this problem. Indeed, even if the new member states are not yet fully integrated in the Schengen system, their citizens benefit already from the principle of the freedom of movement within the whole EU and this will be the case for the Bulgarians and Romanians in the future. Historic relationships, such as those between France and the Arab or African states, between Britain and the countries of South Asia, Africa and the Caribbean, between Germany and Turkey, and so on, also exist between central and eastern Europe, and the CIS countries and Vietnam. In addition, for some years, we have also noted a flow of migrants, both legal and illegal, into these countries, from North East China and South Asia. What will the future of these special relations be, within the great and unified European market?

Shutting the gates by means of the Schengen Agreement does not really tackle this issue, since the generalization of the visas already disturbs transborder relations which are indispensable for the financial survival of the western border regions of the new member States of the EU. We should also note that in western Europe, Romanian, Russian, Ukrainian, Armenian and other workers compete in the legal and particularly the illegal labour market with the immigrants coming from the countries which have recently joined the EU or the countries of the South, the traditional suppliers of immigrants to the EU. We should also note that the employers of central and eastern Europe are not short of ideas either. Budapest has become an important base for Chinese financial activities today, organized by recent immigrants benefiting from the support of the Chinese authorities and the networks of the Communist Party (labour party) in Hungary. We note a similar phenomenon in Poland. In an article in the newspaper *'Gazeta wyborcza'*, has indeed noted the presence of North Korean welders in the shipyards of Gdansk and agricultural labourers in the country's fruit and vegetable plantations. The welders' salaries are purported to be directly paid by the Polish entrepreneurs to the North Korean authorities who repay their labourers around 60 zlotys per month, in other words ¤75, for more than ten hours of work per day. The Polish Labour Inspection has made some enquiries but the trade union *Solidarnosc* has rejected the findings of the article, arguing that this is nothing but propaganda against the government from the political factions which are now in opposition.[8]. We mention this argument to stress to what extent the labour market in the new member states of the EU must also be taken into account so as to develop the global vision required for our thinking and to show to what extent EU enlargement opens a veritable Pandora's Box. When reading these different facts, we see that international negotiations regarding the regulation of migrations must be framed on the basis of the mutual interest and dissociated from the universal principle of the freedom of movement to which the societies of the former Eastern bloc are particularly attached.

Winners and Losers?

Despite numerous apprehensions at the beginning, Britain and Ireland declare that they have not noted any substantial increase of the number of the unemployed in their countries since the opening of the labour market to workers coming from the new member states of the EU in 2004. This seems to be true as regards the middle classes who do not seem to be in competition with the new immigrants, and seem in the contrary, to benefit from the increase of consumption and the development of the services resulting from this phenomenon. On the contrary, the local labourers, confined in low-class jobs, seem to have been affected by a recent increase of unemployment in their job sectors. Therefore, the immigration from the East could in the long run prolong the effects of the neoliberal politics applied in Britain, which lead to the development of a service economy and deindustrialization.

On the other hand, in emigration countries, unemployment rates are globally decreasing since 2004. In Poland, a country where this has been more noteworthy, this rate has even decreased from almost 20 per cent to less than 15 per cent with the number of emigrants having exceeded the number of new unemployed persons by 3 per cent.[9] The same holds true for Slovakia, Lithuania and Latvia.

In this context, one cannot deny, for the moment, that emigration contributed to the relaxation of the climate in the labour markets, without leading to catastrophic consequences for the country of reception. However, the results must be judged in the medium and long term. It is all the more worrying that should the slightest recession occur, the local workers of the country of reception will have to compete for jobs which are already occupied by immigrants. For the country of emigration, according to a number of economists in the country of origin, the advantages of emigration are first noted in the short term: relaxation on the local labour market, transfers of capital to the families remaining in the country,[10] training in new professions for seasonal workers returning to their country, pressure for the increase of salaries in the country of origin when there begins to be a shortage in a number of specialties. But we should not exclude that the lack of qualified workers in central and eastern Europe will be filled by the immigration of labour coming from countries located further East who will accept even lower salaries. We may assume that shortly, the countries of central and eastern Europe within the heart of the enlarged EU could exercise pressure to lessen the immigration laws in Schengen Europe so as to facilitate the immigration from the East or the South in their countries and to revitalize their eastern border regions. This is all the more true as their increasingly ageing populations will make them wish to rejuvenate them, through immigration, as immigrants pay taxes, which pay pensions, the financing of which is causing a lot of problems even today.

Among other problems linked to new migrations, we note the multiplication of immigration agencies of a criminal nature, which promise employment contracts that are not respected in western countries, or are involved in organizing exploitative gang work, which is incompatible with the labour legislation of the countries of reception or origin. In this context, we should mention in particular the phenomena of prostitution and mafia operations which have been moving into western Europe from eastern and, more recently, central Europe since 1989. These illegal activities developed following the structural disintegration of the east European police forces after 1989, which allows them today to bypass even the most stringent immigration laws. Warsaw, Vienna and Berlin are today the logistical bases for the so called Russian mafias, which maintain illegal labour networks.[11]

The leaders of central Asian countries already stress the extent to which their countries and Russia constitute natural transit routes for drugs coming from Afghanistan, which is in principle pacified by the armies of NATO. In this context, the corruption resulting from poverty that we also encounter in the new member States of the EU can only contribute to the preservation of parallel migration flows, which exist already.

From the point of view of central Europe, the consequences of the migration phenomena of the last two years are also contradictory. According to the Polish Employers' Organisations for example, the balance sheet of the migrations post 2004 is moderated. They note the existence of new 'tensions' in the labour market for certain jobs such as the shortage of workers in the construction industry, qualified workers in general, nurses, doctors, plumbers, train drivers, IT programmers and translators.[12]

It was mainly the young and relatively educated who left the country, even if it was to find jobs requiring lower qualifications elsewhere, which were nevertheless better paying. The 'brain drain' phenomenon, which is very well known in Third World countries, is repeating itself today in the East, as had happened earlier, before 1939. Contrary to the French Minister of the Interior, Nicolas Sarkozy who in Paris calls this phenomenon 'chosen immigration', many researchers and politicians from the eastern areas of Europe increasingly call this procedure 'Assistance to development' from poor countries in favour of richer countries, given that it is the latter that benefit from the training expenses made by the emigration countries. Moreover, British employers stress on their behalf, that the immigrants coming from all the countries of the former Eastern bloc have a better level of qualifications for a same post, as compared to immigrants coming from southern Europe and *a fortiori* from Third World countries.[13]

We note however, that graduate migrants tend more than others to return to their country after a lapse of time and we may consider their stay as an enriching factor for their country of origin, in view of the experience and training they obtain. For the countries of reception, the experts stress the significance of the increase of the rate of consumption caused by the flow of new immigrants coming from the East,[14] and the end of the shortage of specialists in some sectors such as the health or information technology sectors. In Britain, we note that 90 per cent of central European immigrants occupy non-qualified jobs in the fields of agriculture, construction, child care, house cleaning, care for the aged, and hotel and catering work. However, these immigrants have also contributed to absorb labour deficits in qualified jobs such as doctors, nurses (where they are underpaid in comparison with their British work colleagues), or IT programmers.[15] Globally, we may note that both the employers and the British government seem to be satisfied with the consequences of opening their labour market; paradoxically, it is the conservatives who are clearly less satisfied.

We also note that even among those favouring open borders, there is apprehension before the consequences of the accession of Romania and Bulgaria, showing that even in London the condition of 'too many' may have been reached.

Diversified Estimates

The results of the migration phenomenon cannot be objectively analysed unless the subjective aspects are included therein, which give a broad explanation of the behaviours of immigrants and their progress in the country of origin. Indeed, even if we could establish that many of the immigrants from the new member states are young, active and usually over-qualified for the job that they do in the reception country, they do not usually compare their situation with that of the reception country's inhabitants or the status that they could have had in their country of origin, but to the financial level they've attained, as compared to most of their compatriots. This explains the feeling of superiority that they usually have, which makes them accept their situation, and that is usually a source of envy for their new neighbours. The 'complex of superiority' is usually added to this, to consider oneself 'European' as regards the older immigrants coming from 'post-colonial countries'. This self-satisfaction varies nevertheless widely from one country to the next and in particular in reception countries which were historically linked to the Mediterranean world. In the case of France, for example, the post-colonial populations know the language of the reception country, as well as its culture to which they were associated with as early as their school years, in their country of origin. This proves to be a source of misunderstandings among the various categories of immigrants, but also with the indigenous population. On this issue, we noted to what extent the employers' or the employees' behaviour, for example, varies widely from one case to the other. Some French people think for example that an Algerian worker is more French in his behaviour than a Polish worker, whereas a German will be more inclined to stress the more European behaviour of a Slovak, as compared to that of a Turk. These examples remain nevertheless very schematic and should be refined, since all these behaviours vary widely from one place to the other and constantly evolve – something that we empirically noted in France when discussing with workers and employers in the construction sector and the scientific research field. Let's also note that the 'information campaigns' organized under the aegis of the EU and taking place in the East since 1989 served to influence thinking in these societies, often convincing them of the attractiveness of 'western' Europe, and the opportunities offered by the more open and flexible global culture. These feelings often compensate for the very many cases of over-exploitation of the work force or the fact that there are many immigrants from the new member States among the homeless, who dare not return to their country, as they are ashamed to face their compatriots who are convinced that anyone who doesn't succeed in the West is *a priori* incapable.[16]

We also note that the proliferation in Germany, France, Britain, Ireland, and so on, of newspapers, radio programmes, magazines, meeting places, trades, and so on, created by and for the new immigrants; all these contribute to their inclusion in the local dynamics, as well as weakening their feelings of disorientation, and creating information and assistance networks which moderate those factors that cause frustration. We also cannot neglect the fact that religious and ethnically based organizations play a significant role in the welcoming of new arrivals. Is this the reason why we see more emigrants in the countries which have a strong Catholic tradition and national churches (Poland, Slovakia, Lithuania)? For the Poles and the Czechs, the engineers' organization established in Britain since the immigration wave of the Second World War cooperate with their counterparts in their countries of origin to welcome new arrivals, find jobs and maintain and constant flow of exchanges and scientific cooperation between the two countries.

We therefore establish, as has been already the case for the countries of the South, that the presence or not of a previous immigration from the same country, can facilitate, or on the contrary, render the integration of new migrants more difficult, durable, or temporary. And this is all the more true if we observe in parallel the misunderstandings between the different generations of immigrants, reinforced by the habits or requirements inherited from socialism and appreciated differently from one another. We also note that the freedom of movement for the new member States of the EU facilitates the travelling from and to and the regular exchanges between the countries whereas the closure of the borders before 2004 for the central European countries and since 2004, for the countries of the CIS or the South, favour the embedding of illegal immigration, which is marginalizing and destabilizing.

Migrations, Social Stability and Intolerance

For central and eastern Europe, whose enthusiastic adoption of neo-liberalism after 1989 was later followed by more sceptical trends, the partial opening of the labour market of the EU in 2004 is an incontestable safety valve for those in power. For example, a recent headline in the Polish trade union magazine announced 'The Revolutionaries escaped to Ireland',[17] whereas at the end of the 1990s, the local media frequently announced that major social tensions were imminent and that anti-capitalist trends or opinions were burgeoning. This potential for revolt has decreased for the time being, at least in the short term, because of emigration. But the example of the strikes organized by Polish 'chosen immigrants' sub-contracting in the port of Saint-Nazaire, which was backed by the French trade union CGT, or the strikes organized by immigrant workers in some Irish supermarkets, seem to show that the phenomenon described above is not necessarily a one-way street. We also note that some trade union militants, more radical than others, have recently become more active in Britain and France and sometimes play a role in the development of trade union trends in their countries of origin.[18]

It's worth noting here that despite the fact that British film maker Ken Loach is now making a film in Britain and Poland on emigrants, the phenomena of over-exploitation, disrespect for labour laws, the new form of slavery in the West, and so on, this issue is not reflected in the efforts of the film makers of the new EU member states, who have long been in a crisis situation themselves. For many eastern Europeans, Europe is a substitute for the powerful traditional patriotism which existed in their countries and which is currently in a relative crisis situation. This form of 'Europeanism' may sometimes connect with racism against the rest of the world, since the European biological colonialism of the nineteenth century was not fundamentally dismantled in the eastern European consciousness during the socialist era; the subsequent dismantling of the Marxist ideology managed to invoke the reappearance of primitive prejudices; this was noted in particular with the launch of the US war against Iraq, to which they noisily associated themselves as belonging the elite of the new Europe. Confronted in the West by the massive and unknown presence in their country of immigrants from the South, and in particular from Arab-Muslim countries, this situation may provoke racist reflexes and Islamophobia in some new immigrants to the West. The extent of these feelings varies widely between social classes and countries.

We also stress the notable difference between the behaviours exhibited by immigrants working in isolation and those engaging in activities which put them in direct contact with other classes of workers. For example, there is little contact between eastern European immigrants working in the agricultural sector in Spain and their competitors, that is, immigrants from Arab or African countries; indeed, the east Europeans seem to be inclined to display their Europeanism as an asset against their competitors in the labour market. These workers in turn continue to be marked by prejudices that appear much less often in the construction industry, for example, where there are more frequent contacts and where the prejudice is replaced more often with cooperation and mutual interests. It seems also that the integration of various immigrants is slower in countries where communities have been traditionally divided residentially, such as in Britain than in countries which are socially less divided and thus more ethnically mixed, such as France, where some sort of homogenization is already under way, despite the existing obstacles. Studies investigating the turmoil in the suburbs in 2005 have also shown that their cause was mostly social at the end of the day and that despite the fact that the rioters were mainly young men of Arab and Muslim origin in some neighbourhoods, elsewhere, depending on other parameters, they could be originating from the Antilles, Africa, or even be of Gallic origin, in particular in Pas de Calais, a region of endemic unemployment. The 'Islamic signs' displayed side by side with portraits of Che Guevara are often a unifying symbol among all the young, whether Muslims or not, who are impressed by what they think to be their 'suburbs' intifada' or their new 'Latin' guerrilla.[19]

We've seen how numerous are the elements that must be considered to understand the migration processes that are taking place and their possible consequences. Moreover, the importance of the traditional links between the various societies, families, groups, cultural habits and the importance of the social differences both in the country of origin and in the country of reception must also be considered in the future. The various migration phenomena generate contradictory feelings and take part in processes which lead to very different perspectives. This explains why the evolution of opinions on social, political, ideological and religious issues varies widely from one place to another, from one time to the next. The factors of competition between the South and East or, on the contrary, the integration of migrants, are both taking place in parallel and no one can fully control them. Therefore, in order to avoid occasional and unforeseeable developments we must hope that systematic research is engaged, with a long-term perspective, on the migration and movement phenomena, and that the public authorities of the countries concerned in the West, East and South undertake together the task of conduct open debate and even regular and considered negotiations to manage these phenomena for their mutual interest. We are only at the beginning of this exploration, which is still very spontaneous and divided; it is this that permits the fishers of troubled waters to formulate all kinds of unverifiable assertions and awakens fear, which, as history has always proven, is the worst counsellor.

Bruno Drweski is a lecturer at the National Institute of Oriental Languages and Civilizations (INALCO) in Paris.

NOTES

1. Janine Ponty, *The Ignored Polish – History of the Immigrant Workers in France between the Two World Wars* (Paris: Sorbonne, 2005).
2. *Biuletyn migracyjny*, 8 (2006), p. 4.
3. Exodus Polaków <http://www.antoranz.net/CURIOSA/ZBIOR6/C0602/20060225-QZD08072_emigracja>.
4. 'Immigration : Fever in Britain', 23 August 2006 <http://www.euractiv.com/fr/elargissement/immigration-poussee-fievre-royaume-uni/article-157205>.
5. Exodus Polaków.
6. Bozena Klos, 'Migracje zarobkowe Polaków do krajów Unii Europejskiej', *Infos – Biuro analiz sejmowych*, 2 (23 October 2006).
7. Patrycja Maciejewicz, 'Ciemna liczba szukajacych pracy w UE', *Gazeta Wyborcza*, 27 July 2006 <http://praca.gazeta.pl/gazetapraca/1,67527,3446910.html>.
8. Mikolaj Chrzan and Marcin Kowalski, 'Koreanskich niewolników jest w Polsce wiecej', *Gazeta Wyborcza*, 27 March 2006.
9. 'The new Britons'.
10. According to information from the National Bank of Poland, an increase of 4 billion zlotys was recorded in relation to 2003, transfers of capital from abroad for a total amount of 22 billion zlotys which corresponds very little in fact, for an internal consumption amounting to 607 billion zlotys. However, transfers of capital also occurred mainly by the furnishing of capital in cash when travelling to the country. The Bank of Poland did nothing but confirmed, without giving numbers, that an increase in the purchase of British sterling was recorded, Maciejewicz, 'Ciemna liczba szukajacych pracy w UE'.
11. B. Drweski, 'The mafias of the East – an heterogenous and original phenomenon', *La Pensée*, 324 (October–December 2000), pp. 27–38.
12. IAR, 'Emigracja zarobkowa szkodzi polskiej gospodarce'. <http://www.money.pl/gospodarka/wiadomosci/artykul/emigracja;zarobkowa;szkodzi;polskiej;gospodarce,217,0,162009.html>. Contrary to what we might believe, the wave of migrant Polish plumbers, which was openly welcomed by the French elite on the eve of the discussion on the European Constitution, was barely a trickle; most of them went to Britain.
13. The new Britons'.
14. The British authorities estimate that the GDP in Britain increased by £280 million in 2004 thanks to the contribution of the work of Polish immigrants alone; Klos, 'Migracje zarobkowe Polaków do krajów Unii Europejskiej'.
15. Pawel Jagiello, 'Polacy na europejskim rynku pracy', 1 September 2006 <http://wiadomosci.polska.pl/specdlapolski/article.htm?id=207906>.
16. Fiona Barton, 'Are young Polish workers robbing their country of its future?', *Daily Mail*, 19 May 2006 <http://www.dailymail.co.uk/pages/live/articles/news/news.html?in_article_id=386931&in_page_id=1770>.
17. Ryszard Górski, 'Eksport nadziei – Wstep do dyskusji – Rewolucjonisci wyjechali do Irlandii?', *Trybuna Robotnicza*, 7 (16 November 2006).
18. Ryszard Górski, 'The Polish of Saint-Nazaire', *L'Humanité*, 30 July 2005.
19. Haydée Saberan, 'The crisis in the suburbs – in the north, at court, the rioters away from clichés – The young whites coming from deprived neighbourhoods appear before justice', *Libération*, 18 November 2005; Jean-Emmanuel Ducoin, Éditorial: 'Discriminations', *L'Humanité*, 26 November 2005.

Albanian Organized Crime

XAVIER RAUFER

A Typical Case: The Koçiu–Rapiki Affair

In March 2001, Sokol Koçiu, the former head of the judicial police of Albania, was arrested with the Kosovar drug trafficker, Mentor Hadergjonaj, for dealing in cocaine. The pro-governmental press presented Koçiu as being close to Ramush Haradinaj, the former Kosovo Liberation Army commander and President of the Alliance for the Future of Kosovo. (In 2004, Haradinaj was elected Prime Minister of Kosovo before being tried for war crimes in 2007.) The opposition press, by contrast, rejected this allegation and instead emphasized the role of Hashim Thaçi, the leader of the Democratic Party of Kosovo (PDK). Hadergjonaj himself called into question the Prosecutor General, Albert Rakipi, in particular by producing photographs of a Christmas party at which the Prosecutor's wife was in the company of two known drug traffickers. The magistrate recognized that the Prosecutor was indeed friends with them.

Two brothers, Frederik Durda and Arben Berballa, were implicated with Hadergjonaj and Koçiu in a network which was said to have imported forty tonnes of cocaine into Albania for sale in western Europe. Durda left Albania in 1976 after being convicted twice for theft. He settled in New York where he was arrested for drug trafficking, and returned to Albania to settle there permanently in the 1990s. Berbella has always lived in Albania, working first as a veterinary surgeon and later as an army officer. At the same time, the two started up their business in Tirana using Greek cargo ships registered in Malta. One of Durda's companies installed the IT systems of various prosecuting authorities, including the one in Tirana. An operation by the police forces of seven South American countries, the United States and European states (including France, Greece, Russia, the Netherlands and Italy) led to the arrest of eighteen people and the seizure of two boats, weapons, a million dollars and eight tonnes of heroin and cocaine. In the deal, two Albanian drug dealers, Vehbi Hysi, aged 60, and Alfred Nina, 32, had even travelled to Colombia as voluntary hostages in order to guarantee payment for the deliveries. Therefore there is no denying the fact that the organizations which grew out of the Kosovo Liberation Army are involved in organized crime.¹

The armed activities of the Albanian minority in Serbia and Macedonia started up again in 2000 and 2001. In January 2000, the Army for the Liberation of Presevo, Medvedja and Bujanovac (UCPMB) was created. This group was particularly active along the heroin routes, particularly the Gnjilane–Bujanovac route where there are heroin-producing laboratories. According to some observers, the main goal of the guerrillas in these Serb and Macedonian zones was to force the international community to send in troops to take over from the local police. An impotent international force is much better for criminal 'biznes' – smuggling drugs, arms, cigarettes, oil, women, and so on. Furthermore, the fall of the Milosevic regime in Belgrade had destabilized the trade controlled by this clan, especially oil and cigarette smuggling. Two rival clans therefore emerged. Hashim Thaçi's Democratic Party of Kosovo, allied itself to the Democratic Party of Albanians (DPA) in Macedonia, which was headed by Arben Xhafcri – whose deputy was implicated in cigarette smuggling in 2000 – and to some clans within the Albanian secret services. The other clan, Ramush Haradinaj's Alliance for the Future of Kosovo (AAK), allied to warlords like Sabit Geçi (who was arrested in October 2000 in Pristina for various Mafia activities including pimping) and Rrustem 'Remi' Mustafa ('Commander Remi', indicted with Haradinaj in August 2002 for kidnap and torture). Ekrem Lluka, a Kosovar businessman who bankrolls the AAK, was implicated when forty tonnes of contraband cigarettes were seized at Pec. KFOR, the Kosovo Protection Force, suspects Lluka of also being involved in drug trafficking.

Within the UCPMB, the following persons are to be noted: Muhamed Xhemajli, a drugs trafficker in Switzerland who joined the KLA in 1998, and Xhavit Hasani who fled Macedonia in 1998 because of his involvement in organized crime and who joined the KLA. In 1999, he controlled the Vitina region and the criminal activities which go on there.

Western intelligence services believe that Albanian extremists (the PDK and the AAK) finance the purchase of their weapons with the money they earn from trafficking in heroin. They buy SA-7 anti-aircraft missiles and SA-18u anti-tank missiles as well as grenade-launchers, other anti-tank weapons, machine guns and precision guns. In July 2000, the Swiss authorities seized a shipment of arms destined for Kosovo. Several Kosovars resident in Switzerland were arrested in connection with this seizure. The traffickers bought anti-tank weapons in eastern Europe which were officially destined for Africa.

More generally in the Albanian-speaking areas of Albania itself, Serbia and Macedonia, Albanian clans are involved in numerous forms of contraband: cigarettes, oil, pirate DVDs, false telephone cards, forged currency, and so on.

In May 2002, for instance, the UN police in Kosovo discovered a secret cigarette factory near Gnjilane, a place already known for its heroin laboratories, where they seized twenty tonnes of contraband tobacco.

The Production and Trafficking of Drugs in Albania, Macedonia and Kosovo: Facts and Evidence

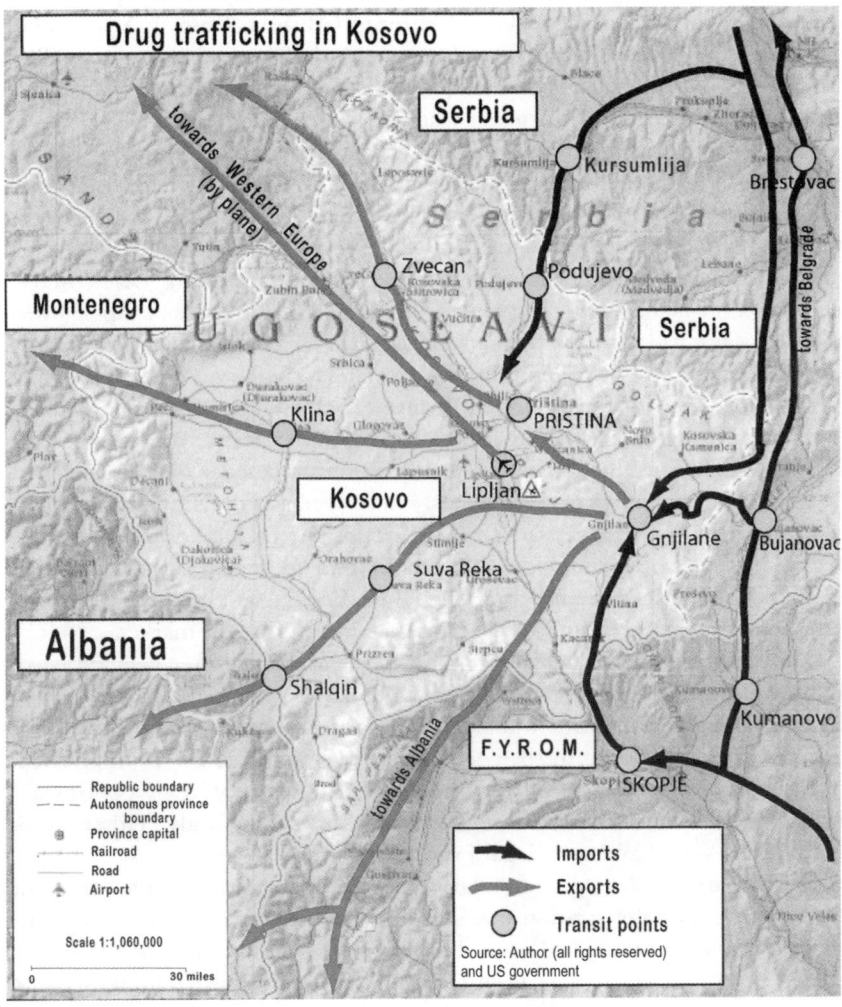

Cannabis is widely produced in the region under examination. The drug is exported principally to Italy (via networks of people traffickers)[2] and Greece. Some is consumed in Greece and some is re-exported via the port of Piraeus, particularly to Britain. In Italy, the resale of cannabis is in the hands of Albanian or Italian-Albanian gangs. It seems that it is also sold on to other countries, especially to Albanian gangs operating in Belgium. Albanian cannabis contains high levels of tetra-hydrocannabinol, the principal active ingredient, to the extent that it can be traded for Turkish heroin.

Certain sources maintain that there are also heroin and cocaine production factories in the Albanian-speaking territories (Albania, Kosovo, Macedonia), and there have also been attempts to grow opium. There are even rumours of newly constructed ecstasy factories. The *savoir-faire* of the Albanian Mafiosi in Belgium is said to come from gangs who specialize in synthetic drugs. One case seems to confirm that Albanians are involved in ecstasy production. On 18 October 2001, the US Federal Drug Enforcement Agency (DEA) dismantled an ecstasy-producing laboratory in California. It was one of the most sophisticated ever discovered in the US, capable of producing 1.5 million tablets a month. Twenty-nine people were arrested including Derek Mayer Galanis, 29, who had spent five months in Kosovo in 2000. Phone taps showed that this gang had contacts with a general in the KLA as well as with Tommy Gambino, son of Rosario Gambino, an associate of the New York Mafia family of the same name, who was himself involved in the 'Pizza Connection', a huge heroin-dealing network in the 1980s in the United States, run by the Cosa Nostra in Sicily and the US, which used pizzerias to launder the drugs money.

Albania, like the rest of the Balkans, is also used as a transit point by South American criminal organizations. These groups rely on local criminals for the logistics of their drug trafficking. Cocaine shipments are basically destined for western Europe but one anti-drug official in the UN said he though that the drug cartels were intending to penetrate the East European market.

The corruption of regional authorities by the Mafia has often been revealed. In May 2002, Lieutenant-Colonel Shamet Bejko, chief of police at the port of Durrës, and Major Alush Muho, the head of the military province, were arrested. They had been in their jobs for three months and were implicated, along with two other port officials (including a close relative of a politician) in a network of heroin trafficking between Turkey and Italy. Thirty kg of heroin were seized and ten illegal immigrants arrested.

In February 2002, 20 kg of heroin were seized in an Albanian government car carrying false registration plates which was on its way to Kosovo. This showed that a new route had opened for heroin trafficking: the drug no longer passed through Greece or Italy but instead through Kosovo and Serbia, and from there into Central Europe. Several people were arrested including Shkelqim Konci, chauffeur to Adi Shamku, the head of the youth wing of the

Albanian Socialist Party, who was also a senior official in the Ministry of Transport. Konci had been Monika Kryemadhi's driver, also a leader in the youth wing of the Socialist Party and the wife of the former Prime Minister, Ilir Meta. When his mobile phone was examined, it turned out to have the numbers in it of various senior officials and heads of the police.

In February 2004, a policeman in northern Albania was arrested for complicity in arms and drug dealing. His brother, a drug trafficker who lived in Montenegro near the Albanian border, was also arrested; the police seized pistols, guns, 300 kg of explosive and hundreds of rounds of ammunition.

In January 2004, as a group of illegal immigrants were attempting the sea crossing from Albania to Italy, twenty-one Albanians died of cold and fifteen disappeared near the port of Vlorë. Only eleven of the would-be emigrants survived. As a result, several people were arrested in Albania including Gjergli Robaj, Deputy Director of the port of Vlorë and the owner of the boat in question, as well as two policemen. The Prosecutor also opened two investigations into Bardhyl Rrokaj, head of the fight against terrorism in Shkodra, in the northern Albania, and his colleague Ilir Rrokaj, head of the transport police in Vlöre. Both were accused of facilitating people trafficking.

Heroin Trafficking

In November 2001, a Europol report estimated that 40 per cent of the heroin on the streets of Europe was sold by Albanians and that Albanians transported a further 40 per cent. European intelligence services have identified fifteen Mafia clans operating in northern Albania and involved in the heroin trade.

At the end of August 2000, the Albanian police intercepted a shipment of heroin in Elbasan, a town in central Albania, and seized 24 kg of heroin. In June 2001, two Albanians and nine Kosovars were arrested by the United Nations Mission in Kosovo (UNMIK) in Gnjilane in eastern Kosovo: 1.5 kg of heroin was seized. The Albanian port of Durrës remains the principal transit point for shipments of heroin to Italy: 10 kg were seized in May 2003, when five people were arrested including an Albanian resident in Italy; 14 kg in September 2003, when two Albanians living in Germany were arrested; 13.3 and 16 kg in March 2004; 5.7 kg and then 11.2 kg in April 2004; 18 kg in September 2004; 2.3 kg in December 2004. In May 2004, the Italian police targeted a network importing heroin from Durrës to Bari: nine Italians and seven Albanians were arrested including the head of the gang, Nako Huta, known as 'Nasho'. The UN police in Kosovo has done its work there too: 18 kg of heroin were seized in July 2003, when three people were arrested; 55 kg were seized and six arrested in January 2005; 100 kg and seven arrests in September 2004.

One notorious drug trafficker is Daut Kadriovski, an Albanian from Macedonia. Born in 1949, and based in Turkey, he is suspected of exporting heroin to western Europe and the United States. He was arrested in Tirana in September 2001: he was in possession of ten different passports and he had had cosmetic surgery to alter his appearance. He had been sentenced *in absentia* to twelve years in prison in Italy in 1996 and he had been also wanted in Hungary: Kadriovski had been arrested in Germany in 1998 but had escaped.

Albanian criminal gangs also operate out of Bulgaria, between the Albanian-speaking countries and Turkey. In June 2002, a Kosovar who was wanted for the murder of three policemen in Albania was arrested in Sofia. He was wanted in Germany, the Netherlands (where he had escaped from prison in 1993), Serbia and Austria, for heroin trafficking and for currency counterfeiting. He was extradited to the Netherlands and prosecuted for various acts of revenge perpetrated in 2000.

Although they are linked to the Turkish *maffya*, the Albanians have supplanted them in the heroin market, especially in Switzerland, Sweden and Norway. In April 2004, the Danish police estimated that these groups import between 80 and 90 per cent of the heroin consumed in Denmark. Heroin increasingly transits through Kosovo, then Serbia and eastern Europe.

People Trafficking

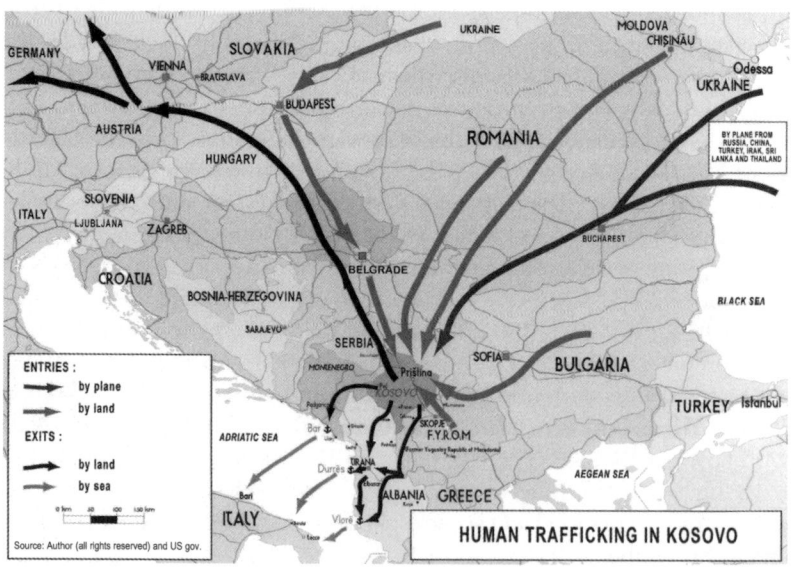

The Albanian Mafia is heavily involved in people trafficking. Albania plays a major role in smuggling Kurdish, Iraqi, Pakistani, Afghan, Chinese, Albanian and African immigrants to Italy, in collaboration with other criminal organizations including the Turkish *maffya*, Chinese triads and above all the *Sacra corona unita*, the Mafia in Puglia. Illegal immigrants and their people smugglers are regularly caught in Italy and Albania. In 2002 alone, the French police in Pristina seized 1,500 false passports at the airport. These documents were mainly used for people smuggling.

In summer 2002, the Italian authorities opened an investigation on the basis of rumours that children were being smuggled. An increasing number of illegal immigrants were aged between 12 and 17 and 20 per cent of them were girls. Some were illegally adopted, some were subject to acts of paedophilia, some were involved in organized gangs of beggars, others in organ trafficking. In 2002–03, the Albanian secret service (SHIK) and the Italian police investigated a network involved in trafficking human organs which was based in Albania, Macedonia, Greece, Italy and Germany. The Albanian SHIK identified a clinic at the centre of this trade. The disappearance of several children were attributed to this network, whether for the trade in organs or for illegal adoption. Five people were sentenced to between fifteen and twenty years in prison. The clinic's doctor involved is still at large. According to a report published by an Italian organization in November 2004, at least 2,000 children have disappeared in Albania over the last ten years. Kidneys are sold for €25,000, livers for €10,000. The trade in human organs is said to amount to €1.2 billion a year.

Other Albanian gangs smuggle immigrants into Britain, whether through Italy or France (where the hand-over points are Menton and Vienne) or via Austria and Germany. These illegal immigrants are often taken to northern France or Belgium. In Belgium, the people smugglers force the immigrants to engage in criminal activites themselves to work their passage. The Albanian Mafia also supplies false passports to enable people to enter Britain. In April 2001, the Belgian police undertook twenty-six searches which led to ten arrests, including the two Albanian leaders of the gang. They seized false passports, false identity cards and the means for producing these.

According to one Albanian daily, illegal immigrants must pay between €2,500 and €3,000 to get to western Europe, €6,500 to get to Britain or Ireland, and between €13,000 and €15,000 to get to the US or Canada.

In its prostitution rackets, the Albanian Mafia has extended its 'supply zones'. Previously, women and girls came from Albanian-speaking countries. Now, the Albanians use women and girls from Bulgaria, Ukraine, Russia, Romania and Moldova, as well as from western Europe.

Not All Albanian Criminals are Mafiosi

As in all countries, there are in the Albanian-speaking countries and among the Albanian diaspora small local gangs which operate independently of the gangs and families which constitute the real Mafia. Some émigrés agree to do one or two deliveries of heroin, a little job to earn some money when they need it. However, such occasional dealers are not Mafiosi as such.

But There Really is an Albanian Mafia[2]

The Albanian Mafia emerged in the 1960s among the Catholic Albanian diaspora. Originally from northern Albania, émigrés had settled on the east coast of the United States where they lived in close contact with Italian-Americans to whom they felt culturally linked, Albania having been occupied by Italy under Mussolini. They went to the same churches. By definition, a Mafia is invisible and anonymous – the mere revelation of its existence can be punished by death – and so its existence can be discovered only by empirical analysis based on the observation of precise facts. These include:

- A hierarchy with respected bosses, escorts and protégés

- An effective law of silence or *omertà*, which means that accused persons deny everything while witnesses of Albanian origin (prostitutes for example) clam up during any investigation or trial[3]

- Extremely secretively practices including hiding criminal clans behind folk or cultural clubs, with members residing in one European country while carrying out operations in others; multiple pseudonyms and surnames; multiples operational cells which know nothing about each other's existence; a systematic refusal to accept the identity attributed to individuals by the police after their investigations (even brothers of whom it has been proved by DNA tests that they are siblings deny even knowing each other and ignore one another during arrests)

- Symbiosis between politics and crime. In Albania, every criminal who counts is protected by a senior police officer who is in turn himself 'covered' by an important politician. In March 2002, for instance, the chief of police and the head of the port authority in Durrës were arrested for heroin trafficking even though they were in charge of the town's fight against drugs. Such scandals are common in Albania: without exception, the subsequent trials end in acquittals 'for lack of evidence'.

An important symptom of Mafia activity is the fact that each gang remains closely linked to its territory of origin, in this case Albania, Kosovo, western

Macedonia and southern Montenegro, the territories where Albanian is spoken.

Characteristics of the 'Export' of the Albanian Mafia

The Belgian, Swiss, Italian, British and Polish examples display characteristics which are common to Albanian-Kosovar Mafia organizations. These include:

- Marriage with local women
- A refusal to live in 'ghettoes'
- A desire to obtain proper residence permits or even citizenship (especially in Switzerland)
- The purchase of small businesses, particularly bars, followed by the takeover of other businesses through intimidation
- Agreements with other foreign gangs, especially Romanian, Bulgarian and North African ones, or with local gangs including Italian Mafias, local pègres, Hell's Angels in Belgium, Chinese triads or Jamaican Yardies in Britain
- A tendency to move around a lot (so-called 'criminal nomadism')
- Networks based on families and clans
- Involvement in numerous different criminal activities
- *Omertà*, maintained by acts of violence or threats, especially against family members who have remained in their home country
- The use of aliases and false papers
- The 'conspiratorial' use of mobile phones, involving the frequent change of SIM cards
- 'Investments' in their country of origin.

A novel aspect is the significant activity of Albanian criminal gangs on the Internet. Sites showing acts of violence including torture and other atrocities are run by Albanians. The females shown on these sites are probably Albanian prostitutes. Albanian pimps in Britain also use Internet sites to advertise their 'escort services'.

Such criminal activity requires extreme fluidity and mobility. There are no Albanian ghettoes in Europe, even though there are some 700,000–800,000

Albanians living in the European Union. Albanian Mafiosi have no desire to put down roots in the areas where they practise their criminal activities. In general, they repatriate the proceeds from their activities to their country of origin. To take the example of Belgium, since the Belgium police has the greatest expertise in Europe on the Albanian Mafia,[4] there are Albanian Mafia gangs in Antwerp, Brussels (in de Schaerbeek, Molenbeek and Saint-Gilles), Liège (where the Celepija and Toma clans operate), Namur (the Krasniqi clan), and Verviers-Mons (the Murati clan). These clans are linked to one another and they work together at the European level, including with other gangs in Aachen or Cologne. For instance, they 'barter' eastern European prostitutes for drugs.

Albanian Mafia clans also have criminal links to North African Maghrebins in France, where they are involved in the trade in stolen luxury cars, with gangs of bikies like the Hell's Angels in Antwerp, teams of Romanian burglars who work for the Albanian gangs, and the Turkish *maffya*, already mentioned, which deals in drugs particularly in Aachen.

Xavier Raufer is a professor at the Institute of Criminology in Paris.[5]

NOTES

1 See *Lettre Internationale des Drogues*, October 2001.
2 See Stéphane Quéré and Xavier Raufer, *Le crime organisé* (Paris: Presses universitaires de France, collection *Que Sais-Je?*, 2005), no. 3, p. 538.
3 See, for example, 'Le procès de six Albanais soupçonnés d'être les responsables d'un réseau de prostitution illustre les difficultés de la justice face à ce type d'affaires', *Le Monde*, 23 June 2002.
4 See 'Note d'Alerte' on the Albanian Mafia: <www.drmcc.org>.
5 The part of this article dealing with the Albanian Mafia has benefited from the expertise and archives of General Bujar Ramaj, the second-in-command of the Albanian security service (SHIK). General Ramaj is now President of the Albanian Centre for the Fight against Organized Crime with whose work the author has been collaborating for several years. The author is also Director of Studies at the Department of Research into Criminal Threats, Université-Paris II. See <www.drmcc.org> and <www.xavier-raufer.com>.

Transnational Networks: The Case of the Chinese of Zhejiang

VÉRONIQUE POISSON

France is home to one of the oldest and largest Chinese diasporas in Europe. Originating mainly from the southern province of Zhejiang, they come from the suburbs of the port town of Wenzhou, located around 500 km south of Shanghai.

Between 200,000 and 300,000 Zhejiang live in France. Among them are a number of merchants and entrepreneurs, who have the ability to mobilize both human and financial resources to establish networks based on an ethnic-entrepreneurial organization. These men operate outside the boundaries of nation-states by means of transational financial networks.

Following the development of commercial exchanges between China and Europe, the movement of goods and the accompanying financial flows have particularly intensified, as has the flow of migrants. However, and paradoxically, both China and Europe have maintained the control of their borders, as far as migration is concerned. In such a context, the management of the movement of migration passed into the hands of private organizations. Although the diaspora area is not necessarily an area where no rights exist, and the practices that have developed therein are not always contrary to the legislation of the reception countries, we will discuss the consequences of this paradox and in particular, the relation between migration flow, the tightening of border restrictions and the development of organized crime (an exponential activity of which is illegal immigration). Two aspects enable the development of this analysis on the transnational networks of the Chinese of Zhejiang: the first one is historical, the second is financial.

Rooted in a migration tradition more than a hundred years old, we will show that some clans (settled in their own particular villages) created networks through many generations and in many continents, with China as their nexus. Their geographical mobility is characterized by a steady movement back and forth between between different continents, articulated through very flexible entrepreneurial networks. Despite the fact that official Chinese historiography emphasizes the financial factors of emigration, the ethnographic works show that the political factors in the migration dynamics have been a constant feature both before, and after, 1949.[1] Against a backdrop of extreme violence, emigration would be part of a strategy in which earning money would enable people to attain some kind of autonomy. This financial autonomy would aid the migrant in opposing those circumstances which would limit their freedom of activity and movement.

After a brief historical overview, we shall concentrate on a more recent era, characterized by the intensification of commercial exchanges and focusing on the entry of China into the World Trade Organisation. We shall examine the development of organized crime and the contemporary forms of slavery in the heart of this transnational area. Based on an on-site study conducted recently by the International Labour Organisation in France,[2] we will show that Chinese migrants do not always choose to belong to ethnic entrepreneurial networks and that this is the consequence of a financial process based on the exploitation of labour.

To conclude, we shall present some joint actions undertaken by France and Italy with the Chinese associations within the framework of a European Programme called 'Equal: The Chinese of Europe and Integration' implemented between 2001 and 2006 so as to mitigate extreme situations.

Historical Factors of Emigration over a Hundred Years

Issues of domination and exclusion lie at the heart of the emigration movements from the province of Zhejiang, which have been taking place for over a century and which are based on a political separation that stands to this day. In fact, this region's emigration towards Europe started as early as the 1930–1940s, centering around the districts of Wencheng and Qingtian, both frontline posts of the Guomindang troops. Wencheng was even the site of an American Air Force base. Moreover, this micro-region was characterized by a shortage of arable land due to the very hilly landscape. These lands belonged to landlords who controlled commercial societies in the prosperous ports of Shanghai and Ningbo. Some of the tribes in Wencheng and Qingtian belong to classes of Mandarins who had close relations with the imperial court of the Qing before 1911, and are linked to a pro-Guomindang class of merchants and industrialists who initiated financial exchanges particularly in the Asia-Pacific zone (Japan and South East Asia, following the Sino-Japanese War of 1937–54). However, from 1932 onwards, trade relations ceased and entrepreneurs were forced to turn to other markets: during this period, between 20,000 and 30,000 Qingtian emigrated to Taiwan, the United States and Europe. Others who remained were marginalized as the local authorities no longer invested in the region. Upon the arrival in France of migrants originating from Qingtian, one of the most important pro-Guomindang movements in Europe was formed, which had around six hundred members. Some of the merchants of Qingtian who became wealthy either abroad or elsewhere in China, repurchased the most arable lands in the neighbouring planes of Wenzhou (at Ouhai and Li'ao in particular). Originating from village communities organized on the line of tribal alliances, the Qingtian spread abroad and in China, establishing entrepreneurial networks as early as the 1920s and 1930s. During our ethnographic work on the plains, we noted that the pioneers of this emigration had relations with the tribes of Qingtian who came from the hills.

The Organization Overseas

Taken from an article on the history of the Chinese abroad in a local newspaper, the biography of Hu Xizhen illustrates the portrait of a candidate for emigration at that time. In 1925, Hu decided not to enlist in the Guomindang Army and left China to make a living in Japan. That same year, he went to Italy:

> From there, he went to France where he knew some compatriots who were itinerant merchants. Following the enactment of new legislation between 1931 and 1932, the French government suppressed the Chinese pedlars and restricted the hiring of immigrant workers in French factories. He returned to Milan and set up a business making leather goods. In 1937, he exhibited at the World's Fair in Paris and one of his leather purses was a huge success. Following this good fortune, he brought his family over to work in the various workshops that he opened.[3]

The international trade fairs enabled the establishment of trade relations abroad, as pointed out by the French Consul in Ghent (Belgium) in 1935:

> We note that around two hundred Chinese left China to try their luck at the Brussels Exhibition. The majority have travelled via San Francisco thanks to the low fare offered today by the Pacific shipping companies. It seems that only a few of them shall arrive here through Marseille or Hamburg. It is feared that many of these Asians will establish themselves as pedlars and trinket sellers.[4]

Acting as a logistical support to emigration, the trade societies also played a key role. Indeed, many Chinese societies producing fancy goods and those of Chinese manufacture in Europe supplied the hundreds of newly arrived itinerant merchants.

It is again the diplomatic sources which illustrate this organization: a letter from the French Minister of the Interior dated 2 March 1931, about a group of Chinese from Zhejiang, states that

> ... these foreigners arrived in Marseille on 21 January 1931 on board the liner *Angers* with transit visas via France for Portugal or Switzerland. Instead of going to these countries they headed for Paris to trade their exotic bibelots. By examining their correspondence, it seems that these foreign traders, who all had relations with Kune He Chong Cie company in Shanghai belonged to an international organization with connections all over: Paris, Manchester, Warsaw, Moscow, Milan, Brussels. They had visas in their passports to enter and leave Germany, which they obtained thanks to the intervention of the Chinese at Ravensburg, Stuttgart, Passau and the German embassy in Copenhagen.

Being the initiators and the driving engine of the first migratory trends towards Europe, the Qingtian, confined in an area surrounded by hills where arable land was not enough, had no other choice but leave the land of their ancestors and spread over a vast territory, aided by their compatriots.

Following the political turmoil before and after 1949, emigration abroad became one of the few choices for supporting the family.

The Hukou Glass Ceiling

More recently, other political factors have contributed to the increase in emigration: mainly, the anti-capitalist and anti-entrepreneurial ideology of the Maoist years, the implementation of an internal passport (the hukou), and the affects of enforced family planning.

In the 1980s, the government brought in new legislation permitting only one child per family. This enforced birth control motivated many clans to go abroad: emigration served, in this particular case, as an escape from the political restrictions which threatened the basis of the social structure of the clan.

As regards population movements inside China, these have been controlled since the 1950s by the internal passport, the hukou. It is not always easy for a internal passport holder to leave and establish themselves elsewhere and the children of migrants are prohibited from joining their parents. This internal border is a source of discrimination, given that town dwellers have better social protection, a more diversified access to the labour market and better paid jobs, and their children benefit from better education in town schools. In such a context, some internal passport holders consider it more beneficial to emigrate abroad rather than establish themselves in towns where they will not have the same rights as established town residents (unless they pay a punitive tax).

Moreover, until the 1990s, entrepreneurial activity was viewed with suspicion in a society whose overwhelming communist ideology disapproved of such enterprise; most merchants held 'pocket passports' which would enable them to leave the country at any time. Both officially and through bribes, they created an autonomous space in which they could carry out their business with relative freedom, avoiding taxation from the central government, fiscal office or other control administrative authorities. However, having achieved a certain level of financial prosperity, it would become difficult to conceal their wealth and they preferred to go abroad with their profits. Those who had close family ties abroad would make financial transfers for investment overseas, using family members as intermediaries. From the 1980s, as China began to open up to the outside world, the networks mobilized for emigration extended themselves to village communities other than those of the Qingtian. At the beginning of the 1990s, at the peak of emigration of the Zhejiang towards France, even the town dwellers of Wenzhou emigrated. For emigration, it is no longer critical to have family contacts abroad: having enough money to travel is sufficient. Upon arrival in France, there are many jobs in the clothing industry and the catering sector. The initial debt can be repaid quickly after two or three years of work.[5]

Though the sociological profile of migrants has changed over decades, emigration as a survival strategy remains the same, both before and after 1949. Indeed, in the country of origin, China, the existence of internal borders is the origin of the discriminatory situations that power the dynamics of migration. But whatever the form of political pressure, the migratory movements of the Chinese of Zhejiang are characterized by returns, 're-migrations' and 'rebound migrations', in which neither the country of origin nor the reception country function as an anchor: more often, the first country of establishment is a bridge to the next migration towards another destination.

Contrary to other diasporas, the Zhejiang one is defined by the inventiveness of those who transcend their difficult social condition, that is, the racism they experience within their national society. They organize themselves, resist opposition, and thrive on the financial and business networks that they weave in the diaspora area, exceeding the borders of nation states to form a transnational area. The choice of country of destination is unimportant, as long as it allows them to work; and it is possible to leave at any time if circumstances seem more promising elsewhere.

From the Movement of Goods to the Trafficking of Human Beings

According to Peter Nolan's thesis,[6] the illegal and informal trade practices that were put in practice in the Wenzhou region from the 1970s proved that the central government lacked control on a local level. The underground economy had organized itself around small private banks and tightly-knit family businesses. In 1984, the government established the 'Wenzhou economic model', famous throughout China, which only confirmed that the government was now acknowledging an already existing system, of which it had never control, in particular, the control exercised by local officials.[7]

By making financial liberalization official, which had been previously been pursued feverishly by the private local entrepreneurs, the central government exerted some control by making them accountable to the government finance officials. This is how M.X. Fulin,[8] an industrialist from Wenzhou, described the situation prevailing at the time:[9]

> Why do so many people in Wenzhou go abroad? ... Because of the controls. We want to carry out a little trade. This is capitalism and we risk ending up in prison .... Those who trade with the outside world need some assistance ... I am one of them ... I thought of going abroad myself in 1987. Now, it is better. If something is not going well with the officials of the Party – and there is trouble sometimes – we have a pocket passport and we are ready to go.

What the Chinese of Zhejiang in Beijing, in Italy and in France have in common is that their shared experience in their home village was characterized by a reluctance to deal with, and a suspicion of, the local government officials. Xiang Biao[10] compares this experience to a similar movement in the Taiji region,[11] which arose from the establishment of professional and sometimes transnational networks, fostering a kind of autonomy as opposed to State control. In France, the 1999 census found that the proportion of Chinese businessmen is above the national average, that is, 2.7 per cent of the active population in France overall, but 10.6 per cent in the case of the Chinese working population.[12] The number of Chinese businesses employing more than ten employees is less than the national average, as they are usually small family-run enterprises. In France, 76.7 per cent of Chinese immigrants work in the production of consumer goods, and in services to private individuals (that is, cleaning, child care, and so on), out of which 30.2 per cent are in the catering industry and 23.8 per cent in trade.[13]

Chinese businesses mainly employ Chinese workers, and conversely, most Chinese workers work in Chinese businesses. The rapid increase in the number of businesses created a demand for labour that must be satisfied by immigration, as the local labour supply is not sufficient. Indeed, data compiled by the International Labour Organisation between 1992 and 1998[14] indicate that the major infraction as far as as illegal work was concerned was the non-declared worker. The proportion of offences regarding the employment of foreigners of all nationalities who are not in possession of labour documents decreased from 13 per cent to 3 per cent beween 1992 and 1998, with the exception of Chinese employers who frequently use people without residence documents or who are in the country illegally.

The need for labour abroad is one of the major drives of emigration from China. A candidate for immigration chooses his destination according to his employability (this determines his ability to repay his debt). To emigrate, he must contact a smuggler; if he tried to go through legal channels, he would not be able to obtain authorization to emigrate to live in France. To collect the money for the trip – between €20,000 and €30,000[15] – he will have to borrow money from his family, given that in most cases, he would not have the full fee even if he sold all his belongings. Upon arrival in France, a part of his salary is earmarked for the travel loan repayment. It is because of this that illegal migrants furnish a cheap, vulnerable, dependant and flexible labour force to the underground market. In France, as in Italy, this ethnic economy only functions when it is integrated with the local labour market, which consists of a chain of sub-contractors (from principals to manufacturers) who are not Chinese themselves. They may even be large and well-known commercial outfits.[16] In a field of activity marked by intense competition, one can only make a significant profit margin thanks to illegal networks and work conditions, a situation in everyone (client, seller and principal) is complicit. From the on-site study conducted in 1996 in Italy by François Brun and Ren Kelong, it is found that the Chinese managed to adapt to the clothing manufacture market in the province of Prato (Tuscany) which was in crisis:

> Chinese businesses are increasingly smaller in general but more and more numerous (479 in 1997, 1,559 in 2002, 1,724 in 2003). Recently, the Chinese learned how to survive the requirements of the economic situation. Taking the lead in the production of ready-to-wear fashion and benefiting from the 'Made in Italy' label, they now have a major share of the market that had become difficult, based on the quick execution of orders, guaranteed by the work of Chinese sub-contractors![17]

According to the latest statistics[18] on the struggle against illegal immigration and related networks between August 2005 and August 2006, even though the removal of illegal foreigners increased by more than 16 per cent, the arrests of network members and employers of illegal workers decreased by 11.71 per cent. The on-site studies on migrations show that the majority of the new migrants who are deported, come back to France some time later, after having been returned to their country of origin by force. The migrants must pay for the journey again but they actually are more likely to be able to repay double the journey-price (that is, €20,000 x 2) by going abroad than they would be able to pay the price of one journey by staying in China. The 2004 International Labour Organisation study on the trafficking and exploitation of Chinese immigrants in France shows that the increase in deportations and the tightening of borders led to the development of illegal immigration channels and made the smuggler networks even more complex.

The waves of regularization facilitated circulation within the European area: migrants were able to get round the dichotomy between administrative status (possible in some areas) and the opportunities on the labour market (which do not always coincide with the country where the residence permit was issued). Indeed, during the period of regularization, they would go to a European country to obtain administrative status and then were able to follow the labour market. Thus when they return to the first country of establishment (for example, France), should they be deported, they will not be sent back to China but to the country in which they have a residence permit.

Upon the repayment of the journey debt, migrants' dependence on their ethnic business networks remains important, for example, if the immigrant worker wishes to set up their own business, they know they can borrow again. The money may also be borrowed from the family in China. It is important to remember that between 1985 and 1990, many government businesses went bankrupt in China, following the outflow of capital that the managers reinvested either in China or abroad to set up their own businesses. The initial investment capital may vary between €30,000 in the case of a sweatshop to €230,000 in the case of purchasing a business.

In France, the issuing of a merchant's licence is subject to possession of a residence permit. Those migrants who cannot obtain this permit, even after living in France for five or ten years therefore must turn to trustworthy people (usually their family and close friends) to serve as a front for the business. Children who emigrated at a young age or who were born in France are also key to the development of a family business strategy, given that in addition to having a legal status, unliked their parents, they also very often have mastered the French language. The interdependence within these networks is both financial and familial: it concerns both the new migrant who must repay the journey debt and the new businessman who has just borrowed his initial capital in China or in France. Belonging to these networks is closely linked to the development of a business strategy: indeed, on the one side, the commercial success of businesspeople is an indicator of integration, on the other, the gap between the low status of an illegal immigrant and the 'omnipotent' businessperson inhibits social advancement. In this context, as R. Rastrelli stresses, 'Reducing the situation to a forced relation of a master-slave using violence, oppression and isolation, minimizes the significance of the cultural identity and the rules defied by the local market itself.' The ILO report stresses that the ethnic mixing and the diversity of the professional activities of the Chinese contribute to the enlargement and open up the niches where the Chinese are concentrated (such as the garment industry, leather goods manufacture and Asian catering).

Recognizing this trend, the European Franco-Italian programme entitled 'The Chinese of Europe and Integration' was concluded in 2006: one of its goals was to propose training programmes for young people between 16 and 25 years of age and for women in various professional careers. Paradoxically, the economic area is structured like an autonomous area enabled to operate freely and at the same time, like a no-rights area where the exploitation through labour and organized crime takes place. The new migrants are integrated in economic terms, by working as the labour force for their compatriots who are already established. The power of the transnational professional networks is fuelled by the financial flows and merchants between China and Europe, perpetuating an unequal and hierarchical order. The debate remains open and heated as to how to develop in our reception countries, and societies, the means to fight within those areas where the law is not functioning.

Véronique Poisson is a member of the Migrinter Laboratory in Poitiers.

NOTES

1 The Peoples' Republic of China was established in 1949.
2 The study was carried out between 2003 and 2004 and was commissioned by the SPL (Special Action Programme on Forced Labour of the International Labour Organisation). This study's finding were published in C. Gao Yun and Véronique Poisson, *Le trafic et l'exploitation des immigrants chinois en France* (Brussels: Bureau international du travail (BIT), 2005).
3 Pan Hongsong, 'Liyi Huaqiao Hu Xizhen', *Ziliao Huijuan, Zhejiang Huqiao Lishi Yanjiu*, 1 (1983), pp. 34–42.
4 AD Series (diplomatic archives): *Asia 1930–1940*, sub category 'Common Affairs', 103, p. 113.
5 Since 2000 and following the collapse of textile production in France, it is more and more difficult to repay the debt after some years of work and some migrants even become insolvable as a result of unemployment.
6 Peter Nolan and Furen Dong, *Market forces in China: Competition and Small Business – The Wenzhou Debate* (London and New York: Zed Books Ltd, 1990).
7 When the control officials came to inspect a business, no mention was made of the illegal trade practices against rake-offs. This way of functioning was developed on such a scale that we can talk about institutionalized racketeering (on behalf of the representatives of the State) against private trades.
8 M. Xu at the time of the interview was the director of one of the biggest eyeglass factories in the town of Wenzhou, employing more than 1,500 persons. Originating from Qingtian, some of his family members which emigrated to Hong Kong and Europe.
9 Interview number 7, conducted on 21 March 1998 at Wenzhou, annex of the thesis of V. Poisson, p. 183.

10 Xiang Biao, 'Zhejiang village in Beijing : Creating a visible non-state space through migration and marketized networks', in Frank N. Pieke and Hein Mallee (eds), *Internal and International Migration* (London: Curzon, 1999), p. 244.
11 The comparison with the movements of Taiji enables Xiang Biao to establish the fact that this autonomous area (created from the establishment of transregional financial networks) is not of an anti-State nature, but in a position of a movement of withdrawal, ready to establish links with the government authorities at the right moment.
12 Emmanuel Ma Mung, 'Immigration and Ethnic Labour Market', in F. Hillmann, E. Spaan and J. Van Naerssen (eds), *Asian Migration and Labour Market Integration in Europe* (London: Routeledge, 2004).
13 Ibid.
14 Yun and Poisson, *Le trafic et l'exploitation des immigrants chinois en France*, p. 66.
15 This amount corresponds to around 20–25 years of an average salary (calculated on the basis of €100 per month)
16 Interview with OCRIEST (Central Office for the Repression of Illegal Immigration and the Employment of Foreigners without Documents) and the Inspection of Labour, BIT 2004
17 François Brun and Ren Kelong, 'Europe: the area of the mobility of migrants: The Chinese of Italy', *Migrations Sociétés*, 18, 107 (September–October 2006), p. 151.
18 According to the National Crime Watch.
19 Referred to in Yun and Poisson, *Le trafic et l'exploitation des immigrants chinois en France*, p. 66, p 134): R. Rastrelli, *Chinese Immigration in Prato* (Prato, 2001), p. 134

PART VIII
Psychoanalysis and Geopolitics

Islam Wrestling with the 'West': Reflections on the Foundations

PHILIPPE RÉFABERT

I have long wondered about the value of the paternity attributed to God in the Jewish tradition. I considered that the expression 'Our Father', so often repeated in liturgical texts, lent itself too easily to mockery by certain atheists who use it to express their impatience with religion.[1] 'Infantile neurosis' has been the most common invective levelled at the phrase since Freud, who saw the figure of God only under the imaginary, idolatrous aspect it doubtless sometimes presents, in Judaism as elsewhere. Indeed, I did not see the usefulness of this designation and I tended to prefer, in this context, the Islamic tradition where the use of such an attribute is unthinkable and condemned as a sign of idolatry.

But suddenly I realized that the sacred prohibition against ascribing this attribute to Allah could have serious consequences. At the same time, I was discovering the richness it brought to Judaism — as well as to Christianity. Behind this seemingly infantile, innocent attribute, I glimpsed the possibility that designating God as 'father' opens the way to 'parricide' in the sense given this term by Socrates when he opposed 'Father Parmenides'. This designation introduces a fundamental aspect of the process of becoming-a-subject, at the end of which the 'imaginary' father is divested of his all-powerful character — a process fraught with risk and suffering, in which the child holds their own, affirms their identity and says 'I'. Using the attribute 'Father' to describe God opens the possibility of saying 'no' to Him, as Moses does when God wants to destroy the people of Israel — the possibility of opposing the Father without disavowing Him — and, having done so, of counting on Him as the support, the foundation on which to construct oneself as a responsible subject. Thus, my interest in Islam led me to perceive that this paternal designation makes it possible to do battle with God, to choose Him as an *adversary*.

In Islam, God who is the Unknowable, merciful Benefactor, Lord of the Worlds, All-Hearing, All-Knowing, All-Powerful, the Powerful and the Wise can under no circumstances be described as the Father of Believers. Thus, revolt against God is pre-empted from the start and Allah does not mention wrestling with the Angel when He inspires the Prophet concerning Jacob, who is mentioned numerous times in the Koran and who is accepted posthumously by Allah among the Muslims, as are so many other biblical figures.

How could the faithful oppose Allah who is Lord of All-the-Worlds, who has no name ('Al llâh' means 'the God'), Allah who comprises all determinations and, as such, tends to absorb alterity? It is easy to see why Islam is an object of fascination in the West, where the 'One' remains *fundamentally* in conflict with itself.

The philosophy of being has tried to do away with this internal conflict, but has never totally succeeded in doing so. A worldly consequence of this division is seen in the West in the fruitful battle that has been raging between secularism and religion for a millennium, between science and religion: the science of 'what' and the science of 'who' (G. Hansel), domains that have been clearly distinct since Newton.

Because Islam pushes the conflict outside its Unity, it remains a sublime horizon for Jews and Christians. Once the conflict between the 'One' and Himself had been expelled, Islam could not separate the political from the scientific. On the other hand, Islam has been the source of the most beautiful love poems, the inspiration of the greatest mystics who, in a seemingly paradoxical manner – as exemplified by Ibn al-Arabi – are almost the only ones in Islam to be able to reconcile its contradictions and descend from the sublime heights to the banality of everyday life.

Given this expulsion of conflict, the individual himself – who is the summit of determination, the very incarnation of the principle of determination itself – is an object that the Lord of All-the-Worlds is reluctant to recognize and tends to assimilate, to absorb into Himself. This is the move that a secular person, or a person with secular values like myself, must understand in order to appreciate the extent of the misunderstanding about human rights that separates the Islamic world from nations founded on a Judeo-Greco-Roman-Christian ethic.

When Islam is not inert, it is engaged in a frantic search for non-being, not because it is not acquainted with it, but because it has disposed of it with sublime carelessness. The Muslim is restricted to a dual possibility: the inside or the outside. He must choose and cannot attain perplexity, a term evocative of weaving (from *plectere*, to weave), composed of the weft threads and the warp threads of the spiritual and the civic; he is expected to choose between partaking in the Whole or not being. As a result, the obligation of being at once a Muslim *and* the citizen of a secular State places him in a position where he is pulled in different directions.

When civilization concerns the organization of citizens in a secular State, the Muslim, the Believer, finds himself in an impasse because, in the Islamic world, 'secular' means 'against the religion', and because the structural absence of a third element – that could be created by the 'no' – prevents the separation of religion from political power. This flaw of Islam did not appear in Egypt with the Muslim Brothers who, at the end of the nineteenth century, translated 'hukm' as 'power' instead of 'judgement' or 'illumination', as suggested by Abdelwahab Meddeb in *La Maladie de l'Islam* (The sickness of Islam), a highly instructive book published in 2002. No, this flaw could have been present in the foundations, as Daniel Sibony claims in his book *Nom de Dieu* (Name of God), also published in 2002.[2]

The expulsion of discontinuity confers great beauty on the apocalyptic style of the Koran, a beauty that, in Rilke's words, 'is nothing but the beginning of terror'. This beauty depends on an affinity with the limitless, with death, with the oceanic feeling produced by some musical compositions and by the inebriating incantation of the suras. History, human rights, or psychoanalytic psychology fail to approach and understand this universe because in it the fundamental principles of discourse are not the same. In the first case, the One is in conflict with itself; in the second, the One, the *Umma* – surrendered and pacified like the Whole – is in conflict with the rest of the world.

The original expulsion of the intrinsic division causes the Islamic world to be suspended, frozen in the moment of the beginning. From there, it reminds the not-yet-Islamic world of the miracle of this beginning, of the time of the apocalypse, this terrible moment when the divine is revealed, the 'big bang' of the Messianic era. The beauty of Islam is based on the proximity of this beginning to which everything leads it back. This particularity of Islam would be best expressed in the words of Islamic scholar Jacques Berque, were it not for their slight irony: 'What the Arab expects from the future is that it should give him back the past.'

In effect, Islam cannot leave behind a 'past' that remains present, a continued present. At the end of *Moses and Monotheism*, Freud advances the idea that by taking possession of the primal Father, a personification of the Whole, Islam has known periods of spectacular glory that were, however, short-lived precisely because these peoples could not oppose this all-powerful figure, loving and merciful. In more modern terms, what this means is that Islam lacks the 'symbolic father', the one who comes into 'existence' when the sons kill the father. The symbolic father is born of the parricide perpetrated on the head of the tribe, the primal Father. He is a dead father, not a father who 'detached' himself from his own father by going into exile, as Abraham did by leaving his home.[3] No, the head of the tribe was subjected to murder, an act that only becomes parricide in retrospect, since the (symbolic) father is born with the death of the Father. The Law emerges extemporaneously from the murder of the primal Father, an event that gives rise to the advent of the symbolic Father. In common language, 'symbolic father' and 'law' are notions that witness *the inscription of the trace of death in life*. In order for discontinuity, for the trace of death to be inscribed in life, someone has to have said 'no' to the whole, a fore-word that at once inscribes the individual in the whole and creates the conditions for separation, provided that the *subject* reappropriates this fore-word and that he says 'no' when the time comes, on his own behalf. Having done this, the subject will be an individual grounded both inside, and outside. This is the process of subjectivization that started in the West between the first and the second writing of the *Iliad*, according to E.R. Dodds; and with Adam according to the story of creation in the second book of Genesis. To say that Islam does not facilitate this process is an understatement: it actually tries to prevent it.

Islam knows only one place, a place from which the other is excluded. The same is true at the level of temporality. Islam, in which Revelation and Creation are contracted into a single time, as Franz Rosenzweig rightly noted, is held still in the time of miracles, in the Messianic era. This terrible moment, defined in Judaism as Messianic time, is transformed by Islam into ordinary time. Islam seems to be held hostage in this time outside linear time, outside the historical time of days and of hours, in this hyper-dense time of the present that is endlessly arriving. What paralyses Islamic civilization and Islam, what maddens the side of God when it awakens, what empowers the party of God (*Hezbollah*) when the fences fall – be it the Berlin Wall, the computer revolution or the so-called sexual 'revolution' in the West – is the fact that the rest of the world is in motion and that this motion dislodges the tectonic plates of the Islamic world that has remained fixated on a future it hopes will bring back its glorious past.

Ahmad (Mohammed), the Messenger, whose coming was predicted by the voice of Jesus (according to the Othman's Vulgate[4]) also announces the completion of time and the approach of the end of time. Islam is also an eschatology. In it, beginning and end are joined and this joining makes the figure of wholeness even more perfect. Encompassing is lived in temporality.

Everything is organized as if Islam has appointed itself guardian of the sublime, of the original paradox, of the moment when life meets death, when life and death, beginning and end, come together. This 'exquisite' moment, sung by western poets, is, for the Muslim, not his daily lot but the joy he keeps in reserve. He sets up camp at the foot of the sublime. No doubt most Muslims remain at a distance from this camp at the foot of the divine but, secretly they encourage and admire those who create the structure, that is, the fanatics who set it up. These *fanatics* – etymologically, the 'inspired ones' – are, after all, personifying the structure, animating the foundations, keeping them alive.

Like it or not, the secular world – originating from the Jewish and Christian worlds that have a common fate after two millennia during which the Christian world, depending on the era, tolerated or tried to annihilate the other – facilitated the passage of the Muslim nation, the *Umma*, to another type of relation with Transcendence, *invitus invitam*.

Our world, commonly called 'the West', which is recovering from religious and secular totalitarian regimes, is confronted with the emergence of a nation, albeit divided, but comprising, nevertheless, 1.3 billion people united in one faith, with a very old history in some places and a very recent history in others, worshipping a radically unknowable God who – once again, structurally – is *fundamentally* a stranger to discontinuity, to determination, to the singularity of the 'I'; a God who rejects mediation and the judgement of the majority; a God who dictated the Book himself.

These are the fundamental elements that allowed an Islamic scholar to write, in response to the Pope's speech in Regensburg:[5] 'That it is our duty [as Muslims] to listen serenely and in silence, to these words sent to us directly from Transcendence.'[6] These scholarly words imply that the Transcendence, the holy figure unique among all others, superlatively unique, addresses the Believer 'directly', without mediation. This figure is as beautiful as it is terrible. How beautiful and terrible it seems to have the privilege of repeating the verses, of sharing the incantatory fervour of those who chant the sura and bring to light the 'inimitable language of the Koran'! Just as beautiful and terrible is the practice that prefers incantation to study, faith to reason, the exhilaration of belonging to a community of the Faithful who know what it is to feel, in this communion, the *oceanic feeling* that Freud – who remained a stranger to it – left to Romain Rolland, the pacifist, who developed on it with delight.

Forgetting the intrinsic division is the dream of every human being; but when this dream is turned into a reality that is then personified by members of a State, the danger for civilizations is great. The apocalyptic text of the Koran offers the dream of simplicity. Allah gave His people a message free from 'tortuous wording' and 'explicitly clear', which defines the infidel as the one who refuses to surrender to the Kingdom of God, who offends the Whole and the peace guaranteed by the Whole – surrender and peace, let us remember, come from the same root: *s-l-m*.[7] But such a God who is Whole cannot be an '*opponent*'. And a man or a group of men cannot live without combat, without battling an opponent. Today, vast segments of the Muslim world have chosen to see as their *adversary* – the one turned toward them, *ad versum* – the West, the United States of America, Israel and their allies. This new reality requires that we learn to know those who have placed us in this role, for better or worse.

Philippe Réfabert is a psychoanalyst and author whose most recent book is *De Freud à Kafka* (Paris: Calmann-Lévy, 2001). This article was first published in French in *Tribune Juive Canada*, 22, 1 (March 2007). It was translated into English by Agnes Jacob.

NOTES

1. I would be tempted to add 'and in Christian texts' if not for the fact that the place and meaning of 'parricide', the essential theme of this article, take on such different connotations in Judaism and in Christianity.
2. Abdelwahab Meddeb, *La Maladie de l'Islam* (Paris: Seuil, 2002); Daniel Sibony, *Nom de Dieu* (Paris: Seuil, 2002).
3. F. Benslama is offended that the hypothesis of the lack of a symbolic father in Islam can be raised, because, as he writes in *La Psychanalyse à l'Epreuve de l'Islam* (Psychoanalysis and the Test of Islam) (Paris: Aubier, 2002): ' ... either Islam does not recognize the symbolic father, in which case the notions of "success", of "time" and of "monotheism" make no sense ... or the persistence of the *primal Father* is present in all the spiritual systems through a constant antagonism within them toward the symbolic father, a continuous battle between the "God the obscure" and "God the Sublime". In that case, Islam is no different than the rest ... ' (p. 118). What is the figure of the symbolic Father in Islam? According to F. Benslama, that of Abraham showing 'the *genesis of the father* as detachment from the *primal Father*'. The idea of detachment from the primal father is radically different from Freud's notion and, more generally, from psychoanalytic thinking. For Benslama, the symbolic father would be the one extracted from 'fusion' with the primal Father! If this were so, psychoanalysis would be shaken by the test to which the author *submits* it in his book. It would succumb altogether, lending itself to being taught in medical schools, in psychology programs and theological institutions.
4. 'And [*remind them*] when Jesus, son of Mary, says: "Oh, children of Israel! I am the apostle of Allah [sent] to you, declaring true what the Torah said before me, and announcing an Apostle who will come after me and whose name will be Ahmad"', Koran, Sura LX1, 6A.
5. What did the Pope say? He recalled the words of the fourteenth-century Christian emperor who suggested to his Muslim interlocutor that, 'Whoever would lead someone to faith needs the ability to speak well and to reason properly, without violence and threats ... [and that] to convince a reasonable soul, one does not need a strong arm or weapon of any kind, or any other means of threatening a person with death ...'
6. Youssef Seddik in *Le Figaro*, 22 September 2006.
7. Despite the insinuations of those who maintain that *s-l-m* is far from being solely the root of words like 'surrender', and that this root has other derivatives such as peace, salvation, health and so on, the word 'Muslim' is used in the sense of 'surrender' in the revelation of the Prophet (see Koran, Sura II, verse 136 and III, 19–20).

Immigrants and the West's Culture of Dependency

THEODORE DALRYMPLE

Whatever you may think of the social, economic and cultural effects of illegal immigration into Western European countries, one fact, in my view, is indisputable: that many of the migrants, whether they be fleeing persecution or merely seeking a better life elsewhere, show considerable determination, enterprise and courage, in short character, in risking the journey.

It is never easy to abandon one country, one language or one culture for another; but when you add the hazards of the journey itself to the clandestinity and precariousness in which the migrants expect to live, it is clear than only those who are highly motivated, either by fear or hope, will take the risk. Asylum seekers and illegal immigrants are often remarkable people.

In my work as a doctor in a British inner city hospital (and the prison next door) I met a large number of asylum seekers and illegal immigrants. I also produced for the courts quite a number of medico-legal reports on those claiming asylum. Very occasionally I was able to confirm their stories by detective work on the internet; at other times, I was able to disprove their stories by their gross internal inconsistencies. Most often, however, it was impossible for me to tell with any degree of certainty whether what they were saying was true, wholly or partially, though I believed that most of them had had horrible experiences of one kind or another, not necessarily political in nature. But when someone tells you that, at the age of eighteen, she saw her father tied to the tree, doused with petrol and burnt to death by government troops, it seems almost indecent to doubt it. Certainly, the consequences of not believing it when it is true are far worse, psychologically speaking, than those of believing it when it is not true. I pitied the poor bureaucrats deputed to sort the asylum sheep from the economic migrant goats, whose job it was to determine what actually happened months ago in a village 7,000 miles from their office, in a country of which they knew nothing, and whose decisions were of such vital importance for real, breathing human beings.

However, one thing struck me in my work almost more forcibly than anything else. I certainly did not subscribe to the view that most migrants – at any rate, the ones I came across – were seeking to become wards of the state. They did not want to be parasites. On the contrary, most of them were desperate to work and the relative flexibility of the labour market in Britain compared to that of other countries was what attracted them to Britain in the first place.

But those asylum seekers who decided to play by the rules as laid down by the authorities found themselves in a situation that they had not really anticipated, and far worse than that of their fellow asylum-seekers who did not play by the rules. Having claimed asylum, they were sent by the authorities to a town or a city at some distance from where arrived in the country, allocated to a hostel for asylum seekers, often with others with whom they shared no common language, given food and a tiny amount of pocket money. They were expressly forbidden to work, and were told they risked having their applications for asylum turned down if they did.

A few weeks of this regime were sufficient to turn active, intelligent, hopeful young people into listless and spiritless beings, whose only sign of life was a certain querulousness about such matters as the size of the TV screen in the hostel sitting room. Their utter dependence destroyed them psychologically in a way in which their terrible experiences, whatever they were, but up to and including ill-treatment by the police and outright torture, had not. Several of them told me that they could not attend the English classes that had been laid on for them, and which took place less than half a mile away, because they had no money for the bus fare and sometimes it rained and they had no umbrellas or raincoats. These were people, remember, who had travelled thousands of miles in containers in lorries, crossing very dangerous frontiers, baking by day and freezing by night, hungry always, and who had often had to relieve themselves where they sat or stood.

Perhaps the most remarkable case with which I had to deal was a man who had already been granted asylum, and had married a local woman. He had been drafted into the army of his country, whose government was notorious for its ruthlessness and brutality. He had taken part in the suppression of an insurrection and in the aftermath had been trained as a torturer. He gave me the impression that he took to the work with some enthusiasm, but unfortunately (at least for him) he fell from grace and found himself in the place of the tortured. Certainly he bore the scars of some kind of maltreatment.

He had managed to escape to Britain after his release from custody and successfully applied for asylum. He started a shop and soon did quite well. Unfortunately, he then had a slight car accident: he had pulled up at a roundabout when a car, travelling at 5 miles per hour, crashed into his rear.

He claimed that the resultant injury to his neck, caused by whiplash, was so severe that his life was now ruined, he could never hope to work again, though he was still only in his thirties. Thus a man had been able to survive both being a torturer and being tortured, but a minor accident at a roundabout in England was too much for him.

It is well known that whiplash injuries do not occur in countries in which the legal system does not recognise them as a basis for claims for compensation. This man, of course, was hoping for compensation on a very large scale.

In a way, this story is very reassuring. It suggests that, contrary to the gloomy prognostications of many, immigrants from widely divergent countries are capable of learning the local culture, and seeking long-term payments at the expense of others.

Theodore Dalrymple is a retired psychiatrist and prison doctor in Britain.